Copper Amine Oxidases

Structures, Catalytic Mechanisms, and Role in Pathophysiology

Copper Amine Oxidases

Structures, Catalytic Mechanisms, and Role in Pathophysiology

Edited by

Giovanni Floris

Bruno Mondovì

CRC Press
Taylor & Francis Group
Boca Raton London New York

CRC Press is an imprint of the
Taylor & Francis Group, an **informa** business

CRC Press
Taylor & Francis Group
6000 Broken Sound Parkway NW, Suite 300
Boca Raton, FL 33487-2742

First issued in paperback 2017

ISBN 13: 978-1-138-11175-2 (pbk)
ISBN 13: 978-1-4200-7680-6 (hbk)

Library of Congress Cataloging-in-Publication Data

Copper amine oxidases : structures, catalytic mechanisms, and role in pathophysiology / edited by Giovanni Floris and Bruno Mondovì.
p. ; cm.
Includes bibliographical references and index.
ISBN-13: 978-1-4200-7680-6 (hardcover : alk. paper)
ISBN-10: 1-4200-7680-9 (hardcover : alk. paper)
1. Amine oxidase. 2. Copper enzymes. I. Floris, Giovanni, 1945- II. Mondovì, Bruno.
[DNLM: 1. Amine Oxidase (Copper-Containing)--metabolism. 2. Amine Oxidase (Copper-Containing)--pharmacology. QU 140 C785 2009]

QP603.M6C67 2009
572'.548--dc22 2008038892

Visit the Taylor & Francis Web site at
http://www.taylorandfrancis.com

and the CRC Press Web site at
http://www.crcpress.com

Contents

Acknowledgments

The editors wish to thank Drs. Francesca Pintus and Delia Spanò of the Department of Applied Sciences in Biosystems, Cagliari University, Italy, for their valuable help in the preparation of this book.

Introduction: The Dawn of Pink Enzymes

Our interest in enzymes capable of inactivating primary amines has been always keen due to the biological and pharmacological activities of these compounds. Much less interest focused on the so-called diamines that were considered for decades to be only products of body decomposition as testified by the trivial names of the most popular ones, i.e., *cadaverine* and *putrescine*. The interest in the enzymes specific to diamines was only marginal until the 1960s. In fact, diamine oxidases were considered no more than biochemical curiosities unworthy of investigation even though they were known to be present in several mammalian tissues since they were described more than 70 years ago in a seminal paper by German biochemist E.A. Zeller titled Zur Kenntnis der Diamine-oxydase (*Helv Chim Acta* 21: 1645, 1938).

An incidental shaker of the then stagnant field of investigation was the late Professor Doriano Cavallini whose inexhaustible curiosity and ingenuity in investigating the metabolism of sulfur compounds led him to search for enzymes able to oxidize cystamine, an intriguing metabolite of cysteine in the 1960s. Thus he targeted diamine oxidases and committed me and other "youngsters" in his laboratory to purifying diamine oxidase from pig kidneys.

The task was very difficult at a time before HPLC and affinity chromatography were not yet available and recombinant DNA technologies were far in the future. We eventually succeeded and obtained a few milligrams of homogeneous pig kidney diamine oxidase, perhaps the first enzyme of this family to be purified and characterized. To our surprise, the enzyme solution looked pinkish. We already knew that the enzyme contained copper and in fact we were able to quantify this metal by the then revolutionary technique of atomic absorption spectroscopy. However, the amount of copper contained could not have accounted for the color that obviously came from another source.

This simple observation opened the chase for the unknown chromophore. Other groups had already demonstrated that diamine oxidases, at variance with the much more popular monoamine oxidase, were inhibited by semicarbazide and related compounds, much like pyridoxal phosphate (PLP)-containing enzymes that were then becoming extremely fashionable as a result of joint efforts of American and Soviet Union scientists.

Even our biochemistry department was involved in investigating PLP-dependent enzymes. Our first guess was that an unusually bound PLP molecule was indeed responsible for both the color and the semicarbazide sensitivity of diamine oxidase since the visible absorption and the enzymic activity disappeared at once when phenylhydrazine was added to a sample. Thus, we started a fruitless search for PLP by every technique then available, including fluorescence, limited proteolysis by chemical and/or enzymatic methods, and others.

In the meantime, we continued to work on several copper proteins such as azurin, plastocyanin, ceruloplasmin, laccase, ascorbate oxidase, and superoxide dismutase. We also had to tackle the issue of the role of the metal in the activity of these proteins. They contained at least two types of copper, namely blue and non-blue, of which only the blue copper was observable by optical spectroscopy. In some compounds, the copper was totally or partially silent and we were induced to try different spectroscopic approaches. One of the brightest researchers of the group, Joe Rotilio, went to the University of Wisconsin to learn from a leading authority in the field, Professor Helmut Beinert, how to use and apply electron paramagnetic resonance (EPR) spectroscopy. Upon his return to Italy, Rotilio was able to start studying all the different copper proteins by the new technique that soon allowed us to distinguish the divalent copper moieties as Type 1 (blue) and Type 2 (non-blue) copper. We took advantage of meeting many outstanding scientists who contributed substantially to the understanding of copper biochemistry including Harry Gray, Jack Peisach, Israel Pecht, Irwin Fridovich, James A. Fee, and the late William E. Blumberg and Bo Malmstrom.

The group headed by Professor Bruno Mondovì split into two halves, one that worked with Joe Rotilio on superoxide dismutase and the flood of oxygen radical issues that emerged from their efforts. I joined the other group in studying the beautiful single and multicopper blue proteins.

Pig kidney diamine oxidase, beef plasma amine oxidase, and lentil seedling diamine oxidase were studied by both groups and by newcomers like Rinaldi, Floris, and their colleagues in Cagliari who added another member to the copper amine oxidase family, namely the one produced by the Mediterranean shrub *Euphorbia characias*. Despite our heroic efforts, no PLP was found in any of these proteins.

Eventually a group in Delft led by J.A. Duine described for the first time a new organic cofactor in enzymes, namely pyrroloquinoline quinone (PQQ), the spectroscopic features of which (absorption in the visible spectrum and redox behavior) seemed to match those of copper diamine oxidase that was found to be independent of copper. This new cofactor was widespread in nature and was even claimed to be a new vitamin. However, its presence in copper amine oxidases was never confirmed. Finally, the question was settled by Judith Klinman who demonstrated conclusively that the pink color of the enzymes and their sensitivity to carbonyl reagents such as semicarbazide was due to trioxyphenylalanine quinone (TOPA or TPQ). Since then all the copper-containing, hydrazine-sensitive amine oxidases have been shown to contain TPQ derived from postsynthetic modification of a particular tyrosinyl residue of these enzymes.

Several amine oxidases (AOs) have been crystallized and their tridimensional structures revealed, thanks particularly to the efforts of Peter Knowles. The new technique of EPR spectroscopy enabled us more than 30 years ago to demonstrate that the mechanism of action of AOs involved the formation of a radical centered at the TPQ moiety but no change of copper valence. Many more observations of the spectroscopy, mechanism, substrate specificity, steady state, and rapid kinetics of AOs were performed, but the physiological role of these enzymes despite their distribution in all living species and in many different tissues remains uncertain.

Three main hypotheses have been put forward concerning the biological importance of AOs. First, their role is to remove biologically active molecules like primary amines including histamine and many diamines including putrescine, a key precursor of polyamines, essential for many functions of cells. Second, their true importance may be the formation of the corresponding aldehydes that in some cases were found to affect dramatically cell differentiation, proliferation, and survival. The third possibility may be the "on-site" production of hydrogen peroxide that many authors, including ourselves, consider a signaling molecule rather than a harmful side product.

Of course, as is often the case in biology, it is reasonable that each hypothesis is valid under a particular set of conditions. However, at least in some cases, the true substrates of AOs have not been yet identified. Very recently, research activities focusing on methylated arginines has given rise to very unusual products like methylurea. Finally, the identification of vascular adhesion protein-1 (VAP-1) as a semicarbazide-sensitive amine oxidase (SSAO) and of human histaminase as amiloride-binding protein-1 (ABP-1) has further broadened the spectrum of biological roles of these enzymes.

In this context it is worth mentioning that more than 30 years ago, Professor Cavallini, using our purified diamine oxidase as a reagent, was able to demonstrate for the first time that some aldehydes derived from aminothiols like cystamine after cyclization acted as possible endogenous ligands for the then "orphan receptors" that bound drugs like the diazepines. It appears therefore that the field of the pink enzymes is far from exhausted and remains fascinating and worthy of further research.

The present book will give the AO fans a valuable reference and general readers a thorough introduction to these enzymes.

ALESSANDRO FINAZZI-AGRÒ

Editors

Giovanni Floris was born in Camerino, Italy. He began his career at the University of Cagliari in 1974 as a researcher in biochemistry. After serving as an associate professor from 1985 to 2000, he was named a full professor of biochemistry and continues in that capacity today. He was the director of the Department of Sciences in Applied Biosystems at the University of Cagliari from 2000 through 2006.

Dr. Floris served as president of biological sciences at the University of Cagliari from 2001 to 2006 and has been the president of biotechnology at the university since 2007. He is also a member of the Italian Society of Biochemistry and Molecular Biology.

Dr. Floris has written more than 100 papers and has authored or co-authored several books. His current research interests are structures and functions of copper proteins and peroxidase and the role of amine oxidases in plant physiology.

He can be reached at florisg@unica.it.

Bruno Mondovì was born in Verona, Italy. He earned an M.D. and a Ph.D. from the University of Rome. After joining the university as an assistant professor in 1951, he advanced to associate professor in both biochemistry and applied biochemistry, then full professor, and was named emeritus professor in 2004. In addition to his teaching activities at the University of Rome, he served as a visiting professor at Albert Einstein College of Medicine of Yeshiva University in New York, University of Quebec, Montreal, and the Polish Academy of Sciences in Lodz.

Dr. Mondovì was the recipient of the Wilhelm Konrad Roentgen International Prize for Oncology in 1985 and the Feltrinelli Prize for Medicine in 1989, both awarded by the Accademia Nazionale dei Lincei.

Among other professional activities, Dr. Mondovì was the director of the Institute of Applied Biochemistry at the University of Rome for 16 years and president of the Pharmacy Course at the university for 17 years. He organized several International Manziana Conferences on Copper in Biological Systems. In addition to serving as an organizing member and treasurer of the Scientific Committee of the European

Biochemical Societies, Dr. Mondovì is or has been the treasurer and a member of the Italian Biochemical Society, president of the European Histamine Society Congress, and a member of the American Chemical Society, New York Academy of Sciences, and the Italian Society of Biochemistry and Molecular Biology.

Dr. Mondovì has written almost 400 papers, edited and wrote several books, and holds a number of patents. He has served as an invited speaker at 60 international meetings. He continues to perform research on the structures and functions of copper proteins, biochemical mechanisms of heat sensitivity of cancer cells, and the role of amine oxidases in pathophysiology.

Dr. Mondovì's e-mail address is bruno.mondovi@uniroma1.it.

Contributors

Andrea Bellelli
Department of Biochemical Sciences
University of Rome La Sapienza
Rome, Italy

Frans Boomsma
Department of Internal Medicine
Erasmus University Medical Center
Rotterdam, The Netherlands

Peter Boor
Department of Clinical and Experimental Pharmacology
Slovak Medical University
Bratislava, Slovakia
and
Division of Nephrology
RWTH University of Aachen
Aachen, Germany

Sinéad Boyce
Department of Biochemistry
Trinity College
Dublin, Ireland

Franca Buffoni
Department of Preclinical and Clinical Pharmacology
University of Florence
Florence, Italy

Christian Carpéné
French National Institute of Health
and
Université Paul Sabatier
Institut Louis Bugnard
Toulouse, France

Alessandra Cona
Department of Biology
University of Rome III
Rome, Italy

Gavin P. Davey
School of Biochemistry and Immunology
Trinity College
Dublin, Ireland

Maria Luisa Di Paolo
Department of Biological Chemistry
University of Padua
Padua, Italy

David M. Dooley
Department of Chemistry and Biochemistry
Montana State University
Bozeman, Montana, USA

Rodolfo Federico
Department of Biology
University of Rome III
Rome, Italy

Alessandro Finazzi-Agrò
Department of Experimental Medicine
University of Rome Tor Vergata
Rome, Italy

Giovanni Floris
Department of Applied Sciences in Biosystems
Cagliari University
Cagliari, Italy

Wieslawa Agnieszka Fogel
Department of Hormone Biochemistry
Medical University of Lodz
Lodz, Poland

Ivo Frébort
Department of Biochemistry
Palacky University
Olomouc, Czech Republic

Martha Motherway Gildea
School of Biochemistry and Immunology
Trinity College
Dublin, Ireland

J. Mitchell Guss
School of Molecular and Microbial Biosciences
University of Sydney
Sydney, New South Wales, Australia

Andrew Holt
Department of Pharmacology
Faculty of Medicine and Dentistry
University of Alberta
Edmonton, Alberta, Canada

Sirpa Jalkanen
University of Turku
and
National Public Health Institute
Turku, Finland

Mircea Alexandru Mateescu
Department of Chemistry and Biochemistry
Université du Québec á Montréal
Montreal, Quebec, Canada

Emanuela Masini
Department of Preclinical and Clinical Pharmacology
University of Florence
Florence, Italy

Andrew G. McDonald
School of Biochemistry and Immunology
Trinity College
Dublin, Ireland

Rosaria Medda
Department of Applied Sciences in Biosystems
Cagliari University
Cagliari, Italy

Bruno Mondovì
Department of Biochemical Sciences
University of Rome La Sapienza
Rome, Italy

Laura Morpurgo
Department of Biochemical Sciences
University of Rome La Sapienza
Rome, Italy

Minae Mure
Department of Chemistry
University of Kansas
Lawrence, Kansas, USA

Réginald Nadeau
Department of Physiology
Université de Montréal
and
Sacré-Coeur Hospital
Montreal, Quebec, Canada

Toshihide Okajima
Institute of Scientific and Industrial Research
Osaka University
Osaka, Japan

Aldo Olivieri
School of Biochemistry and Immunology
Trinity College
Dublin, Ireland

Jeff O'Sullivan
Dublin Dental School and Hospital
Trinity College
Dublin, Ireland

Michael I. O'Sullivan
Dublin Dental School and Hospital
Trinity College
Dublin, Ireland

Pavel Peč
Department of Biochemistry
Palacky University
Olomouc, Czech Republic

Paola Pietrangeli
Department of Biochemical Sciences
Rome University La Sapienza
Rome, Italy

Laura Raimondi
Department of Preclinical and Clinical Pharmacology
University of Florence
Florence, Italy

Adelio Rigo
Department of Biological Chemistry
University of Padua
Padua, Italy

Marko Salmi
Department of Bacterial and Inflammatory Diseases
National Public Health Institute
University of Turku
Turku, Finland

Tiina A. Salminen
Department of Biochemistry and Pharmacy
Åbo Akademi University
Turku, Finland

Lawrence M. Sayre
Department of Chemistry
Case Western Reserve University
Cleveland, Ohio, USA

Hubert G. Schwelberger
Department of General and Transplant Surgery
Medical University Innsbruck
Innsbruck, Austria

Marek Šebela
Department of Biochemistry
Palacky University
Olomouc, Czech Republic

Anna Stasiak
Department of Hormone Biochemistry
Medical University of Lodz
Lodz, Poland

Roberto Stevanato
Department of Physical Chemistry
University of Venice
Venice, Italy

Shinnichiro Suzuki
Graduate School of Science
Osaka University
Osaka, Japan

Katsuyuki Tanizawa
Institute of Scientific and Industrial Research
Osaka University
Osaka, Japan

Keith F. Tipton
School of Biochemistry and Immunology
Trinity College
Dublin, Ireland

Antonio Toninello
Department of Biological Chemistry
University of Padua
Padua, Italy

Ewa Toporowska-Kowalska
Department of Pediatric Allergy, Gastroenterology, and Nutrition
Medical University of Lodz
Lodz, Poland

Mercedes Unzeta
Faculty of Medicine
Universitat Autònoma de Barcelona
Barcelona, Spain

Anton H. van den Meiracker
Department of Internal Medicine
Erasmus University Medical Center
Rotterdam, The Netherlands

Giuseppe Zanotti
Department of Chemistry
University of Padua
Padua, Italy

Abbreviations

AGAO *Arthrobacter globiformis* amine oxidase
AO amine oxidase
BSAO bovine serum/plasma amine oxidase
BzAO benzylamine oxidase
CAO copper-containing amine oxidase
DAO diamine oxidase
ECAO *Escherichia coli* amine oxidase
ELAO *Euphorbia characias* latex amine oxidase
HPAO *Hansenula polymorpha* amine oxidase
LOX lysyl oxidase
LSAO lentil (*Lens esculenta*) seedling amine oxidase
MAO monoamine oxidase
PKAO pig kidney amine oxidase
PPLO *Pichia pastoris* lysyl oxidase
PSAO pea (*Pisum sativum*) seedling amine oxidase
TPQ 2,4,5-trihydroxylphenylalanine quinone
TPQ_{red} reduced form of TPQ
TPQ_{ox} oxidized form TPQ
TPQ_{aq} aminoquinol form of reduced TPQ (TPQred)
TPQ_{imq} iminoquinone form of TPQ
TPQ_{sq} semiquinone radical form of TPQ
TPQ_{ssb} substrate Schiff base of TPQ
SAO serum/plasma amine oxidase
SSAO semicarbazide-sensitive amine oxidase
VAP-1 vascular adhesion protein-1

1 History

Franca Buffoni

As a pupil of Professor H. Blaschko whose research greatly contributed to our knowledge of the copper-containing amine oxidases (Blaschko, 1962 and 1974), I am pleased to introduce this book describing the most recent results of research focusing on these enzymes.

The copper-containing amine oxidases (CAOs; EC1.4.3.6, amine:oxygen oxidoreductase [deaminating]) comprise a large class of enzymes that catalyze the oxidation of primary amines to the corresponding aldehyde concomitant with the reduction of dioxygen to hydrogen peroxide. They are widely distributed in nature, found in prokaryotic and eukaryotic organisms (Zeller, 1963; Blaschko, 1974). Interest in these enzymes has greatly increased in recent years, resulting in the discovery that they play certain important physiopathological roles.

CAOs allow prokaryotes and fungi to use primary amines as sole sources of nitrogen for growth. In plants they are important for the synthesis of alkaloids (Tabor and Tabor, 1964), cell wall formation, and wound healing (Davidson, 1993). Four enzymes found in mammals belong to class EC 1.4.3.6:

1. Intracellular AOs, also called diamine oxidases (DAOs)
2. Tissue copper-containing amine oxidase or semicarbazide-sensitive amine oxidase (SSAO), a somewhat misleading name because all Cu/TPQ AOs, unlike FAD-dependent enzymes, are inhibited by semicarbazide and other carbonyl reagents
3. Blood plasma spermine oxidase (SAO)
4. Blood plasma benzylamine oxidase generally termed plasma and serum AOs or benzylamine oxidases (BzAOs)

Best discovered DAOs in the tissues of mammals in 1929. The enzymes were called histaminases because they oxidized histamine. Zeller introduced the diamine oxidase name in 1938 on the basis that pig kidney histaminase acted also on short aliphatic diamines such as 1,4-diaminobutane (putrescine) and 1,5-diaminopentane (cadaverine).

SAO was discovered by Hirsch (1953) in the blood plasma of sheep. BzAO was first described by Bergeret et al. (1957) in the blood plasma of horses. Two years later Blaschko et al. (1959) described an amine oxidase found in the blood plasma of many mammals that differed from SAO in that it exerted no significant action on spermine and spermidine. Blaschko called this enzyme benzylamine oxidase (BzAO) because benzylamine was the best substrate. BzAO has the same substrate specificity (aging only on primary amines) and inhibitor sensitivity as SAO. SAO occurs in ruminants (Blaschko, 1962) whereas BzAO occurs in the blood plasma of nonruminant mammals, in ungulates not possessing SAO, in elephants, in some carnivores and primates (e.g. baboons; Blaschko and Bonney, 1962), and in humans (McEwen, 1965a).

CAOs share many common biochemical properties and clearly differ from mitochondrial monoamine oxidases (MAOs) although they share some overlap in substrate specificity. CAO enzymes also share a similarity with lysyloxidase (LOX), the enzyme of the extracellular matrix that oxidatively deaminates the ε-amino group of specific lysine and hydroxylysine residues of collagen (Pinnell and Martin, 1968).

Copper deficiency has been shown to produce defects in the structural stabilization of the fibrous protein of connective tissue with a consequent pathological manifestation in several species of animals. The diminished tensile strength of elastin and collagen is the result of impaired cross linkages of their polypeptide chains (Carnes, 1971). This is dependent on a decrease of LOX activity but according to Carnes, also of BzAO activity of connective tissue that may play a role in the formation of elastin cross links. Pure crystalline BzAO of pig plasma was able to oxidize the purified proelastin of copper-deficient pig aorta (Carnes, 1971).

Evidence has been obtained via immunofluorescence histochemistry of the presence in the connective tissues of all pig organs of a protein possessing the same immunological determinant as pig plasma BzAO (Buffoni et al., 1977a). An intense fluorescence was observed in the blood vessels of all organs of the pig (Banchelli and Buffoni, 1981). In copper-deficient pigs, the BzAO activity was absent or very low; after copper therapy, the enzymatic activity made its appearance. Copper-deficient animals died as consequences of extensive vascular damage (Buffoni and Blaschko, 1965), and similar observations were made in chicks (Bird et al., 1966; Chow et al., 1968).

The presence of BzAO activity in blood vessels and in the connective tissue explains the biochemical discovery of the SSAO enzymatic activity in different tissues of mammals. The first demonstration of this activity was obtained by use of irreversible inhibitors of mitochondrial MAO. When MAO A and B activities were completely inhibited by an irreversible inhibitor, an enzymatic activity able to oxidize primary amines such as benzylamine and phenylethylamine and inhibited by semicarbazide remained in tissue (Lyles and Callingham, 1975; Roth and Gillis, 1975; Dial and Clarke, 1978; Lewinsohn et al., 1978). Roth and Gills suggested that the amine oxidase present in lungs and inhibited by semicarbazide was similar to the enzyme found in plasma and probably associated with vascular endothelium.

Göktürk et al. (2007) recently described macrovascular changes in mice overexpressing human SSAO (or tissue BzAO) in smooth muscle cells. Elevated pulse pressure together with an abnormal elastin structure in the aorta suggests a rigidity of large arteries in these animals as result of elevated SSAO activity and a physiological role of SSAO (or BzAO) in elastin maturation.

Chronic SSAO inhibition in rats produced lesions consisting of striking disorganization of elastin architecture within the aortic media accompanied by degenerative medial changes in vascular smooth muscle cells (Langford et al., 1999). An increase in the plasma levels of BzAO has been observed in many diseases such as liver cirrhosis that involve fibrosis development (Buffoni et al., 1977b). Recent evidence indicates that BzAO activity is present also in the blood plasma of ruminant mammals. Fresh bovine serum of adult animals contains two amine oxidase activities, one with high affinity for benzylamine and one with lower affinity. The enzyme with lower affinity for benzylamine was identified as SAO (Buffoni et al., 2000). Similar observations were made in sheep (Callingham et al., 1988). BzAO appears in the blood plasma of all mammals and this suggests that this enzyme has a relevant physiological role in addition to its activity in the formation of elastin cross links in connective tissues.

This book presents the most recent research contributions that shed further light on the pathophysiological roles of these enzymes.

2 Nomenclature and Potential Functions of Copper Amine Oxidases

Sinéad Boyce, Keith F. Tipton, Michael I. O'Sullivan, Gavin P. Davey, Martha Motherway Gildea, Andrew G. McDonald, Aldo Olivieri, and Jeff O'Sullivan

CONTENTS

2.1 INTRODUCTION

The properties and behaviors of copper amine oxidase enzymes are reviewed in detail elsewhere in this volume and this account will concentrate on their "identity crisis" and their more intriguing functional diversity. Although some politicians and pundits may disagree, it is generally accepted that it is helpful to be clear what one is actually talking about. This has proven to be more difficult than it should be for the copper amine oxidases because the term refers to a group of enzymes with diverse activities and functions. These include diamine oxidase (DAO), lysyl oxidase (LOX), and an enzyme that acts toward primary monoamines—often called semicarbazide-sensitive amine oxidase (SSAO).

As shown in Table 2.1, some of the enzymes in this group exhibit overlapping substrate specificities, and apparent significant specificity differences among tissues and species make it difficult to define their specificities with certainty. For example, pig plasma SSAO has been shown to oxidize mescaline more efficiently than benzylamine (Buffoni and Della Corte, 1972) but the human enzyme has no detectable activity toward this substrate. Although serotonin (5-HT) is not oxidized by SSAO from most sources, it is a good substrate for the enzyme from pig and human dental pulp (Nordqvist et al., 1982; O'Sullivan et al., 2002; O'Sullivan et al., 2003b).

Our unpublished work indicates that splicing differences may account for this novel specificity. Comparisons of the crystal structures of bovine (Lunelli et al., 2005) and human SSAO (Jakobsson et al., 2005) indicate a basis for the species

TABLE 2.1
Reported Substrate Specificities of Mammalian Amine Oxidases

Amine	SSAO	DAO	LOX	MAO
Adrenaline	No	No	No	Yes
Allylamine	Yes	?	?	No
Aminoacetone	Yes	?	?	No
Benzylamine	Yes	Low	Yes	Yes
Cadaverine	±	Yes	Yes	No
1,3-Diaminopropane	No	Yes	?	Very Low
Dopamine	Yes	Yes	?	Yes
Ethylenediamine	No	Yes	?	No
Histamine	±	Yes	±	No
5-Hydroxytryptamine	No*	No	?	Yes
L-Lysine	±	No	?	No
N^1-Acetyl putrescine	±	±	?	±
N^T-Methylhistamine	Yes	Yes	?	Yes
1-Methyl-4-phenyl-1,2,3,6-tetrahydropyridine	No	No	?	Yes
Noradrenaline	±	±	?	Yes
Methylamine	Yes	No	?	No
n-Pentylamine	Yes	Low	?	Yes
Peptidyl-L-lysyl-peptide	±	No	Yes	No
2-Phenylethylamine	Yes	±	?	Yes
Putrescine	±	Yes	?	No
Spermidine	Yes	Yes	?	No
Spermine	Yes	Low	?	No
Tryptamine	±	No	?	Yes
Tyramine	Yes	±	?	Yes

Notes: ± Some studies have reported activity; others could not detect any; ? No data; * SSAO from dental pulp appears active toward this substrate.

Sources: Data compiled from several sources including: Bardsley et al. (1970); De Matteis et al. (1999); Moldes et al. (1999); Seiler (2000); Elmore et al. (2002); Ochiai et al. (2006), and our work with MAOs.

difference. The narrow funnel or "entrance gate" through which substrates must pass to access the active site is narrower in the human enzyme, providing greater restrictions on substrate access (Jakobsson et al., 2005).

This SSAO term is frequently used to describe the group of copper-containing enzymes that are active toward primary monoamines. This distinguishes them from the flavoprotein monoamine oxidases (MAOs) that are not inhibited by semicarbazide. Conversely SSAO, at least in some species, is insensitive to inhibition by acetylenic monoamine oxidase inhibitors such as clorgyline. Because of this, SSAO has sometimes been referred to as clorgyline-resistant amine oxidase (CRAO). However, most copper amine oxidases are inhibited by carbonyl group reagents such as semicarbazide. This is because of the presence of amino-acid based 6-hydroxydopa; 2,4,5-trihydroxyphenylalanine (TOPA or TPQ) quinone in the case of DAO and SSAO and lysine tyrosylquinone in LOX as a redox cofactor. DAO and LOX are also insensitive to inhibition by clorgyline.

2.2 WHAT'S IN A NAME? A CLEAR CASE FOR CONFUSION

The above brief account of names used in attempts to distinguish the copper-containing amine oxidases from monoamine oxidase indicates the need for greater clarity. Although most scientists agree on the need for standardized, universally accepted systems of nomenclature in biochemistry, chemistry, and genetics, the problem lies in the acceptance of standard names. One simple rule of enzyme nomenclature is that the number of complaints about the existing system (and there are many concerning the copper amine oxidases) will always be exceeded by the number of complaints received when attempts are made to improve the system. In the case of the amine oxidases, the need is for clarity rather than the imposition of uniformity. When nomenclature causes researchers to perform sensible experiments with the wrong enzymes, it is necessary to make changes.

The definitive names and classification (Enzyme Commission [EC]) numbers of characterized enzymes appear on the list of the Nomenclature Committee of the International Union of Biochemistry and Molecular Biology (IUBMB). The most recent version (*Enzyme Nomenclature*) was published in 1992. In order to allow frequent updating since them, information is disseminated through the internet at http://www.chem.qmul.ac.uk/iubmb/enzyme/ and http://www.enzyme-database.org/.

The classification of enzymes follows the simple rule that enzymes are categorized and named after the reactions they catalyzes, i.e., they should be classified in terms of their physiological substrates or group of substrates and the types of reactions involved (Boyce and Tipton, 2000). The EC number consists of four components. EC 1 includes all the oxidoreductase enzymes. Group 1.4.x.y includes the enzymes oxidizing CH–NH_2 groups. EC 1.4.3.y includes those that use oxygen as the acceptor. The final (y) number specifies the individual enzyme within the group. The correct use of this system should mean that the enzyme or group of enzymes cited should be perfectly described by the "accepted name," EC number, and source. The difficulty is sometimes knowing the correct substrate(s) to choose in cases of enzymes that have broad specificities.

Table 2.2 lists information about each of the enzymes currently recognized as amine oxidases. The list excludes the enzymes that catalyze the oxidative

TABLE 2.2
Amine Oxidases Recognized on IUBMB Enzyme List

Accepted Name	EC	Other Names	Reaction	Comments
Amine oxidase, flavin-containing	1.4.3.4	Monoamine oxidase; tyramine oxidase; tyraminase; amine oxidase; adrenalin oxidase; epinephrine oxidase; MAO; polyamine oxidase; serotonin deaminase; adrenaline oxidase; spermidine oxidase; spermine oxidase ***Systematic:*** amine:oxygen oxidoreductase (deaminating) (flavin-containing)	$RCH_2NH_2 + H_2O + O_2 =$ $RCHO + NH_3 + H_2O_2$	Flavoprotein (FAD) acting on primary amines and on some secondary and tertiary amines
Amine oxidase, copper-containing	1.4.3.6	Diamine oxidase; diamino oxhydrase; histaminase; amine oxidase; monoamine oxidase; amine oxidase (pyridoxal containing); benzylamine oxidase; histamine deaminase; histamine oxidase; Cu-amine oxidase; semicarbazide-sensitive amine oxidase; SSAO ***Systematic:*** amine:oxygen oxidoreductase (deaminating) (copper-containing)	$RCHNH_2 + H_2O + O_2 =$ $RCHNH_2 + H_2O + O_2 =$	Group of enzymes including those oxidizing primary monoamines, diamines, and histamine; copper quinoproteins
Ethanolamine oxidase*	1.4.3.8	***Systematic:*** ethanolamine:oxygen oxidoreductase (deaminating)	Ethanolamine + $H_2O + O_2 =$ glycolaldehyde + $NH_3 + H_2O_2$	Cobamide protein
Putrescine oxidase	1.4.3.10	***Systematic:*** putrescine:oxygen oxidoreductase (deaminating)	Putrescine + $H_2O + O_2 =$ 4-aminobutanal + $NH_3 + H_2O_2$	Flavoprotein (FAD); 4-aminobutanal condenses nonenzymically to 1-pyrroline

Cyclohexylamine oxidase*	1.4.3.12	***Systematic:*** cyclohexylamine:oxygen oxidoreductase (deaminating)	Cyclohexylamine + H_2O + O_2 = cyclohexanone + NH_3 + H_2O_2	Flavoprotein (FAD); other cyclic amines can act instead of cyclohexylamine, but not simple aliphatic and aromatic amides
Protein-lysine 6-oxidase	1.4.3.13	Lysyl oxidase ***Systematic:*** protein-L-lysine:oxygen 6-oxidoreductase (deaminating)	Peptidyl-L-lysyl-peptide + H_2O + O_2 = peptidyl-allysyl-peptide + NH_3 + H_2O_2	Also acts on protein 5-hydroxylysine.
Polyamine oxidase	1.5.3.11	***Systematic:*** N^1-acetylspermidine: oxygen oxidoreductase (deaminating)	N^1-acetylspermine + H_2O + O_2 = N^1-acetylspermidine + 3-aminopropanal + H_2O_2	Flavoprotein requiring Fe^{2+}; also acts on N^1-acetylspermidine and N^1,N^{12}-diacetylspermine
Hydroxylamine oxidase*	1.7.3.4	Hydroxylamine oxidoreductase (HAO) ***Systematic:*** hydroxylamine:oxygen oxidoreductase	Hydroxylamine + O_2 = nitrite + H_2O	Hemoprotein with seven *c*-type hemes and one *P*-460-type heme per subunit

Note: * Not reported to occur in mammalian systems.

deamination of free amino acids. The published version of the enzyme list also includes selected references for each enzyme but these have been omitted from Table 2.2 and subsequent tables. The classification of MAOs (EC 1.4.3.4) and the copper-containing amine oxidases (EC 1.4.3.6) in terms of their cofactor requirements, does not follow the usual rules and provides ample scope for confusion because many researchers are interested in what an enzyme does, not the cofactors it uses. For example, several publications claiming to focus on MAOs actually concern SSAOs. One leading supplier marketed SSAO as MAO for many years. Sometimes reference to the source tissue can help, for example, papers about plasma monoamine oxidase probably concern SSAO or DAO because MAO is absent from blood plasma.

The group of copper-containing amine oxidases (EC 1.4.3.6) includes at least two distinct mammalian enzymes: the enzyme that oxidizes primary monoamines (SSAO) and diamine oxidase (DAO). Again, the tissue studied can be a guide since DAO is present in high levels in the intestine, kidney, thymus gland and placenta and relatively high activities of SSAO are associated with cardiovascular smooth muscle cells, adipose tissue, lung, and cartilage. However, these distributions are far from exclusive and both enzymes may be found in blood plasma in variable amounts. The inhibition by carbonyl group reagents such as semicarbazide cannot be used to distinguish these enzymes.

It is reasonable to ask how this confusion arose in the first place. The reason for this departure from the rules is that EC 1.4.3.4 and EC 1.4.3.6 were originally distinguished as monoamine oxidase (also known as tyramine oxidase or tyraminase) and diamine oxidase (DAO, also known as histaminase). Those names were changed after the discoveries that MAO could catalyze the deamination of some diamines and that the DAO from some sources was active toward some monoamines but exhibited little activity toward histamine. Resistance to change may be the main reason for the persistence of this unsatisfactory state, but confusion on the part of both students and researchers about which results refer to what enzyme indicates it is clearly time to clarify the issue.

It is relatively simple to deal with the reclassification of DAO because it has a clear specificity for some amines that serve as poor or nonfunctioning substrates for the other two enzymes. The problem is distinguishing SSAO from MAO because they both catalyze the oxidation of primary amines. The problem is compounded by the species differences in their activities and the specificities of SSAO, along with difficulties in discerning whether publications about semicarbazide-sensitive amine oxidases actually discuss SSAO or DAO.

The nonphysiological amine benzylamine was shown to be a good substrate for mammalian enzymes. Because of this, plasma SSAO has been called benzylamine oxidase. However, benzylamine is also a substrate for DAO and one of the two isoenzymes of MAO. Substrates that have been shown to be oxidized by SSAO and not by MAO include methylamine and aminoacetone. Thus, one of those might be used in a name, but that would tend to obscure the broad substrate specificity. Furthermore, a specific methylamine oxidase, also a copper-containing quinoprotein, is present in some bacteria and fungi (Zhang et al., 1993).

The most consistent difference between MAO and SSAO appears to be the ability of the former to catalyze the oxidation of secondary amines such as adrenaline and some tertiary amines. This can be used as a clear specificity difference to distinguish the enzymes. Therefore, the proposal is to divide EC 1.4.3.6 into (1) diamine oxidase and (2) primary amine oxidase, and assign new numbers to both. Assignment of the EC 1.4.3.6 number to one of them would simply cause more confusion.

The accepted name of an enzyme in the IUBMB enzyme system is usually the name most commonly cited in the literature, provided that the name is not incorrect or ambiguous. As discussed *semicarbazide-sensitive amine oxidase* is clearly an ambiguous name and cannot be used without adding to the confusion that already exists. *Primary amine oxidase* has been proposed as the accepted name, but alternative suggestions are welcome. One obvious issue related to the accepted name is that the logical PAO abbreviation is commonly used for polyamine oxidase. However, many commonly used abbreviations are used for more than one enzyme. For example, DAO is also commonly used as an abbreviation for D-amino acid oxidase and LOX has been used to indicate lipoxygenases. The IUBMB does not concern itself with abbreviations, taking the view that it would be fruitless to try to guide authors about trivial abbreviations of enzyme names. Providing that enzymes are adequately defined, authors should be free to choose whatever abbreviations suit their purposes. In this chapter, we will use PrAO to indicate primary amine oxidase.

The proposed changes to the present EC classifications of DAO and PrAO (SSAO) are listed in Table 2.3. These also necessitate some revisions to the description of MAO; see Table 2.4. These changes will be available for public review and comment at http://www.enzyme-database.org/newenz.php. This is far from the end of the story since additional revisions will be required for the entries of other amine oxidases. Suggestions for new enzyme names, corrections, and amendments to existing entries may be submitted via forms available at http://www.enzyme-database.org/forms.php.

2.3 OTHER MAMMALIAN COPPER-CONTAINING AMINE OXIDASES

In addition to the known amine oxidases, human pregnancy plasma has been reported to contain a semicarbazide-sensitive amine oxidase with a somewhat different specificity than DAO or PrAO (Bardsley et al., 1974a; Seiler, 2000). However, little else is known about the activity. The major copper-containing protein ceruloplasmin has several functions including an amine oxidase activity (Hellman and Gitlin, 2002; Healy and Tipton, 2007). The mechanism of this process appears complex and produces water rather than H_2O_2. However, the activity level is relatively low, with an optimum pH some two units below neutrality. Convincing evidence that it makes a significant contribution to amine oxidation under physiological conditions is lacking.

2.4 PRACTICAL DISCRIMINATION

It is relatively simple to distinguish between MAO and the copper-containing enzymes by their sensitivity to inhibition by semicarbazide. Clorgyline, will inhibit both isoenzymes of MAO at sufficient concentrations and preincubation with 1 *mM*

TABLE 2.3
Proposed Changes to Entries Covering Mammalian Copper-Containing Amine Oxidases

EC 1.4.3.6

Deleted entry

Amine oxidase (copper-containing). This was classified on the basis of cofactor content rather than reaction catalyzed and is now known to contain enzymes with different activities. It has been replaced by two enzymes, EC 1.4.3.21, primary amine oxidase, and EC 1.4.3.22, diamine oxidase. [EC 1.4.3.6 created 1961, modified 1983, modified 1989, deleted 2008.]

EC 1.4.3.21

New entry

Accepted name: primary amine oxidase.

Reaction: $RCH_2NH_2 + H_2O + O_2 = RCHO + NH_3 + H_2O_2$.

Other name(s): amine oxidase; amine oxidase (copper-containing); *benzylamine oxidase*; copper amine oxidase; Cu-amine oxidase; semicarbazide-sensitive amine oxidase; *SSAO*; *polyamine oxidase*; *amine oxidase (pyridoxal containing)*; *diamine oxidase*; *diamino oxhydrase*; monoamine oxidase; *histamine deaminase*; *histamine oxidase*; plasma monoamine oxidase.

Systematic name: primary amine:oxygen oxidoreductase (deaminating and copper-containing).

Comments: A group of enzymes oxidizing primary monoamines with little or no activity toward diamines such as histamine and secondary and tertiary amines. They are copper quinoproteins (2,4,5-trihydroxyphenylalanine quinone) and sensitive to inhibition by carbonyl-group reagents, such as semicarbazide. They differ from EC 1.4.3.4, monoamine oxidase, in their inhibition by semicarbazide and inactivity toward secondary and tertiary amines. In some mammalian tissues the enzyme also functions as a vascular adhesion protein (VAP-1).

EC 1.4.3.22

New entry

Accepted name: diamine oxidase.

Reaction: histamine + H_2O + O_2 = imidazole-4-acetaldehyde + NH_3 + H_2O_2.

Other name(s): diamine oxidase; diamino oxhydrase; *histaminase*; *histamine deaminase*; *semicarbazide-sensitive amine oxidase*; *SSAO*.

Systematic name: histamine:oxygen oxidoreductase (deaminating) (copper-containing).

Comments: A group of enzymes oxidizing diamines such as histamine and also some primary monoamines but with little or no activity toward secondary and tertiary amines. They are copper quinoproteins (2,4,5-trihydroxyphenylalanine quinone) and sensitive to inhibition by carbonyl-group reagents such as semicarbazide.

Note: To facilitate searching, all other names used for an enzyme entry are retained in revised entries. Italics are used here to indicate names that are misleading; underlined names are ambiguous.

clorgyline for 30 min at 37°C is sufficient to inhibit MAO in tissue samples (Tipton et al., 2000). Furthermore, adrenaline appears to be entirely specific as a substrate for MAO. Methylamine appears to be a suitable discriminatory substrate for PrAO. Cadaverine has been used as a specific substrate for DAO, since PrAO does not generally oxidize it. Cadaverine has also been used for the assay of LOX (Langford

TABLE 2.4
Proposed Modifications to Nomenclature of Monoamine Oxidase

EC 1.4.3.4

Accepted name: monoamine oxidase.

Reaction: $RCH_2NHR' + H_2O + O_2 = RCHO + R'NH + H_2O_2$.

Other name(s): monoamine oxidase; tyramine oxidase; tyraminase; amine oxidase; adrenalin oxidase; epinephrine oxidase; MAO; serotonin deaminase; adrenaline oxidase; polyamine oxidase; spermidine oxidase; spermine oxidase.

Systematic name: amine:oxygen oxidoreductase (deaminating) (flavin-containing).

Comments: A flavoprotein (FAD). Acts on primary amines, and also on some secondary and tertiary amines. It differs from EC 1.4.3.x, primary-amine oxidase, in oxidizing secondary and tertiary amines and inactivity toward methylamine. Unlike EC 1.4.3.21 and 1.4.3.22, it is not inhibited by semicarbazide. [EC 1.4.3.4 created 1961, modified 1983 (EC 1.4.3.9 created 1972, incorporated 1984), modified 2008.]

et al., 1999). Putrescine may be a more appropriate DAO substrate to use if LOX is also present.

A plethora of amine oxidase assays have been reported (Tipton et al., 2000). Those based on oxygen utilization or H_2O_2 formation should be effective for all substrates. Others such as the spectrophotometric determination of benzylamine oxidation (Tabor et al., 1954) and the coupled spectrophotometric assay for methylamine oxidation (Lizcano et al., 2000a) are restricted to use with a single substrate. Although the spermine and spermidine polyamines are substrates for PrAO and DAO as well as polyamine oxidase (PAO), the cleavage sites differ, with DAO and PrAO oxidatively deaminating at the amino termini and PAO cleaving at the internal secondary amine positions (Lee and Sayre, 1998; Seiler, 2000). Thus, any assay that determines deaminated products would reveal the involvement of PAO but would not discriminate DAO and PrAO.

2.5 FUNCTIONS OF AMINE OXIDASES

In addition to their amine oxidase functions that include activities toward a number of xenobiotics (Strolin-Benedetti et al., 2007), the copper-containing amine oxidases appear to be involved in several cellular processes as shown in Table 2.5. We now know it is very common for enzymes to have alternative functions unconnected to their catalytic roles (Jeffery, 2003; Tipton et al., 2003). Although, at first sight, PrAO seems particularly busy in this respect (O'Sullivan et al., 2004), several of these alternative activities appear to be linked to the formation of H_2O_2 as a product of amine oxidation.

2.6 OTHER FUNCTIONS INVOLVING H_2O_2

2.6.1 Promotion of Glucose Uptake

Hydrogen peroxide is known to mimic the effects of insulin by causing a recruitment of the intracellular GLUT4 glucose transporter to cell surfaces, although relatively

TABLE 2.5
Alternative Functions of Copper-Containing Amine Oxidases

Function	DAO	LOX	SSAO	Role of H_2O_2
Amine oxidation	Yes	Yes	Yes	Reaction product
Vascular adhesion	No	No	Yes (endothelial)	Yes
Chemotaxis	?	Yes	?	Yes
Glucose transport	?	?	Yes	Yes
Adipocyte maturation	?	?	Yes	Yes
Extracellular matrix	?	Yes	Yes	Yes
Imidazoline binding	Yes	?	Yes	No
Cell death	Yes	Yes	Yes	Yes

high concentrations are necessary and the addition of vanadate is also required in some systems. Substrate oxidation by PrAO has been shown to stimulate glucose uptake by adipocytes and this effect is inhibited by both semicarbazide and catalase, consistent with the involvement of PrAO activity (Enrique-Tarancón et al., 2000). This stimulation of glucose transport is mediated through the phosphatidylinositol 3-kinase (PI3K) signalling pathway, after tyrosine phosphorylation of the insulin receptor substrate proteins IRS-1 and IRS-3 (Enrique-Tarancón et al., 2000).

Thus, it appears that the generation of H_2O_2 during the oxidation of substrates by PrAO stimulates glucose uptake by recruiting GLUT4 to cell surfaces. Similar effects have been reported in cells derived from some other insulin-sensitive tissues (El Hadri et al., 2002) and the stimulation of glucose uptake is paralleled by an insulin-like antilipolytic effect. It has been suggested that methylamine, formed from MAO-catalyzed breakdown of the stress hormone adrenaline may serve as the physiological substrate for this process (McDonald et al., 2007).

This ability to stimulate glucose uptake is not a specific function of PrAO, since the hydrogen peroxide formed in the MAO-catalyzed reaction can do the same and we have no reason to suppose that other H_2O_2-forming enzymes may not do likewise. However, since PrAO is a cell membrane-bound enzyme, it is not surprising to find a small proportion associated with the endosomal vesicles that contain GLUT4 and this may suggest a more specific function for this enzyme (McDonald et al., 2007). It is interesting that down-regulation of PrAO expression by treatment of mice with TNF-α, which is synthesized in fat cells and known to be involved in obesity-linked insulin resistance, resulted in a decrease in amine stimulated glucose uptake by adipocytes (Mercier et al., 2003).

2.6.2 Cell Death

All these enzymes produce H_2O_2 as a product of amine oxidation. Although it is an important cellular signaling molecule, H_2O_2 is toxic at higher concentrations and linking it to cell death under pathophysiological conditions has become a popular pastime. As shown in Table 2.2, several other oxidative enzymes also produce H_2O_2. Under certain conditions, these amine oxidases can generate sufficient H_2O_2 to cause apoptosis in model systems (Houen, 1999; Kunduzova et al., 2002; Toninello et al., 2006; Hernandez et al., 2006; Averill-Bates et al., 2008) but their relative importance and possible significance to human diseases remain to be established.

A further factor is that the aldehyde products are potentially toxic and may also contribute to cell toxicity (Deng et al., 1998; Houen, 1999; Bonneau and Poulin, 2000). All aldehydes are capable of reacting with free amino groups, such as the ε-amino of lysine in protein, to form a Schiff's base, subsequent methylene bridges, and then produce irreversible covalently cross-linked complexes between proteins, as well as between proteins and single-stranded DNA (Bolt, 1987; O'Brien et al., 2005). Formaldehyde, rather than H_2O_2 formation from methylamine oxidation has been reported as the dominant factor in apoptosis in cell lines (Hernandez et al., 2006; Solé et al., 2008) and methylglyoxal, formed from aminoacetone, is known to be highly toxic in this respect (Hiraku et al., 1999). It should be emphasized that copper amine oxidases exhibit nothing peculiar in this respect and the aldehydes derived from flavin-containing amine oxidases are also toxic (Burke et al., 2004; Li et al., 2003).

2.6.3 Vascular Adhesion

Endothelial PrAO also functions as a vascular-adhesion protein (VAP-1). VAP-1 adhesion has, perhaps not surprisingly, been studied only in some tissues such as endothelial cells. However, with the exception of retinal PrAO, humans and rodents appear to carry only a single gene for PrAO (Schwelberger, 2007), which suggests further complexities in the processing and functions of this enzyme. Certainly the name SSAO-VAP-1 that is sometimes used for this compound should not be applied to the enzyme from other tissues that have not been shown to have vascular adhesion functions.

The natural components of the lymphocyte cell surfaces to which VAP-1 binds remain uncertain. However, L-lysine and amino sugars, studied as model cell surface components, bind reversibly to VAP-1. The binding is saturable and specific, since D-lysine and *N*-acetyl amino sugars do not bind in this way. Hydrogen peroxide is necessary for this adhesion process and the H_2O_2 formed during enzyme-catalyzed substrate oxidation is much more effective in promoting binding than that added externally (O'Sullivan et al., 2003a; O'Sullivan et al., 2007; Olivieri et al., 2007). Kinetic studies suggest that these model compounds bind to the PrAO–H_2O_2 complex rather than the free enzyme. It is possible that the conformational changes reported during PrAO action may be involved in this process. If the H_2O_2 necessary for the VAP-1 adhesion process is derived from substrate oxidation, the amines involved remain to be identified, since neither the amino sugars nor L-lysine are oxidized by the enzyme to a detectable extent. However, methylamine resulting from metabolism of the adrenaline stress hormone adrenaline (McDonald et al., 2007) may be a reasonable candidate.

2.6.4 Cellular Maturation and Extracellular Matrix Formation

The role of LOX in the deposition and maintenance of the extracellular matrix via collagen and elastin cross linking is well established. LOX is also present in cell nuclei in amounts that vary with the stage of cell development; one of its actions appears to be the stimulation of mRNA for collagen and elastin mRNA transcription (Kagan and Li, 2003). PrAO also appears to be necessary in extracellular matrix development and maintenance, specifically in the development of normal elastin in vascular smooth muscle (Langford et al., 2002; Göktürk et al., 2007). It is also important for adipocyte maturation (Mercier et al., 2001; Fontana et al., 2001). The H_2O_2, formed as a result of substrate oxidation by PrAO, appears to be necessary for this process, whereas that formed by MAO-catalyzed substrate oxidation was much less effective (Mercier et al., 2001). This may reflect the differences in the locations of the enzymes.

2.6.5 Signalling

LOX, associated with the extracellular matrix, also has a chemotactic function in inducing the directional migration of vascular smooth muscle cells. The process requires active LOX and is inhibited by catalase, indicating the involvement of H_2O_2 (Li et al., 2000). Since PrAO is also an extracellular enzyme, this suggests that it may also be involved in H_2O_2-mediated signalling processes.

2.6.6 Self-Regulation

Under some conditions the H_2O_2 formed during PrAO-catalyzed amine oxidation may lead to its irreversible inhibition. It has been suggested that this may result from the oxidation of sulfydryl groups in the enzyme to form a vicinal disulfide bond (Jakobsson et al., 2005), although oxidation of the cofactor may also occur (Lee at al., 2001a). It has been suggested that this may play an important role in modulating its actions (Jakobsson et al., 2005; Pietrangeli et al., 2004), but the extent to which such inactivation may occur *in vivo* is unknown.

2.7 DRUG BINDING: ACTIVITY NOT INVOLVING H_2O_2

DAO, PrAO, and MAO are reported to be binding proteins for several imidazoline and guanidine compounds (Carpéné et al., 1995; Holt et al., 2004) that are inhibitors of sodium-transporting systems and have been used as diuretic and antihypertensive agents. The relative contributions of each of these enzymes in this respect is unknown but the plasma and endothelial surface locations of PrAO may make it more important for binding such drugs in the circulation. However, the enzymes from different sources differ considerably in their binding abilities (Raddatz et al., 1997; Lizcano et al., 1998). The significance of this binding, which may either activate or inhibit amine oxidase activity (Holt et al,. 2004; Bour et al., 2006), for those taking drugs of this type remains unclear.

2.8 CONCLUSIONS

Separating the two amine oxidases from their unfortunate classification union as EC 1.4.3.6, amine oxidase (copper-containing) and dispensing with the ambiguous semicarbazide-sensitive amine oxidase (SSAO) name should help to reduce the confusion that now exists in the literature. It would be a great relief to know which amine oxidase was under study simply from reading the title of a paper. That is possible for most other enzymes and there is no reason why amine oxidases should be different.

All the copper-containing amine oxidases form hydrogen peroxide as a product of amine oxidation. This product is an important signalling molecule and it appears that the different behaviors of these enzymes and monoamine oxidase may be related to their intracellular and extracellular localizations. The nature of the amine substrates necessary for H_2O_2 production is in most cases unknown and it is not clear whether changes in their levels or those of the amine oxidases are the most important factors regulating these effects. Since the metabolic functions of these enzymes are important, it will be an interesting challenge to determine the relative importance of each of these processes under different physiological conditions.

The use of inhibitors *in vivo* may help in this respect, but such studies have to date been hampered by the absence of a readily available specific inhibitor. Aminoguanidine was used in several studies but it is known also to inhibit other enzymes including arginine kinase, aspartate aminotransferase, and the inducible form of nitric oxide synthase, and it is therefore difficult to interpret results obtained from its use. Furthermore, although PrAO inhibitors have been reported to be effective anti-inflammatory agents (Salter-Cid et al., 2005; Yraola et al., 2007a), the multiple functions of this enzyme make the side effects and longer-term consequences of their use hard to predict. Indeed, PrAO over-expression (Göktürk et al., 2003a; Stolen et al., 2004a) and knock-out (Mercier et al., 2006) have both been reported to result in vascular damage among other complications.

ACKNOWLEDGMENTS

We are grateful to Science Foundation Ireland for support and to Peter Knowles and Bill McIntire for helpful advice about amine oxidase diversity.

3 Cofactors of Amine Oxidases
Copper Ion and Its Substitution and the 2,4,5-Trihydroxylphenylalanine Quinone

Shinnichiro Suzuki, Toshihide Okajima, Katsuyuki Tanizawa, and Minae Mure

CONTENTS

3.1 COPPER ION AND ITS SUBSTITUTION

3.1.1 PREFACE

Copper amine oxidases [amine:oxygen oxidoreductase (deaminating); EC 1.4.3.6; CAOs] contain a redox-active organic cofactor, 2,4,5-trihydroxyphenylalanine quinone (TPQ) that is essential for catalyzing the oxidation of primary amines by dioxygen to the corresponding aldehydes, ammonia, and hydrogen peroxide. The catalytic mechanism can be separated into two half-reactions: the enzyme reduction by the amines (reductive half-reaction; Equation 3.1) and the subsequent enzyme reoxidation by dioxygen (oxidative half-reaction; Equation 3.2):

$$CAO_{ox}(TPQ_{ox}) + R\text{-}CH_2\text{-}NH_3^+ \rightarrow CAO_{red}\text{-}NH_2(TPQ_{aq}) + R\text{-}CHO \quad (3.1)$$

$$CAO_{red}\text{-}NH_2 + O_2 + H_2O \rightarrow CAO_{ox} + NH_4^+ + H_2O_2 \quad (3.2)$$

Each monomer of a dimeric CAO also encompasses one Cu(II) ion as an inorganic cofactor, which is electron paramagnetic resonance (EPR)-detectable non-blue (or type 2) copper (Lindley, 2001). The yellowish pink color of CAOs stems from an intense absorption band in the 460- to 500-nm region assigned to TPQ_{ox}. The weak d–d bands of the Cu(II) centers are hidden in the intense TPQ_{ox} band, but the two circular dichroism (CD) extrema at 660 (+) and 810 (–) nm suggest the existence of d–d transitions (Suzuki et al., 1980). The EPR signals of CAOs have an axial character with $g_{//} > g_{\perp} > 2$, indicating that the Cu(II) ion has a tetragonal geometry.

The Cu(II) centers show a pentacoordinate square pyramidal geometry having three histidinyl imidazole and two water ligands under x-ray crystal structure analysis (Dawkes and Philips, 2001). To determine the role of Cu(II) in the half-reactions (Equations 3.1 and 3.2), several studies on transitional metal(II) substitution of CAOs have been carried out. In particular, the replacement of Cu(II) with Co(II) or Ni(II) has led to some effective approaches to understanding the metal centers of CAOs. The purpose of this chapter is to summarize the preparation and characterization of Cu(II)-depleted and metal(II)-substituted AOs and the reaction mechanisms of metal(II)-substituted AOs with the structures of their active centers. The effects of metal ions other than Cu(II) on TPQ biogenesis of CAO are dealt with in Chapter 8.

3.1.2 Preparation and Characterization of Cu(II)-Depleted Amine Oxidases

The active site Cu(II) ions in CAOs are very tightly bound, making their removal for metal depletion studies difficult. Moreover, despite considerable care at each stage of the Cu depletion process, some Cu contamination is inevitable due to the high affinity and selectivity of the apo-protein for Cu(II). Five methods of Cu depletion have been reported:

1. Fully Cu-depleted BSAO was anaerobically prepared by dialyzing pale yellow BSAO reduced by treatment of the native enzyme with 50 *mM* sodium dithionite, against 0.2 *M* sodium phosphate buffer (pH 7.2) containing 10 *mM* KCN for 8 to 12 hr at 4°C (Suzuki et al., 1981 and 1983). The dialysis against the KCN-containing buffer solution was carried out three times under dinitrogen atmosphere.
2. Cu depletion by dialysis of dithionite- or substrate-reduced BSAO against cyanide under anaerobic conditions was reported (Agostinelli et al., 1997 and 1998). A protein solution was dialyzed at 5°C against carefully de-aerated 0.1 *M* potassium phosphate buffer (pH 7.2). Added to the solution was solid potassium cyanide (final concentration of 10 *mM*). After about 1 hr, solid sodium dithionite or benzylamine (BZA) was added to a concentration of 1.0 *mM* and dialysis was continued for 3 to 24 hr, followed by final dialysis against three buffer changes.

3. Half-depleted Cu BSAO was prepared by dialyzing BSAO against a solution containing 1.0 *mM* sodium *N,N*-diethyldithiocarbamate (DDC) and 1.0 *mM* sodium sulfide or 1.0 *mM* sodium dithionite in 0.1 *M* potassium phosphate buffer (pH 7.2) for 48 hr at 4°C (Morpurgo et al., 1987; Agostinelli et al., 1994a). The solution previously de-aerated by dinitrogen was kept at minimum flow during dialysis. A brown precipitate of Cu(II)–DDC was removed by centrifugation. Excess DDC, sulfide, and dithionite were removed by dialysis.
4. Fully depleted Cu LSAO was prepared by dialyzing LSAO in bidistilled water against 10 *mM* DDC for 6 hr at 4°C. The resulting solution was centrifuged at 105,000 *g* for 2 hr (Padiglia et al., 1999a). The supernatant was dialyzed against bidistilled water, then centrifuged at 15,000 rpm for 30 min.
5. In an anaerobic chamber, 0.1 *M* potassium phosphate buffer (pH 7.2) containing HPAO and 1 *mM* methylamine•HCl was introduced into a Slide-a-Lyzer cassette (10-kDa cutoff) with a syringe (Mills and Klinman, 2000). The Slide-a-Lyzer cassette was placed in 0.1 *M* potassium phosphate buffer (pH 7.2) containing 1 *mM* methylamine•HCl, 1 *mM* KCN, and 1 *mM* sodium ethylene diaminetetraacetate at 4°C. The dialysis buffer was changed in the anaerobic chamber after 6 to 16 hr. The protein was kept through four changes of buffer. After dialysis against the second buffer, the Cu-depleted HPAO was removed from the Slide-a-Lyzer with a syringe. All buffers were degassed prior to use and phosphate buffer was passed through a column of chelating resin to avoid metal contamination.

The characteristics of Cu-depleted BSAO, LSAO, and HPAO are shown in Table 3.1. The relative activities and residual Cu contents of Cu-depleted AOs indicate the difficulty of achieving complete Cu depletion from enzymes because of tight binding of Cu to the active site and some Cu contamination during the depletion processes. Cu-depleted LSAO was prepared by the treatment of DDC (a copper ion chelator) only (method 4).

On the other hand, the treatment of BSAO with DDC and dithionite or sulfide yielded the half-depleted enzyme (method 3), and its enzyme activity was decreased to 16 to 18% by the treatment with dithionate and to <2% by treatment with sulfide. The Cu full removal from BSAO was carried out by reducing the protein with dithionite (a powerful reducing agent) and dialyzing the reduced protein against cyanide (a cuprous ion chelator) (method 1), but the Cu depletion from BSAO by cyanide and BZA (a substrate for BSAO) gave a residual Cu of 0.35 mol atom/subunit (method 2).

Cu-depleted AGAO prepared by method 1 contained only 0.004 mol atom of Cu per mol of subunit (Kishishita et al., 2003)—significantly lower than the remaining Cu contents obtained in the other experiments. Cu-depleted LSAO, BSAO, and AGAO almost quantitatively recovered the Cu contents and enzyme activities on incubation with Cu(II) ion. Harsh reducing conditions such as dithionite treatment damage HPAO, such that reconstitution with Cu did not restore enzyme activity (Mills and Klinman, 2000). Therefore, methylamine, a substrate for HPAO, was used to reduce HPAO instead of dithionite, but the amount of residual copper was found

TABLE 3.1
Characteristics of Cu-Depleted Amine Oxidases

Cu-Depleted AO	Relative Activity (%)[1]	Cu Content (mol atom/ subunit)	Method
Cu fully-depleted BSAO	< 1	0.01 to 0.02	1[6]
Cu fully-depleted BSAO	< 2	0.05 ± 0.01	1[7]
Cu-depleted BSAO($S_2O_4^{2-}$)[2]	2.5	0.11	2[8]
Cu-depleted BSAO(BZA)[3]	4.2	0.35	2[8]
Cu fully-depleted AGAO	0.06	0.0041 ± 0.0002	1[9]
Cu half-depleted BSAO($S_2O_4^{2-}$)[2]	16 to 18	0.92 to 1.0	3[10]
Cu half-depleted BSAO(Na_2S)[4]	< 2	0.94	3[10]
Cu fully-depleted LSAO	–	0.08	4[11]
Cu-depleted HPAO	5.05	0.112	5[12]
Cu-depleted HPAO + PHZ[5]	0.94	Not detectable	5[12]

Notes: [1] Activities of corresponding native CAOs are taken as 100%; [2] Dithionite ($S_2O_4^{2-}$) was used as a reducing agent; [3] Benzylamine (BZA) was used as a reducing agent;.[4] Na_2S was used as a reducing agent; [5] PHZ was added to a five-fold molar excess over protein; [6] Suzuki et al., 1983; Suzuki et al., 1981; [7] Agostinelli et al., 1994a; [8] Agostinelli et al., 1998; [9]Kishishita et al., 2003; [10] Morpurgo et al., 1987; Agostinelli et al., 1994a; [11] Padiglia et al., 1999a; [12] Mills and Klinman, 2000.

to vary slightly, ranging from 10 to 20% of the total Cu (method 5). In Table 3.1, Cu-depleted HPAO initially shows 5% of its original activity, which is attributed to the residual Cu in the protein. The addition of PHZ to Cu-depleted HPAO inactivates the active sites containing residual Cu and gives a relative activity of 0.94%.

The visible absorption spectrum of pink Cu-depleted LSAO prepared without treatment of dithionite shows a broad absorption peak at 490 nm. Native LSAO also absorbs specifically in the visible region (498 nm) owing to the oxidized TPQ cofactor (Padiglia et al., 1999a). However, Cu fully depleted BSAO treated with dithionite exhibits the absence of any distinct visible absorption band, although yellowish pink native BSAO possesses a visible absorption band at 476 nm ($\varepsilon = 3{,}800\ M^{-1}cm^{-1}$) (Suzuki et al., 1983). BSAO and pink Cu-depleted BSAO containing oxidized TPQ (TPQ_{ox}) yielded deep yellow phenylhydrazone products showing intense absorption bands at 448 ($\varepsilon = 40{,}000\ M^{-1}cm^{-1}$) and 432 nm ($\varepsilon = 28{,}000\ M^{-1}cm^{-1}$), respectively.

The absence of Cu shifts the intense band from 448 to 432 nm. On the other hand, the treatment of Cu fully depleted BSAO containing reduced TPQ (TPQ_{red}) with 10 equiv PHZ did not give the intense phenylhydrazone band. In Cu-depleted HPAO, however, PHZ did not react even with TPQ_{ox}, whose spectrum shows an absorption band around 460 nm (Mills and Klinman, 2000). Since TPQ_{ox} in AGAO was also reduced to TPQ_{red} by the Cu removal from the holo-enzyme under the reduction with dithionite, Cu-depleted AGAO exhibited the loss of the 480-nm absorption band diagnostic of TPQ_{ox} (Kishishita et al., 2003). When Cu fully depleted AGAO was air-oxidized in a weak alkaline buffer (pH 9.0) or unfolded in 8 *M* urea, TPQ_{red} in the Cu-depleted enzyme was oxidized to bring about reappearance of the absorption band of TPQ_{ox}. The absorption spectrum of Cu-depleted HPAO shows a unique absorption band having two peaks around 435 and 465 nm (Mills and Klinman, 2000), which is the same as that reported for the semiquinone form (TPQ_{sq}) of TPQ (Padiglia et al., 1999a; Dooley et al., 1991). The reduction or oxidation of the Cu-depleted enzyme with 1 equiv dithionite or exposure to air, respectively, led to the disappearance of the absorption band of TPQ_{sq}.

3.1.3 Preparation and Characterization of Metal(II)-Substituted Amine Oxidases

Metal-substituted AOs were generally prepared by incubation of Cu-depleted AOs in tris(hydroxymethyl)aminomethane (Tris)–HCl buffers with 0.5 to 2 *mM* metal ion at 4 to 30°C for 3 to 48 hr and successive dialysis against the same buffer to remove excess metal ion (Agostinelli et al., 1994a and 1998; Padiglia et al., 1999a; Kishishita et al., 2003). Another metal reconstitution was carried out by several dialyses of Cu-depleted AO against a Tris–HCl buffer containing 2 *mM* metal ion, followed by dialyses against the same buffer (Suzuki et al., 1981 and 1983). Co(II) reconstitution was carried out under anaerobic conditions to avoid oxidation of Co(II) to Co(III).

Table 3.2 shows the characteristics of metal(II)-substituted BSAO, LSAO, HPAO, and AGAO. The early conclusion that Co(II) BSAO is catalytically competent was questioned by some investigators (Dooley et al., 1993) who correlated the activities

TABLE 3.2
Characteristics of Metal(II)-Substituted Amine Oxidases

Metal(II)-Substituted AO	Relative Activity[7] (%)[8]	Metal Content (mol atom/subunit) [Metal]	Cu Content (mol atom/subunit)
Co SAO[1]	13	0.73 [Co]	–
Ni BSAO[1]	2	0.80 [Ni]	–
Zn BSAO[1]	1	0.40 [Zn]	–
Co BSAO[2]	19 to 22	0.95 to 1.0 [Co]	0.11 to 0.33
Co BSAO[3]	15	1.05 ± 0.1 [Co]	0.025 ± 0.005
Co LSAO[4]	7	0.90 [Co]	–
Ni LSAO[4]	–	0.50 [Ni]	–
Zn LSAO[4]	–	0.50 [Zn]	–
Co HPAO[5]	19[9]	0.93 [Co]	0.162
Ni HPAO[5]	1.7[9]	0.71 [Ni]	0.163
Zn HPAO[5]	1.7[9]	0.86 [Zn]	0.156
Co AGAO[6]	2.2	1.060 ± 0.001 [Co]	0.0063 ± 0.0004
Ni AGAO[6]	0.9	0.800 ± 0.024 [Ni]	0.0088 ± 0.0002
Zn AGAO[6]	0.06	1.390 ± 0.004 [Zn]	0.0084 ± 0.0004

Notes: [1] Suzuki et al., 1983; [2] Agostinelli et al., 1998; [3] Agostinelli et al., 1994[a]; [4] Padiglia et al., 1999a; [5] Mills and Klinman, 2000; [6] Kishishita et al., 2003; [7] See Table 3.1 concerning activities of corresponding Cu-depleted AOs; [8] Activities of corresponding native CAOs are taken as 100%; [9] Cu-depleted HPAO was treated with PHZ before reconstitution with metal ion.

of the Co(II)-substituted enzymes with the presence of residual or adventitious Cu. However, recent papers on Co(II)-substituted AOs reported that Co(II) restores some significant activity to the enzyme (Agostinelli et al., 1998; Padiglia et al., 1999a; Mills and Klinman, 2000; Kishishita et al., 2003), while Zn(II) and Ni(II) were minimally active or inactive. Table 3.3 compares the kinetic parameters for native and corresponding metal(II)-substituted AOs, supporting the amine oxidase activities shown by Co(II) AOs. The major difference between native and Co(II)-substituted AOs is seen in the K_m of HPAO and in the k_{cat} of LSAO and AGAO. However, the reason for the difference observed in these AOs is unclear.

Visible absorption (AB), CD, and magnetic circular dichroism (MCD) spectra of Co(II)-substituted BSAO are shown in Figure 3.1 (Suzuki et al., 1981). The anaerobic incubation of Cu-depleted BSAO with Co(II) produced a colored solution. The pink color of Co(II) BSAO results from an intense absorption band at 470 nm, which is similar to that of native BSAO. The MCD spectrum displaying a small positive band around 480 nm and negative band around 580 nm suggests that the high-spin Co(II) site having a tetrahedral geometry is located in BSAO. The CD peaks at 370 and 440 nm can be assigned to TPQ_{ox} and that around 540 nm is due to the d–d transitions of Co(II). The X-band EPR spectrum of Co(II) HPAO at 8 K shows the $g = 6$ and 2.7 features attributed to high-spin Co(II) ($S = 3/2$), which is 90% detectable by EPR (Mills and Klinman, 2000). This EPR spectrum has been interpreted as arising from Co(II) in an asymmetric tetrahedral or five-coordinate environment having significant rhombicity.

The dithionite treatment for the Cu depletion from AGAO gives TPQ_{red} showing the loss of the 483-nm absorption band. Rapid oxidation of TPQ_{red} linked in

TABLE 3.3
Kinetic Parameters for Native and Metal(II)-Substituted Amine Oxidases

Enzyme	Substrate	K_m / μM	k_{cat} / s^{-1}	pH
BSAO[1]	BZA	$(1.9 \pm 0.1) \times 10^3$	0.82 ± 0.03	7.2
Co(II) BSAO[1]	BZA	$(0.66 \pm 0.04) \times 10^3$	0.14 ± 0.01	7.2
LSAO[2]	Putrescine	200	155	7.0
Co(II) LSAO[2]	Putrescine	300	11	7.0
HPAO[3]	O_2[5]	10 ± 1	2.12 ± 0.04	7.1
Co(II) HPAO[3]	O_2[5]	680 ± 110	2.08 ± 0.2	7.1
AGAO[4]	2-PEA[6]	2.5 ± 0.0	75.7 ± 0.8	6.8
	O_2	20.8 ± 6.0	110 ± 7	–
Co(II) AGAO[4]	2-PEA[6]	1.9 ± 0.1	1.51 ± 0.01	6.8
	O_2	16.3 ± 5.8	1.24 ± 0.07	–
Ni(II) AGAO[4]	2-PEA[6]	3.8 ± 0.2	1.30 ± 0.02	6.8
	O_2	18.3 ± 2.4	1.13 ± 0.04	–

Notes: [1] Agostinelli et al., 1998; [2] Padiglia et al., 1999a; [3] Mills and Klinman, 2000; [4] Kishishita et al., 2003; [5] Determined with 1.6 *mM* methylamine (native) or 1.5 *mM* methylamine (Co HPAO); [6] 2-Phenylethylamine.

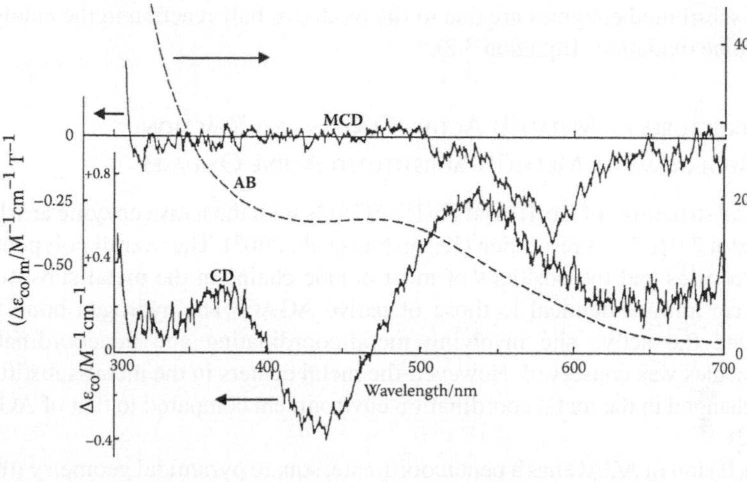

FIGURE 3.1 Absorption (AB), circular dichroism (CD), and magnetic circular dichroism (MCD) spectra of Co(II)-substituted BSAO. (Reproduced with permission from Suzuki et al., 1981.)

Cu-depleted AGAO was observed upon reconstitution with several divalent metal ions (Kishishita et al., 2003). The original metal ion, Cu(II), was the most effective (k_{obs} = 1.0 min^{-1} at 30°C and pH 6.8), and Co(II) and Ni(II) were also found effective, although at lower rates (Co(II), k_{obs} = 0.57 min^{-1}; Ni(II), k_{obs} = 0.065 min^{-1}). Zn(II), Ca(II), Mg(II), and Mn(II) did not support O_2 oxidation of the enzyme-linked TPQ_{red} even after incubation for 24 hr. The reactivities of TPQ_{ox} in Co(II) and Ni(II) AGAOs toward PHZ were similar to that in the native enzyme in which the hydrazone formation was completed within about 3 min. Moreover, upon anaerobic mixing of Co(II) or Ni(II) AGAO with excess 2-PEA at 5°C, the 483-nm absorption band derived from TPQ_{ox} disappeared with rate constants of about 700 and 300 s^{-1} for the Co(II) and Ni(II) enzymes, respectively. The absorption band of the native enzyme disappeared with a rate constant of >1000 s^{-1}. In these reactions, TPQ_{ox} is changed into the aminoquinol form (TPQ_{aq}) of the reduced cofactor (the IUPAC name is aminoresorcinol but aminoquinol has been traditionally used in the CAO field and therefore will be used in this chapter). These results are consistent with the generally accepted mechanism for the reductive half-reaction (Equation 3.1) in which the bound metal ion does not play a direct role in the catalytic reaction. In substrate-reduced AGAO, TPQ_{sq} exhibiting typical absorption bands at about 365, 440, and 470 nm, appears within 30 ms after mixing, resulting from the equilibrium between the Cu(II)/TPQ_{aq} and Cu(I)/TPQ_{sq} states. On the other hand, TPQ_{sq} is not detected in the substrate-reduced Co(II) and Ni(II) AGAOs (Hirota et al., 2001; Kishishita et al., 2003), and TPQ exists probably as TPQ_{aq}.

The one-electron transfer from TPQ_{aq} to the bound metal ion does not occur in the substrate-reduced, metal-substituted enzymes because of the energetically unfavorable Co(I) and Ni(I). Similar behavior was also observed in Co(II) LSAO (Padiglia et al., 1999a). Since the reductive half-reaction proceeds at a much faster rate than k_{cat} in all forms of AGAOs, the differences in the catalytic activities of the native and metal-substituted enzymes are due to the oxidative half-reaction in the catalytic cycle of amine oxidation (Equation 3.2).

3.1.4 Structures of Metal(II) Active Centers and Reaction Mechanisms of Metal(II)-Substituted Amine Oxidases

X-ray crystal structures of Co(II) and Ni(II) AGAOs with the native enzyme at 100 K were solved at 2.0 to 1.8 Å resolution (Kishishita et al., 2003). The overall polypeptide folding structures and the positions of most of side chains in the metal-substituted AGAOs were almost identical to those of native AGAO. The hydrogen bond network around the active site involving metal-coordinating and noncoordinating water molecules was conserved. However, the metal centers in the metal-substituted enzymes changed in the metal coordination environment compared to that of AGAO (Figure 3.2).

The Cu(II) ion in AGAO has a pentacoordinate, square pyramidal geometry (three imidazole groups (N–Cu, 2.2 to 2.3 Å), an equatorial water molecule (O(1)–Cu, 2.0 Å), and an axial water molecule (O(2)–Cu, 2.7 Å)). The Cu(II)-to-Wat2 distance longer than the Cu(II)-to-Wat1 distance is explained by the Jahn–Teller effect. Both metal ions in Co(II) and Ni(II) AGAOs possess an additional water ligand (Wat3) not observed in the native enzyme, showing hexacoordinate, octahedral geometries (N–Co and N–Ni, 2.0 to 2.4 Å; O–Co and O–Ni, 2.0 to 2.3 Å). Although the electron density for the TPQ cofactor determined at an ambient temperature was insufficient to define the orientation of the quinone group (Wilce et al., 1997), the electron densities for TPQ_{ox} in the three enzymes at 100 K clearly defined a single orientation of the quinone group with the C2=O atom closest to the metal ion and the C5=O atom in proximity to the catalytic base Asp 298 (Kishishita et al., 2003). This orientation of the TPQ ring is identical with those in the active forms of *E. coli* (Wilmot et al., 1999; Dawkes and Phillips, 2001) and *Hansenula polymorpha* (Li et al., 1998) CAOs and appears to be held by hydrogen bonds formed between the C2=O atom and Wat2 (2.6 to 2.7 Å). A structural comparison of the active sites of these three enzymes shows that TPQ_{ox} in Co(II) and Ni(II) AGAOs is held in a conformation similar to the TPQ_{ox} cofactor in native AGAO.

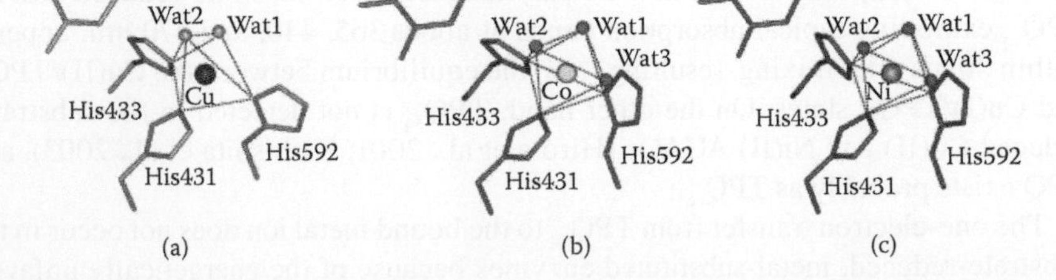

FIGURE 3.2 Metal coordination structures of (a) Cu(II) AGAO, (b) Co(II) AGAO, and (c) Ni(II) AGAO. PDB accession codes: Cu (1IU7), Co (1IQX), and Ni (1IQY). (Reproduced with permission from Kishishita et al., 2003.)

Figure 3.3 represents a proposed mechanism of the oxidative half-reaction by native and metal-substituted AGAOs (Kishishita et al., 2003). The oxidative half-reaction consists of the net transfer of $2e^-$ and $2H^+$ from TPQ_{aq} to dioxygen; the absorption band of TPQ_{aq} at 310 nm (Hirota et al., 2001) results from the $2e^-$ reduction of TPQ_{ox} by a substrate. Following the oxidation of TPQ_{aq}, the products are TPQ_{imq} and hydrogen peroxide. TPO_{sq} was not detected in the absorption spectra of substrate-reduced Co(II) and Ni(II) AGAOs, that is, the $1e^-$ transfer from TPQ_{aq} to these metal ions is unlikely to occur (*vide supra*). It is conceivable that the initial step in the oxidative half-reaction of Co(II) and Ni(II) AGAOs is the direct $1e^-$ reduction of dioxygen from TPQ_{aq} and does not involve a change in the redox state of the metal.

The initial reaction of TPQ_{aq} with O_2 is much slower in Co(II) and Ni(II) AGAOs than in the native enzyme, and most likely is the rate-limiting step in the oxidative half-reaction of the metal-substituted enzymes. In the case of the native enzyme, however, Cu(II) appears to play a key role in the initial step of the oxidative half-reaction, including the possibility of valence change of Cu (Kishishita et al., 2003). The final step in the oxidative half-reaction under single turnover conditions is the hydrolysis of TPQ_{imq} to regenerate TPQ_{ox}. Under the catalytic turnover conditions, this may not be on the reaction pathway, that is, TPQ_{imq} is considered to react directly with a substrate amine to form the substrate Schiff base (trans-imination reaction), as shown in Figure 3.3.

The hydrolyses of TPQ_{imq} to TPQ_{ox} in Co(II) and Ni(II) AGAOs are about 10^3 times as slow as that of the native enzyme (Hirota et al., 2001). These findings are

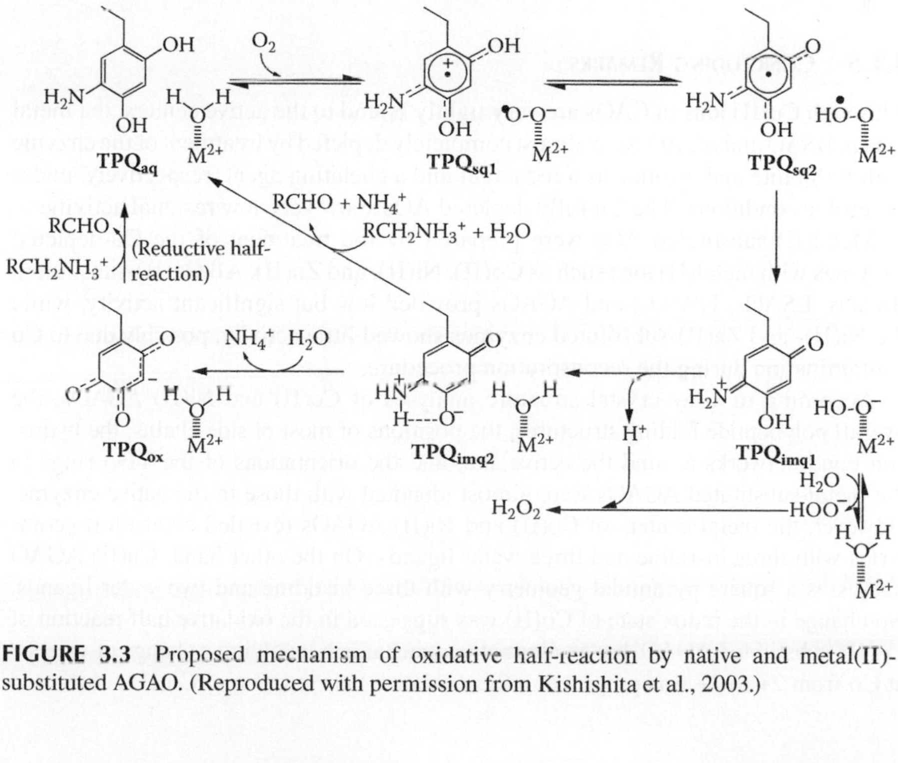

FIGURE 3.3 Proposed mechanism of oxidative half-reaction by native and metal(II)-substituted AGAO. (Reproduced with permission from Kishishita et al., 2003.)

consistent with the observation of a back reaction between the product aldehyde and TPQ_{aq} observed in the metal-substituted enzymes; no back reaction occurs in the native enzyme. The results from metal(II)-substituted AGAOs suggest that the native Cu(II) ion plays an essential role of catalysis in the electron transfer from TPQ_{aq} to dioxygen by providing a binding site for $1e^-$- and $2e^-$-reduced dioxygen and preventing the back reaction between the product aldehyde and TPQ_{aq}.

An alternative mechanism that involves a change of valence at Co(II) in the oxidative half-reaction of Co(II) HPAO has been reported (Mills and Klinman, 2000). In the active site containing TPQ_{aq} and Co(II) close to O_2 (TPQ_{aq}/Co(II)–O_2), the first electron transferred to dioxygen comes from Co(II), forming superoxide ($O_2^{\bullet-}$) bound to Co(III) (TPQ_{aq}/Co(III)–$O_2^{\bullet-}$). A second electron is transferred from TPQ_{aq} to the Co(III)–$O_2^{\bullet-}$ species, forming TPQ_{sq} and Co(III) peroxide (TPQ_{sq}/Co(III)–O_2^{2-}). The last electron transfer is from TPQ_{sq} to Co(III) ion to form TPQ_{ox} and Co(II) ion with release of H_2O_2. This Co(II)/Co(III) mechanism does not involve an initial electron transfer from TPQ_{aq} to the metal and is very different from the possible Cu(II) mechanisms.

A kinetic study of Co(II) HPAO reveals that the major cause of the reduced specific activity compared to the native enzyme was a 68-fold increase in the apparent K_m for oxygen, and saturation of the Co(II) enzyme with oxygen gave a k_{cat} equivalent to the native k_{cat} (Table 3.3). The difference in K_m suggests that the metal may play a role in the formation of a viable O_2 binding site and electrons are passed directly from TPQ_{aq} into pre-bound oxygen without the need for a prior reduction of the metal (i.e., Co(II)/Co(I) and Cu(II)/Cu(I) valence changes) (Su and Klinman, 1998).

3.1.5 Concluding Remarks

Although Cu(II) ions in CAOs are very tightly bound to the active centers, the metal ions in BSAO and AGAO were almost completely depleted by treatment of the enzyme with dithionite and cyanide as a reductant and a chelating agent, respectively, under anaerobic conditions. The Cu fully depleted AOs show very low residual activities.

Metal(II)-substituted AOs were prepared by the treatment of the Cu-depleted enzymes with metal(II) ions such as Co(II), Ni(II), and Zn(II). All Co(II)-substituted BSAOs, LSAOs, HPAOs, and AGAOs provided low but significant activity, while the Ni(II)- and Zn(II)-substituted enzymes showed little activity, possibly due to Cu contamination during the reconstitution procedure.

According to x-ray crystal structure analyses of Co(II) and Ni(II) AGAOs, the overall polypeptide folding structures, the positions of most of side chains, the hydrogen-bond networks around the active site, and the orientations of the TPQ rings in the metal-substituted AGAOs were almost identical with those in the native enzyme. However, the metal centers of Co(II) and Ni(II) AGAOs revealed octahedral geometries with three histidine and three water ligands. On the other hand, Cu(II) AGAO possesses a square pyramidal geometry with three histidine and two water ligands. No change in the redox state of Co(II) was suggested in the oxidative half-reaction of Co(II)-substituted AGAO, but an alternative mechanism involving a change of valence at Co from 2+ to 3+ was proposed in the oxidative half-reaction of Co(II)-substituted

HPAO. On the basis of the results of Co(II)-substituted AOs, it can be presumed that the Cu ion in CAOs catalyzes a direct electron transfer from TPQ_{aq} to dioxygen without its valence change and provides an electrostatically stabilized binding site for $1e^-$-reduced dioxygen, superoxide anion.

3.2 2,4,5-TRIHYDROXYLPHENYLALANINE QUINONE (TPQ)

Topaquinone or TPQ (2,4,5-trihydroxylphenylalanine quinone) is the organic cofactor of CAOs (Janes et al., 1990). TPQ is derived from a conserved tyrosine residue (T-X-N-Y-D/E) by post-translational modifications requiring only Cu^{2+} and dioxygen (Janes et al., 1992). The mechanistic details of the biogenesis of the TPQ cofactor will be discussed in Chapter 8. This chapter focuses on the chemistry of TPQ and how the active site environments of CAOs modulate their reactivity to be optimal for catalysis.

The 4-hydroxyl group of TPQ (Scheme 3.1, TPQ_{ox}; see Figure 3.4 for numbering) has an acidic pK_a (~3 in enzyme, ~4 in solution) due to the resonance delocalization and stabilization of the negative charge between the C2 oxygen and the C4 oxygen (Mure and Klinman 1993). Due to this resonance effect, the C3 proton is exchangeable with deuterium upon incubation with D_2O, as detected by ^1H NMR (Mure and Klinman 1993) as well as resonance Raman spectroscopy (Moenne-Loccoz et al.,

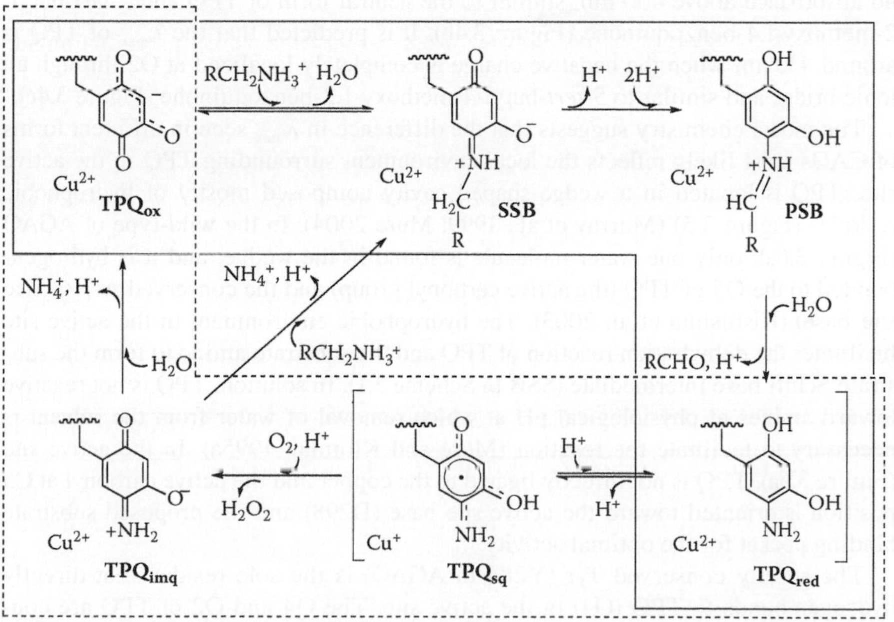

SCHEME 3.1 Proposed reaction mechanisms of CAOs. TPQ_{ox} = resting (oxidized) form of TPQ. SSB = substrate Schiff base intermediate. PSB = product Schiff base intermediate. TPQ_{red} = substrate reduced (aminoquinol) form of TPQ. TPQ_{sq} = semiquinone form of TPQ. TPQ_{imq} = iminoquinone form of TPQ. The reductive and oxidative half-reactions are shown in solid and dotted boxes, respectively.

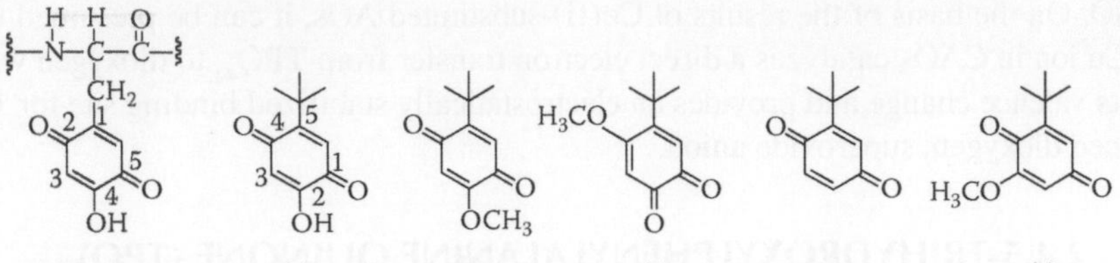

FIGURE 3.4 Structures of TPQ cofactor and model compounds.

1995). The broad absorbance around 480 nm of CAO arises from the resonance stabilized mono-anion form of TPQ. The λ_{max} of this absorption varies from 450 to 500 nm at physiological pH, depending on the origin of CAO and the effects of mutations in the active site.

Solution chemistry suggests an explanation for this observed variation as the λ_{max} of the resonance stabilized mono-anionic TPQ is solvent-dependent (Mure and Klinman 1995a). This effect is ascribed to the extent of charge delocalization between O2 and O4. The fully solvated anion in aqueous solution has a λ_{max} at 484 nm but for a naked anion in an aprotic polar solvent such as acetonitrile, the λ_{max} is at 498 nm. When the negative charge is completely localized at O4 through an ionic bridge (zwitteranionic structure), the λ_{max} is 372 nm (in acetonitrile) and there is no absorbance above 400 nm, similar to the neutral form of TPQ and 5-*tert*-butyl-2-methoxy-1,4-benzoquinone (Figure 3.4b). It is predicted that the λ_{max} of TPQ is around 420 nm when the negative charge is completely localized at O2 through an ionic bridge and similar to 5-*tert*-butyl-4-methoxy-1,2-benzoquinone (Figure 3.4c).

The model chemistry suggests that the difference in λ_{max} seen in different forms of CAOs most likely reflects the local environment surrounding TPQ in the active site. TPQ is located in a wedge-shaped cavity composed mostly of hydrophobic residues (Figure 3.5) (Murray et al., 1999; Mure 2004). In the wild-type of AGAO (Figure 3.6a), only one water molecule is found in the wedge, and it is hydrogen-bonded to the O5 of TPQ (the active carbonyl group) and the conserved Asp (active site base) (Kishishita et al. 2003). The hydrophobic environment in the active site facilitates the dehydration reaction of TPQ and the substrate amine to form the substrate Schiff base intermediate (SSB in Scheme 3.1). In solution, TPQ is not reactive toward amines at physiological pH at which removal of water from the solvent is necessary to facilitate the reaction (Mure and Klinman 1995a). In the active site (Figure 3.6a), TPQ is not directly ligated to the copper and the active carbonyl at C5 position is oriented toward the active site base (D298) and the proposed substrate binding pocket for the optimal activity.

The strictly conserved Tyr (Y284 in AGAO) is the sole residue that directly hydrogen bonds to TPQ (O4) in the active site. The O4 and O2 of TPQ are connected through a hydrogen bonding network including this Tyr and two water molecules, one of which is on the Jahn–Teller axis of copper ion (Figure 3.6a). The partial localization of the negative charge to either C2 or C4 positions results in some blue shift in the λ_{max}. A resonance Raman study on a D298A–AGAO mutant (λ_{max} = 450 nm) revealed that the C2 carbonyl group has more single bond

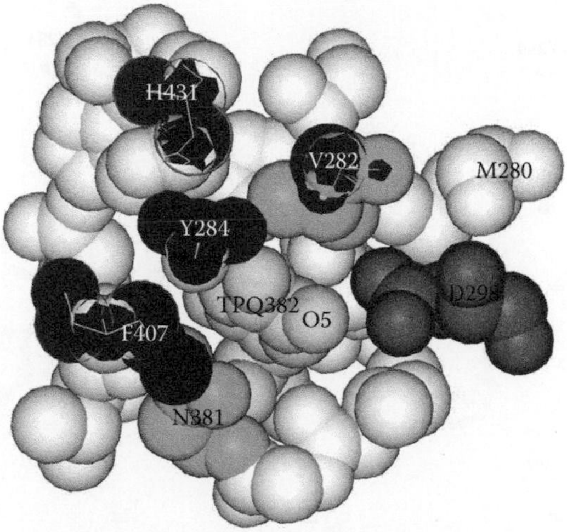

FIGURE 3.5 (See color insert following page 202.) Space-filling representation of region surrounding TPQ_{ox} (yellow) in the "active" conformation of wild-type AGAO. V282 and N381 forming a wedge-shape pocket (green). O5 of TPQ is facing D298 (red).

character than the wild-type, indicating that the electrons are partially localized on O2 (Chiu et al., 2006). In the Y369F mutant (Y364 corresponds to Y284 in AGAO) of ECAO where O4 is not hydrogen bonded to Y369F, the λ_{max} is 20 nm red-shifted from the wild-type (Murray et al., 2001), consistent with the fully delocalized anion.

The deprotonation of the 4-hydroxyl group of TPQ is essential to direct the reaction of amine to C5 carbonyl group. The C2 carbonyl group is more electron-rich due to the delocalization of the negative charge and is not reactive toward nucleophilic addition. It should be noted that the presence of an oxoanion at the C4 position also facilitates the formation of the substrate Schiff base (SSB in Scheme 3.1) by stabilizing the carbinolamine intermediate through hydrogen bonding interaction (Mure et al., 2002). A solution study (Mure and Klinman 1995a) has shown that in 5-*tert*-butyl-1,4-benzoquinone (Figure 3.4d) where such an interaction is expected to be absent, amines do not attack the carbonyl carbons but instead attack the ring carbon. 5-*tert*-Butyl-2-methoxy-1,4-benzoquinone (Figure 3.4b) and 5-*tert*-butyl-3-methoxy-1,4-benzoquinone (Figure 3.4e) both have UV-vis spectroscopic properties very similar to the neutral TPQ (Figure 3.4a). However, only the 2-methoxy group can support the substrate Schiff base formation in which ring amination occurs when the methoxy group is at the C3 position. Although the reaction of 5-*tert*-butyl-2-methoxy-1,4-benzoquinone (Figure 3.4b) with amines is much slower than that with TPQ, it still demonstrates that the hydrogen bond acceptor at C2 position (equivalent to C4 in the TPQ cofactor) is essential for Schiff base formation.

TPQ is connected to the peptide backbone of CAO through the methylene group that originates from the conserved precursor Tyr residue (Y382 in AGAO). The mobility of TPQ is modulated in the active site by a series of H-bonding interactions and by the wedge cavity for optimal activity as mentioned above (Mure 2004). TPQ

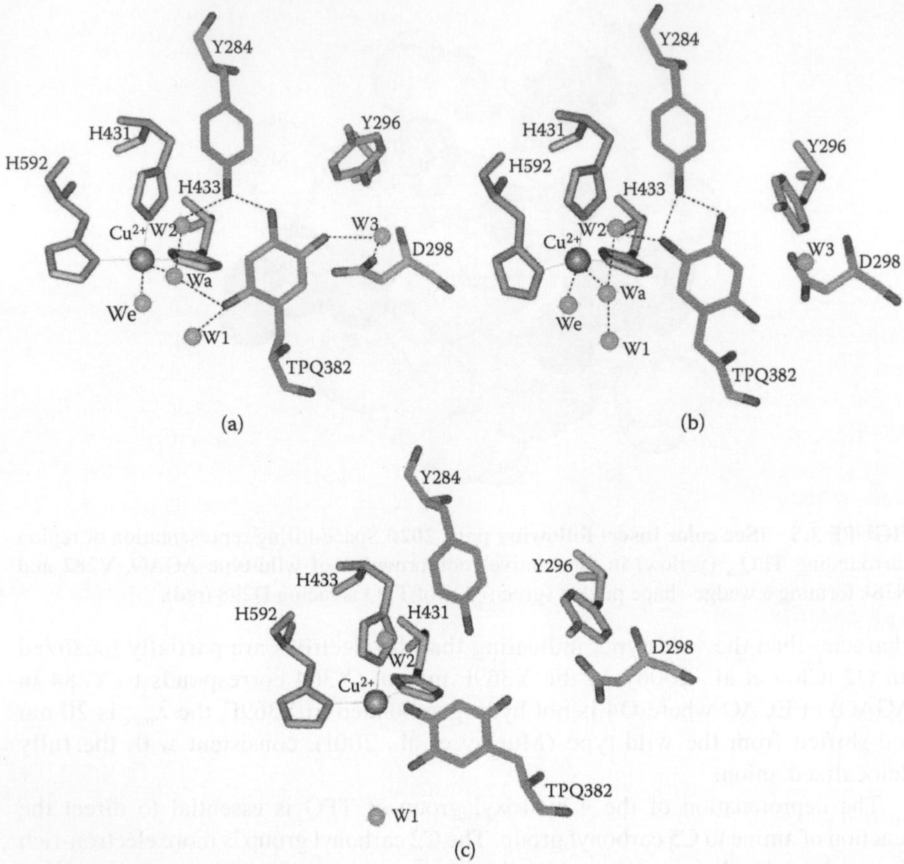

FIGURE 3.6 Active site structures of wild type-AGAO. (a) "Active" conformation of TPQ_{ox}. (b) "Flipped" conformation of TPQ_{ox}. (c) "On-copper" form of TPQ_{ox}. Y296 located at base of proposed substrate channel.

can pivot and rotate in the wedge cavity and can swing out of the wedge when it is freed from the hydrogen bond interaction with the conserved Tyr (Y284 in AGAO).

In the x-ray crystal structure of ECAO at pH 7.0, TPQ is in the active conformation, but the disordered electron density for TPQ suggests some limited mobility (pivoting) (Parsons et al., 1995). When crystals are grown in the presence of ammonium or lithium sulfates (>1.0 *M*) or at acidic pH (<5.0), TPQ is seen in altered conformations in which it has flipped (Figure 3.6b) or ligated to Cu^{2+} through O4, termed the on-copper form (Figure 3.6c) (Wilce et al., 1997). In the flipped form, the TPQ ring has rotated around Cα-Cβ, retaining the hydrogen bonding interaction between O4 and Y284 (Figure 3.6b). The crystals in which TPQ is bound to Cu^{2+} are colorless, indicating charge localization of the O4 anion through ligation to Cu^{2+}.

The mobility of TPQ can be assessed by determining the rates of exchange of O5 with solvent $H_2{}^{18}O$ and of H3 with solvent D_2O, as followed by resonance Raman spectroscopy (Green et al., 2002). In solution, the O5 of the fully delocalized anion exchanges with ^{18}O and H3 exchanges with D within 2 min. The wild-type ECAO and N404A mutant (N404 corresponds to N381 in AGAO, constituting the wedge wall) of HPAO follow this trend, suggesting that the TPQ cofactor in those proteins is relatively mobile. In contrast, the TPQ cofactors in wild-type AGAO, HPAO, and PSAO (pea seedling copper amine oxidase) are much less mobile. H3 showed very slow or no exchange, but O5 still exchanged rapidly. The TPQ in the wild-type PSAO at pH 4.6 (with LiCl) did not show an exchange of H3 and the O5 exchanged very slowly (4.5 hr) with the solvent water, supporting the assignment of TPQ in the flipped orientation as seen in the crystal structure (Wilce et al., 1997).

When TPQ is in the active conformation (Figure 3.6a), it can react with substrate amines to form the substrate Schiff base (SSB in Scheme 3.1). A model study (Mure and Klinman 1995a) indicated that the substrate Schiff base intermediate has a λ_{max} at 350 nm and exists predominantly as a *para*-quinone imine when the net charge is neutral. The negative charge of the 4-oxoanion is localized at O4 either by a zwitteranionic interaction with the neighboring positive charge on the nitrogen at C5 or with an amine salt as in the case of the neutral form. When the net charge is –1, the negative charge will be delocalized, resulting in a red shift in λ_{max} to 450 nm.

The substrate Schiff base intermediate detected in the catalytic cycle of BSAO (bovine serum amine oxidase) has a λ_{max} at ~350 nm (Hartmann et al., 1993), suggesting that it has a *para*-quinone imine structure where the negative charge is localized at O4. The conversion of the substrate Schiff base (SSB in Scheme 3.1) to the product Schiff base (PSB in Scheme 3.1) occurs via active site base (D298)-catalyzed abstraction of a proton at the Cα position of the substrate Schiff base. This deprotonation leads to the reduction of the quinone ring.

A model study has shown that the abstraction of the Cα proton from a *para*-iminoquinone to form a product Schiff base analog occurs via a 1,3-prototropic shift and requires base catalysis (Mure and Klinman 1995a, b). When the acidity of the C proton of the substrate Schiff base is lowered, for example, by alkylation with α-methylbenzylamine or cyclohexylamine, the proton abstraction step is completely inhibited. If the substrate Schiff base had an *ortho*-iminoquinone character in which the negative charge of 4-oxoanion is delocalized, the abstraction of the Cα proton would occur in a 1,5 sigmatropic fashion; the acidity of the Cα proton would not have much effect in reactivity.

In a model study (Mure and Klinman 1995a), the *ortho*-quinone analog of TPQ (Figure 3.4c) oxidized cyclohexylamine readily, but the corresponding substrate Schiff base was not detected. In AGAO, removal of the active site base (D298A) leads to a 1000-fold reduction in catalytic turnover (Chiu et al., 2006). Based on model studies, it is not too surprising that the mutant reacts so slowly but the result clearly demonstrates that the substrate Schiff base has some intrinsic reactivity that must be supported by the active site environment.

The product Schiff base (PSB in Scheme 3.1) in the catalytic cycle is mono-protonated at the C5 imino-nitrogen, facilitating the rapid hydrolysis to yield the substrate-reduced form aminoquinol of TPQ (TPQ_{red} in Scheme 3.1). This protonation state of the product Schiff base is essential for optimal activity. A model study (Mure and Klinman 1995b) has shown that the protonated product Schiff base formed with an alkyl amine, only detectable in the anhydrous acetonitrile, has a λ_{max} at ~390 nm that is ~40 nm red shifted compared to the neutral form. The model compound for the protonated product Schiff base formed with a benzylamine has a λ_{max} at 406 nm and the neutral form absorbs at 368 nm. The protonated product Schiff base has never been observed in CAOs. However, a stable 350-nm species assigned as a neutral product Schiff base has been observed in Co- and Ni-substituted AGAOs (Kishishita et al., 2003). A similar stable species absorbing at 380 nm has been seen in an active site mutant of HPAO (Cai et al., 1997a).

2-Hydrazinopyridine (2HP) is a suicide inhibitor of CAOs and it forms a stable hydrazone adduct mimicking the substrate Schiff base. In the crystal structure of 2-HP inhibited wild-type ECAO (Wilmot 1997), the active site base (D383 corresponds to D298 in AGAO) is in a hydrogen bonding interaction with the nitrogens of the 2HP moiety of the adduct, and the O4 of the adduct is in hydrogen bond interaction with the conserved Tyr (Y369) (Figure 3.7a). The former interactions keep the TPQ ring and the pyridine ring of 2HP moiety non-coplanar, preventing the tautomerization of the hydrazone form to the thermodynamically favored azo tautomer.

When these interactions are disrupted by mutation in the active site, e.g., at higher pH or in the presence of urea, the azo tautomer becomes dominant as is the case of the corresponding model compound in solution (Scheme 3.2) (Mure et al., 2005a). Phenylhydrazine also inhibits CAOs to form a stable adduct that is a tautomeric mixture of hydrazone and azo. The equilibrium is dominantly shifted to the azo form in CAOs, as is the case of the model compound. Both 2-HP and phenylhydrazine adducts have been shown to be mono-anionic at neutral pH in model studies. The 2-hydroxyl group is deprotonated and the 4-hydroxyl group is hydrogen bonded to the azo nitrogen. In the enzyme active site, the 4-hydroxyl group is most likely deprotonated in mono-anionic form and the negative charge is stabilized by hydrogen bond with the conserved Tyr. When fully deprotonated (deprotonation of the 4-hydroxyl group), both adducts of TPQ show a 100-nm red shifted λ_{max} in the enzyme and in solution.

The idea that TPQ in the resting form and also the reaction intermediates must be tightly controlled for optimal activity was introduced by a series of mutation

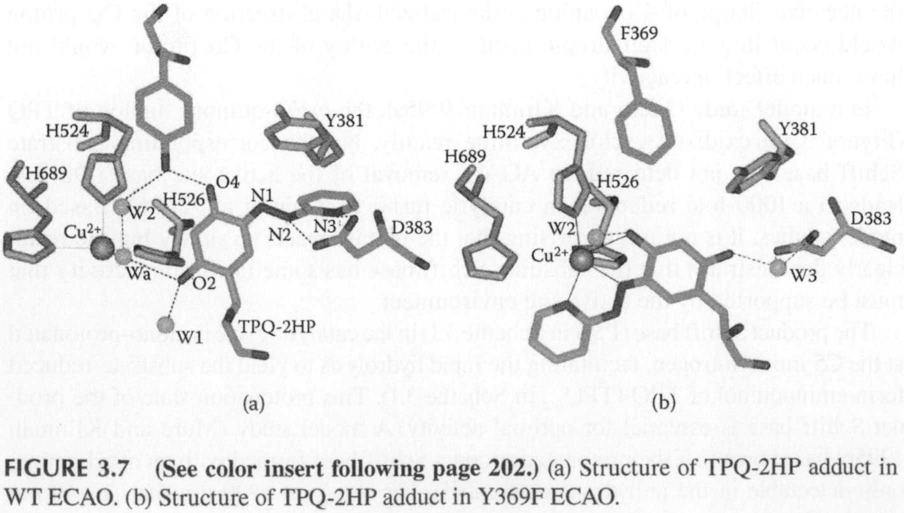

FIGURE 3.7 (See color insert following page 202.) (a) Structure of TPQ-2HP adduct in WT ECAO. (b) Structure of TPQ-2HP adduct in Y369F ECAO.

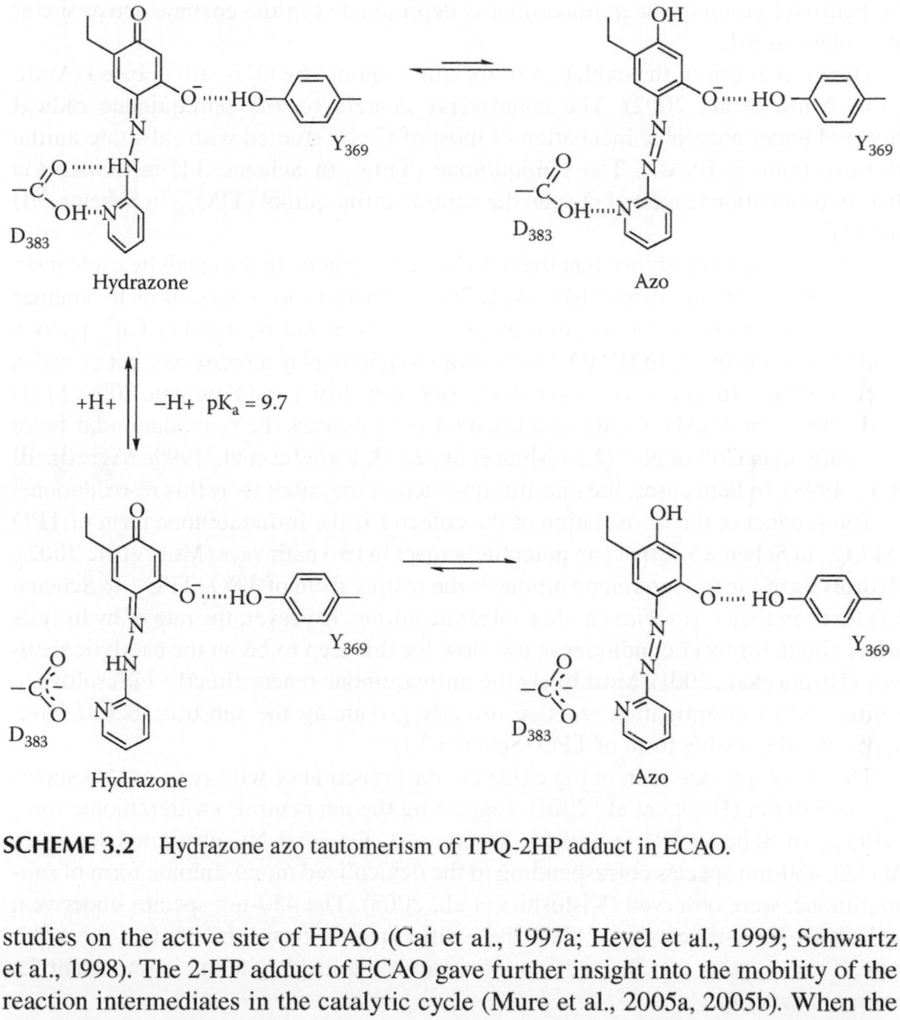

SCHEME 3.2 Hydrazone azo tautomerism of TPQ-2HP adduct in ECAO.

studies on the active site of HPAO (Cai et al., 1997a; Hevel et al., 1999; Schwartz et al., 1998). The 2-HP adduct of ECAO gave further insight into the mobility of the reaction intermediates in the catalytic cycle (Mure et al., 2005a, 2005b). When the conserved Tyr (Y369 in ECAO, corresponding to Y284 in AGAO) is mutated to Phe, the TPQ–2HP adducts gained mobility and swung out the wedge and were trapped on the Cu^{2+} (Figure 3.7b). This clearly shows that the Tyr plays a key role in keeping TPQ and its reactive intermediates within the wedge to facilitate catalytic turnover. Supporting this observation, Y369F-ECAO is much less active than the wild-type (Murray et al., 2001).

The substrate-reduced form of TPQ (TPQ_{red} in Scheme 3.1) is neutral in the catalytic cycle and can be observed following anaerobic reduction of Co and Ni AGAOs with substrate amines (Kishishita et al., 2003). The neutral form of the aminoquinol has a λ_{max} at 296 nm in solution and is ~10-nm red shifted in CAOs when compared to the model compound in solution (Mure and Klinman 1995a). The mono-cationic form of the aminoquinol has a λ_{max} at 278 nm and no absorption above 300 nm. The pK_a of the amino group of aminoquinol is 5.9 in solution and the values of the two hydroxyl groups exceed 9.6 (Mure and Klinman 1993). It is less likely that the hydroxyl group of the aminoquinol is deprotonated in the enzyme active site at physiological pH.

The mechanism of the oxidation of the aminoquinol by O_2 is still debated (Mure 2004; Mure et al., 2002). The controversy centers on the semiquinone radical detected under anaerobic incubation of most of CAOs studied with substrate amine (the exception is BSAO). The semiquinone (TPQ_{sq} in Scheme 3.1) is formed via disproportionation reaction between the neutral aminoquinol (TPQ_{red} in Scheme 3.1) and Cu^{2+}.

A model study has shown that the reactive semiquinone in the catalytic cycle must be the neutral semiquinone (Bisby et al., 2000). There is no consensus as to whether the semiquinone is on the reaction pathway, in other words, whether Cu^{2+} plays a redox role in catalysis. In HPAO, Cu^{2+} is not thought to play a redox role but provides an environment that facilitates the re-oxidation step (Mills and Klinman, 2000; Mills et al., 2002). In AGAO, BSAO, and LSAO, Cu^{2+} catalyzes the re-oxidation far more efficiently than Co^{2+} or Ni^{2+} (Kishishita et al., 2003, Padiglia et al., 1999a, Agostinelli et al., 1998). In both cases, the rate-limiting step in the catalysis is this re-oxidation.

The product of the re-oxidation of the cofactor is the iminoquinone form of TPQ (TPQ_{imq} in Scheme 3.1) that can potentially react in two pathways (Mure et al., 2002). Hydrolysis of the iminoquinone produces the resting form of TPQ (TPQ_{ox} in Scheme 3.1) that can then react with another substrate amine. However, the rate of hydrolysis under single turnover conditions is too slow for this step to be on the catalytic pathway (Hirota et al., 2001). Most likely the iminoquinone reacts directly with substrate amine via a transimination reaction directly producing the substrate Schiff base, bypassing the resting form of TPQ (Scheme 3.1).

The iminoquinone seen in the oxidative half-reaction of wild-type AGAO shows λ_{max} at 340 nm (Hirota et al., 2001), suggesting the net neutral, zwitteranionic form (TPQ_{imq} in Scheme 3.1) is present. In contrast, Co- and Ni-substituted forms of AGAO, 450-nm species corresponding to the delocalized mono-anionic form of iminoquinone, were observed (Kishishita et al., 2003). The 450-nm species underwent hydrolysis 1000-times slower than the zwitteranionic form observed in the wild-type. This suggests that the protonation state of the iminoquinone is important for optimal reactivity.

Although TPQ seems to be compartmentalized in the wedge for optimal activity in the reductive half-reaction, it is possible that the aminoquinol of TPQ has more mobility and can interact with Cu^{2+} site. However, x-ray structure analysis of the substrate-reduced form of ECAO, the aminoquinol, can be seen in the same configuration as the active TPQ (Wilmot et al., 1999). It is proposed that HPAO contains a nonmetal O_2 binding site (Su and Klinman 1998; Goto and Klinman, 2001). Recent studies using Xe binding to the crystals of AGAO, HPAO, and BSAO revealed a hydrophobic pocket near the Cu^{2+} binding site away from the wedge that may accommodate O_2 (Figure 3.6a) (Duff et al., 2004; Johnson et al., 2007; Lunelli et al., 2005). It is still unclear how O_2 reacts with the aminoquinol of TPQ in the presence of Cu^{2+}. It should be noted that the active site of CAO can be divided into two parts, one containing TPQ and the amine substrate binding sites and the other containing Cu^{2+} and the oxygen binding site. Understanding how these two compartments communicate is essential for understanding the oxidative half-reaction of CAO catalysis.

The biogenesis of TPQ has been followed using time-dependent x-ray crystallography (Kim et al., 2002). The conversion of the precursor Tyr to TPQ occurs entirely at the Cu^{2+} site where the 4-hydroxyl group is ligated directly to Cu^{2+}. It is clear that Cu^{2+} catalyzes the biogenesis and activation of the Tyr residue and O_2. However, once biogenesis is complete, TPQ must move off the copper and eventually be captured in the wedge. The driving force for such movement is not clear. The Cu^{2+}-ligated form (Cu-on) of TPQ (Figure 3.6c) has been seen in crystals under certain conditions (high salt, low pH) in which it seems that the active site is dehydrated compared to the active form of CAO (Figure 3.6a). One interpretation for the effect of dehydration caused by high salt under crystallization conditions is that the eliminated water molecules serve as a buffer between Cu^{2+} and TPQ to prevent their direct interaction. A recent study of the D298K–AGAO mutant has clearly shown that the biogenesis intermediates, i.e., dopaquinones, have mobility and can interact with the residues in the wedge (Moore et al., 2007).

Further insights into the chemical properties of TPQ can be gained by a comparison with the related but distinct quinone cofactor found in lysyl oxidase, namely LTQ (lysine tyrosyl quinone) (Mure 2004). At physiological pH, TPQ exists as a resonance-stabilized mono-anion but LTQ is a neutral *ortho*-quinone. The mono-anionic TPQ and the neutral LTQ have very similar visible spectra but the broad absorption of LTQ is slightly red shifted (Mure et al., 2003). They have the exact same redox potential, thereby they can both function as redox cofactors that react with O_2.

LTQ is covalently linked at two positions to the peptidyl backbone, therefore it is predicted that the cofactor will be less mobile than TPQ in CAO. The visible spectra of the phenylhydrazine adducts of TPQ and LTQ at physiological pHs are similar and other methods such as resonance Raman spectroscopy are essential to distinguish the two cofactors. The biogenesis of the LTQ cofactor has been proposed to undergo the similar mechanism as TPQ in which dopaquinone reacts with a lysine side chain in the former and with water in the latter. A recent finding supports this proposal that D298K–AGAO forms an LTQ-like quinone (Moore et al., 2007). Although LTQ and TPQ share many common features, their reactivity patterns reveal clear differences that arise because LTQ is a neutral *ortho*-quinone. For example, LTQ can react with branched alkylamines in lysyl oxidase (Liu et al., 1997) whereas the alkylamines will not react with TPQ in CAO. It is most likely that the mechanism involves a 1,5 sigmatropic shift as seen in other *ortho*-quinones such as 4-methoxy-5-methyl-1,2-benzoquinone.

4 Copper Amine Oxidases from Plants

Rosaria Medda, Andrea Bellelli, Pavel Peč, Rodolfo Federico, Alessandra Cona, and Giovanni Floris

CONTENTS

4.1 INTRODUCTION

The heterogeneous superfamily of amine oxidases (AOs) includes enzymes found in all living systems, ranging from prokaryotes to eukaryotes. These enzymes are involved in cellular and extracellular metabolism of amines (monoamines, diamines, and polyamines) whose oxidation may generate other biologically active substances like aldehydes, ammonia, and hydrogen peroxide. The classification of copper amine oxidases (CAOs) takes into account their origins and substrate specificities and includes mammalian, microbial, and plant amine oxidases.

The enzymes of plant origin especially are subjected to intensive studies of the interface of biochemistry and plant physiology since both sciences play key roles in essential metabolic pathways in plants (Bachrach, 1985). In fact, CAOs generally play a nutritional role in microorganisms, utilizing primary amines as nitrogen or carbon sources, and in mammals are involved in detoxification and metabolic and vascular diseases. Although their physiological role in plants is not precisely known, they are implicated in wound healing, detoxification, and cell growth by regulating intracellular diamine and polyamine levels. The aldehyde products may have a key role in the biosynthesis of some alkaloids and the hydrogen peroxide released may be involved in the formation of lignin and suberin.

Amine oxidase activity in plant extracts was first demonstrated in 1948 and was later found in different tissues (axes, cotyledons, embryos, leaf, latex, seeds, seedlings, shoots, roots, xylem) of various plant species: *Arachis hypogae, Avena sativa, Canavalia ensiformis, Cicer arietinum, Cucumis sativus, Eichornia stricta, Euphorbia characias, Glycine max, Heliantus tuberosus, Hyoscyamus niger, Hordeum vulgare, Lathyrus cicera, Lathyrus sativus, Lens esculenta, Lupinus luteus, Nicotiana tabacum, Onobrychis viciifolia, Oryza sativa, Phaseolus vulgaris, Pisum sativum, Setaria italica, Thea sinensis, Trifolium subterraneum, Trigonella foenum-graecum, Triticum aestivum, Vicia faba,* and *Zea mays.*

Many enzymes exhibiting AO activities have been purified to homogeneity and characterized from these different species, mostly from leguminosae [*Cicer arietinum* (seedlings), *Glycine max* (seedlings), *Lathyrus cicera* (seedlings) and *Lathyrus sativus* (seedlings), *Lens esculenta* (seedlings), *Lupinus luteus* (seedlings), *Onobrychis viciifolia* (leaves), *Phaseolus vulgaris* (seedlings), *Pisum sativum* (seedlings, cotyledons, and embryos), *Trifolium subterraneum* (leaves) *Trigonella foenum-graecum* (seedlings), and *Vicia faba* (seedlings and leaves)] but also from gramineae [*Hordeum vulgare* (seedlings) and *Zea mays* (seedlings)], and from other two species, *Euphorbia characias* (latex) and *Thea sinensis* (leaves) (Medda et al., 1995a; Luhová et al., 1995).

The best studied and characterized are the enzymes from seedlings of pea *Pisum sativum* (PSAO; Kumar et al., 1996), lentil *Lens culinaris* (LSAO; Medda et al., 1998), and latex of the Mediterranean shrub *Euphorbia characias* (ELAO; Padiglia et al., 1998a). To date, the complete amino acid sequences are available for several amine oxidases, and some enzymes have been crystallized and analyzed by x-ray diffraction. Of these, the unique plant CAO crystal structure available is PSAO (Kumar et al., 1996). Moreover, a CAO-encoding gene *ataol* has been isolated and characterized from *Arabidopsis thaliana*, and sequence analysis reveals that it encodes a 668-amino acid polypeptide with 48% identity to CAOs from peas and lentils (Møller and McPherson, 1998).

Generally, plant AOs are homodimeric proteins of about 70 kDa, containing both Cu ions and TPQ that are essential for enzyme function, so that they are irreversibly inhibited by carbonyl reagents and reversibly inhibited by copper chelating agents. While the redox cycling of TPQ was well documented in all AOs examined, a change in the valence of the Cu ions is found only in plant enzymes that are much more active than their animal homologues (Padiglia et al., 2001a). Although their substrate specificity is usually broad, ranging from monoamines (aliphatic and aromatic) to diamines and polyamines, they oxidize putrescine and cadaverine with high affinity.

This chapter deals with the reaction mechanisms, kinetic properties, inhibitors, and physiological roles of plant CAOs.

4.2 CATALYTIC MECHANISM

A crucial decision when describing the catalytic cycle of any enzyme is whether completeness or clarity is to be privileged. In this review, we shall try to follow an intermediate strategy: we shall present a complete qualitative description coupled to a simplified quantitative description. The main reason for this choice is that the catalytic cycle of CAO involves a large number of theoretically expected or experimentally demonstrated intermediates. Some of them are poorly populated or have spectroscopic features very similar to those of more prominent species; thus they are difficult to resolve and, whenever possible, they are neglected in the analysis of experimental data.

Scheme 4.1 illustrates the minimal kinetics of the catalytic cycle of CAOs. It can be divided into the reductive, substrate-dependent and oxidative, oxygen-dependent half cycles. The reductive half cycle leads from the couple of the oxidized enzyme and its amine substrate to the reduced enzyme and the aldehyde product; its intermediates are a (possible) noncovalent complex, and the quinone–ketimine (substrate Schiff-base), quinol–aldimine (product Schiff-base), and aminoquinol derivatives. The oxidative half cycle converts the aminoquinol derivative into the oxidized enzyme and involves a noncovalent complex of the reduced enzyme and O_2, and in CAOs from plants, a characteristic semiquinone radical derivative. Reoxidation releases ammonia and hydrogen peroxide. Although this description applies to CAOs from both plants and from mammals, they differ greatly with respect to rate constants and relative populations of the different intermediates and are best dealt with individually (Table 4.1).

When resting oxidized AO from bovine serum (BSAO) is mixed with benzylamine, the earliest spectroscopic changes recorded are assigned to the reversible formation of the quinone–ketimine intermediate. Hartmann et al. (1993) attributed to the formation of this intermediate an increase in absorption at 340 nm occurring with the apparent rate constant of $k_1 = 12\ s^{-1}$ at 5 *mM* benzylamine, pH = 7.2 and T = 25°C. The quinone–ketimine intermediate is then converted into the aminoquinol–aldimine intermediate in an exponential process that involves the reduction of TPQ, to which Hartmann et al. assigned the apparent rate constant of $k_2 = 2.5\ s^{-1}$. Since reduction of TPQ causes a major absorbance change, it is detected easily in stopped flow experiments. Reactions 1 and 2 are reversible and the apparent rate constants contain contributions from several reactions that cannot be easily deconvoluted. The reduction of TPQ (reaction 2) is associated with abstraction of a proton from the substrate and presents a significant isotopic effect; it is slower by a factor of 15 in the case of dideuterated benzylamine (Hartmann et al., 1993). Substituted benzylamine derivatives present significantly different values of k_2 and may undergo side reactions; they are not dealt with in this chapter.

Under similar conditions, Bellelli et al. (2000) found that the formation of the quinone–ketimine intermediate of BSAO and benzylamine could be described as a second order reversible reaction, with an apparent equilibrium constant of 2.8 *mM*; the reduction of TPQ drained from the equilibrium mixture with the rate constant of $k_2 = 1.2\ s^{-1}$, slightly lower than the value of Hartmann et al. (1993).

The evolution of the quinone–ketimine occurs via dissociation of benzaldehyde that leaves behind the aminoquinol derivative of TPQ. Since benzaldehyde absorbs at 250 nm, this step is usually used to monitor reaction velocity, both in steady state and stopped flow experiments. Bellelli et al. (2000) assigned to the rate constant of this step (k_3) the value of $3.3\ s^{-1}$. Since the reduction of the quinone–ketimine derivative is the slowest step in the cycle, the steady state mixture is mostly composed by oxidized enzyme derivatives.

Reoxidation of the aminoquinol derivative involves two steps: (1) the bimolecular second order formation of a reversible noncovalent complex with oxygen and (2) the

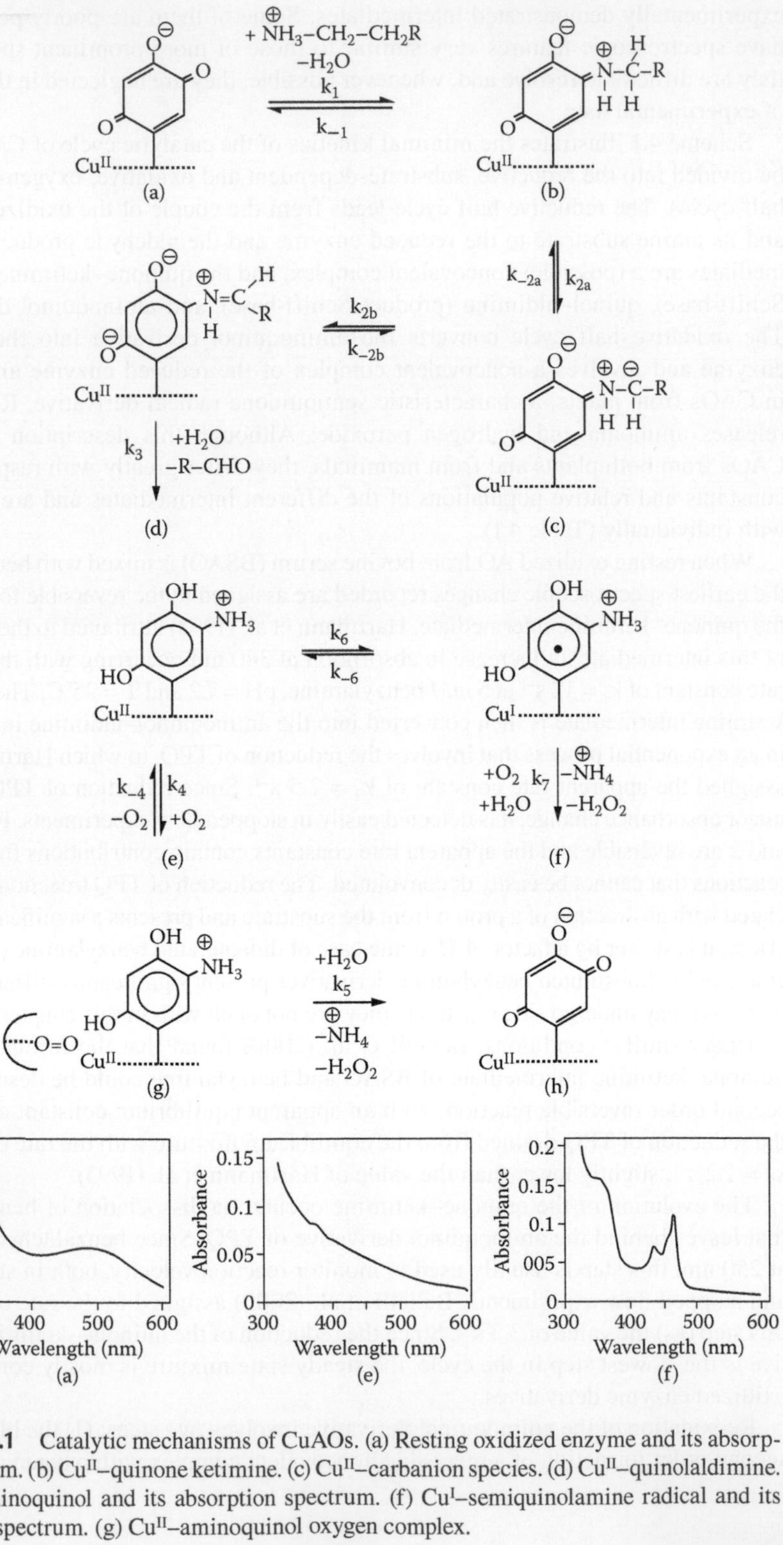

SCHEME 4.1 Catalytic mechanisms of CuAOs. (a) Resting oxidized enzyme and its absorption spectrum. (b) Cu^{II}–quinone ketimine. (c) Cu^{II}–carbanion species. (d) Cu^{II}–quinolaldimine. (e) Cu^{II}–aminoquinol and its absorption spectrum. (f) Cu^{I}–semiquinolamine radical and its absorption spectrum. (g) Cu^{II}–aminoquinol oxygen complex.

TABLE 4.1
Rate Constants of Individual Steps of Catalytic Cycles of BSAO and LSAO

	BSAO + Benzylamine	LSAO + pDABA	LSAO + Putrescine
k_1 ($M^{-1}s^{-1}$)	(> 1000)	2.5×10^5	4×10^5
k_{-1} (s^{-1})	(> 2.8[a])	(3)	(3)
k_2 (s^{-1})	1.2	8	200
k_{-2} (s^{-1})	(0.01)	Negligible	Negligible
k_3 (s^{-1})	3.3	200	200
k_4 ($M^{-1}s^{-1}$)	3.6×10^5	5×10^5	5×10^5
k_{-4} (s^{-1})	33	Negligible	Negligible
k_5 (s^{-1})	20	> 100	> 100
k_6 (s^{-1})	N/A	20,000	20,000
k_{-6} (s^{-1})	N/A	20,000	20,000
k_7 ($M^{-1}s^{-1}$)	N/A	3×10^7	3×10^7

Notes: [a] Value cannot be determined with precision because the pertinent intermediate forms and decays rapidly; the pertinent equilibrium constant is estimated at 2.8 mM, yielding a well-defined ratio of k_{-1}/k_1 (see Bellelli et al. 2000); values in parentheses are uncertain; N/A = not available.

Sources: Bellelli et al., 1985, 1991, 2000; Turowski et al., 1993).

electron transfer coupled to the release of hydrogen peroxide and ammonia (Su and Klinman, 1998; Bellelli et al., 2000). The first reaction is spectroscopically silent, but is clearly demonstrated by the order of the overall reaction between 1 and 2; it has been assigned a rate constant of $k_4 = 3.6 \times 10^5$ $M^{-1}s^{-1}$. The second reaction is irreversible and associated to large changes in the absorption spectrum; it has been assigned a rate constant of $k_5 = 20$ s^{-1} (Bellelli et al., 2000).

Reoxidation is catalyzed by the redox chemistry of copper: indeed, if copper is chelated with cyanide or removed, the aminoquinol derivative cannot be oxidized by oxygen (Agostinelli et al., 1997). If copper is substituted with cobalt, reoxidation is possible but much slower (Agostinelli et al., 1998).

In the case of plant CAOs reacting with their physiological short chain diamine substrates, e.g., LSAO with putrescine, the formation of the quinone–ketimine intermediate may be reasonably approximated as a single second order process with $k_1 = 4 \times 10^5$ $M^{-1}s^{-1}$; this is followed by the monomolecular step leading from the quinone–ketimine to the aminoquinol–aldimine intermediate that is much faster than in BSAO ($k_2 = 200$ s^{-1}). The release of the product and the formation of the aminoquinol derivative occur with a rate constant similar to k_2, i.e., $k_3 = 200$ s^{-1}. Thus, two macroscopic differences exist between the BSAO–benzylamine and LSAO–putrescine couples: (1) the latter has a k_{cat} almost 100-fold higher than the former; and (2) in the latter, the reduction of the quinone–ketimine intermediate is not rate limiting and the steady state mixture contains a significant population of reduced enzyme intermediates (Bellelli et al., 1991); the steady state mixture in the former consists mainly of oxidized derivatives of the enzyme (Bellelli et al., 2000).

LSAO and the other plant amine oxidases also oxidize chromogenic aromatic substrates such as p-dimethylaminomethyl benzylamine (pDABA). In this case, the formation of the quinone–ketimine intermediate can be described with reasonable approximation as a single second order process with $k_1 = 2.5 \times 10^5$ $M^{-1}s^{-1}$ followed by the monomolecular step leading from the quinone–ketimine to the aminoquinol–aldimine intermediate that is faster than in BSAO, but much slower than in the same enzyme reacting with putrescine ($k_2 = 8\ s^{-1}$). The subsequent release of the product is monitored easily because the reduced aminoquinol derivative thus formed equilibrates rapidly with the semiquinolamine radical whose spectrum is characteristic (Scheme 4.1), but is fast and cannot be estimated with precision with this substrate; Bellelli et al. (1985 and 1991) tentatively assigned to this step the same rate constant determined for putrescine, i.e., $k_3 = 200\ s^{-1}$. Since under these experimental conditions k_2 is the rate limiting step, the steady state mixture is mainly composed by oxidized derivatives.

The reoxidation of the aminoquinol derivative in plant AOs may occur via two different pathways: one is in principle similar to that described for CAOs from mammals, except that the deviations from simple bimolecular kinetics are less evident. The other involves a specific semiquinone radical intermediate that forms because of the transfer of one electron from the reduced TPQ to the Cu atom that is reduced to Cu(I) and never exceeds 50% of the total reduced enzyme, possibly because it only occurs for one monomer in the dimer. This intermediate is relatively easy to detect since it accumulates when the enzyme is reduced anaerobically, and has distinguishable absorption and EPR spectra (Dooley et al. 1991); moreover, it is in equilibrium with the aminoquinol derivative and the equilibrium constant depends on temperature, so that T-jump can be used to investigate the rate constants of its formation and decay in anaerobiosis (Turowski et al. 1993).

Since the semiquinone radical equilibrates very rapidly with the aminoquinol derivative, and since its reaction with oxygen is almost 100-fold faster than the reaction of the latter (Bellelli et al., 1991; Medda et al., 1998) at neutral pH, it accounts for most of the reoxidation of the enzyme. As the pH moves away from neutrality, however, the equilibrium population of the radical intermediate diminishes and the other pathway becomes more relevant; this allowed Medda et al. (1998) to determine the rate constants of the other reoxidation pathway and compare them with the corresponding values for BSAO (Table 4.1).

4.3 INHIBITORS

Enzymologists have found useful tools in finding and using specific inhibitors and following their effects on enzyme actions and investigating structure–function relationships. Inhibitors can be divided into reversible, irreversible, and mechanism-based inactivator groups. Several reversible and irreversible inhibitors have been described (Bardsley, 1985; Padiglia et al., 1998b) and used to investigate the structure–function relationships of plant amine oxidases (Šebela et al., 2003). Mechanism-based inactivators of plant CAOs have also been described (Longu et al., 2005a).

4.3.1 Reversible Inhibition

Two main types of reversible inhibitors acting on plant amine oxidases are competitive and noncompetitive. Only sodium azide is known as an noncompetitive inhibitor of PSAO (Peč and Haviger, 1988; Luhová et al., 1996) and of AO from *Trigonella foenum-graecum* (FGAO; Šebela et al., 1997) that blocks the hydrolysis of the enzyme substrate complex. Mixed types of inhibitors have not been encountered.

4.3.1.1 Reversible Competitive Inhibition

Diamino- and monoamino-ketonic compounds were synthesized and evaluated as inhibitors of PSAO (Macholán, 1969). 1,5-Diamino-3-pentanone is the most powerful competitive inhibitor of the enzyme. A magnitude increase of ~1 in the inhibition constant was observed when the carbon chain of the inhibitor was shortened by one methylene group as in 1,4-diamino-2-butanone. A significant decrease in inhibition occurs when one amino group is eliminated as in 1-amino-2-butanone or replaced with a bromine atom as 1-bromo-4-amino-3-butanone; the short-chained 1-amino-2-propanone is almost ineffective.

All these compounds show strong affinities for binding to the TPQ cofactor as normal substrates, but their complexes are hydrolyzed very slowly at a rate about 1% that of putrescine (Macholán, 1974). The other competitive inhibitors of AOs are some alkaloids with piperidine skeletons and piperidine derivatives (PSAO; Peč, 1985); cinchona alkaloids (PSAO; Peč and Macholán, 1976); bivalent transition metal complexes with diamines—the strongest inhibiting effect shown by Cu(1,3-diaminopropane)Br_2, K_i = 9 *mM* (LSAO; Devoto et al., 1986), 1,2-diaminoethane, K_i = 5.2 *mM* and trans-1,2-diaminocyclohexane, K_i = 3.6 *mM* (PSAO; Peč and Frébort, 1992a). Methylglyoxal-bis-guanylhydrazone, an inhibitor of PSAO, shows inhibition depending on preincubation time. The inhibition without preincubation is competitive (K_i = 0.26 *mM*). When the enzyme is preincubated for 1 hr with the inhibitor, the kinetics seems to reflect noncompetitive inhibition (K_i = 0.017 *mM*; Yanagisawa et al., 1981).

Competitive inhibitors of FGAO are 1,5-diamino-3-pentanone, 1,4-diamino-2-butanone (K_i ~0.02 *mM*), aminoguanidine, cinchonine, L-lobeline, β-aminopropionitrile, and pargyline (Šebela et al., 1997). Some physiologically important derivatives of 4,5-dihydroimidazole are competitive inhibitors of PSAO (Peč and Hlídková, 1987). A series of *N,N′*-bis(2-pyridinylmethyl) diamines was synthesized and characterized for their competitive inhibition effects toward PSAO (Stránská et al., 2007). Several compounds synthesised included N^6-(3-aminopropyl)adenine (APAD), N^6-(4-aminobutyl)adenine (ABAD), N^6-(4-amino-*trans*-but-2-enyl)adenine (ATBAD), N^6-(4-amino-*cis*-but-2-enyl)adenine (ACBAD), and N^6-(4-aminobut-2-ynyl) adenine (ABYBAD). Compounds APAD and ACBAD functioned as competitive inhibitors of grass pea AO (GPAO). LSAO was inhibited competitively by both APAD and ACBAD (Lamplot et al., 2005). Plant cytokinins, typically N^6-(D^2-isopentenyl)adenines, are weak competitive inhibitors of PSAO while their behavior as substrates was not observed. Observed K_i values ranged from 0.9 to 3.9 *mM* (Galuszka et al., 1998).

4.3.1.2 Reversible Noncompetitive Inhibition

Noncompetitive inhibition of plant AOs has been observed with chelating agents such as cyanide and azide (Rotilio, 1985). Diethyldithiocarbamate (DDC) causes precipitation of the copper chelate complex, causing a loss of enzymatic activity, and was thus often used to obtain copper-free enzymes (Mondovì et al., 1967a). DDC (K_i = 45 *mM*) was used to obtain copper-free LSAO (Rinaldi et al., 1984) in a number of studies of the active site (Medda et al., 1995b, 1997a, b).

LSAO is inhibited noncompetitively by sodium cyanide and sodium azide. PSAO is inhibited noncompetitively by sodium azide (Peč and Haviger, 1988; Luhová et al., 1996). Imidazole derivatives such as carnosine, anserine, and impromidine behave as noncompetitive inhibitors of PSAO (Bieganski et al., 1982a). The chelating agents 1,10-phenanthroline and 2,2´-bipyridyl are noncompetitive inhibitors of PSAO (Devoto et al., 1986). Diethylentriamine (K_i = 72 *mM*) and triethylentriamine (K_i = 0.57 *mM*) are strong noncompetitive inhibitors of PSAO (Devoto et al., 1986).

In contrast to trans-1,2-diaminocyclohexane, a previously described competitive inhibitor, the vicinal diamine *cis*-1,2-diaminocyclohexane (K_i = 2.9 *mM*) is a weak noncompetitive inhibitor of PSAO (Peč and Frébort, 1992a). ELAO shows a noncompetitive inhibition by N^6-(3-aminopropyl)adenine and N^6-(4-amino-*cis*-but-2-enyl) adenine (Lamplot et al., 2005). Pyridine carbaldoximes and alkyl pyridyl ketoximes act as strong noncompetitive inhibitors of PSAO.

The pyridine moieties in the studied compounds showed remarkable influence on inhibition potency. Other compounds studied include pyridine-2-carbaldoxime, pyridine-3-carbaldoxime pyridine-4-carbaldoxime, methyl-2-pyridyl ketoxime, methyl-3-pyridyl ketoxime, methyl-4-pyridyl ketoxime, propyl-2-pyridyl ketoxime, hexyl-2-pyridyl ketoxime, and octyl-2-pyridyl ketoxime (Mlíčková et al., 2001). FGAO is inhibited by Cu-complexing agents such as triethylenetetramine, diethylenetriamine, o-phenanthroline, 8-hydroxyquinoline, 2,2´-bipyndyl, imidazole, potassium cyanide, aminoacetonitrile, hydroxylamine, acetone oxime, and benzamide oxime (Šebela et al., 1997).

4.3.2 Irreversible Inhibition

Irreversible inhibitors of AOs are mostly carbonyl group reagents that interact with the TPQ cofactor. Typical reaction with phenylhydrazine is characterized by high affinity formation of an adduct that shows strong absorption with a maximum at 415 to 445 nm. The most effective irreversible inhibitors of AOs are substituted hydrazines, phenylhydrazine, semicarbazide, thiosemicarbazide, and phenylsemicarbazide. Carbonyl reagents were used to solve the problem of conflicting data for the number of functional active sites of dimeric AOs (Padiglia et al., 1992).

PSAO is inhibited in a time-dependent manner by the hydrazides of acetic, benzoic, nicotinic, picolinic, and other acids (Peč et al., 1992a). N^6-(4-aminobut-2-ynyl) adenine was found to inhibit GPAO, LSAO, and ELAO in a time-dependent manner (Lamplot et al., 2005). Longu et al. (2005b) studied irreversible interactions of LSAO with NO-derivatized polyamines.

4.3.3 Mechanism-Based Enzyme Inactivators

1,4-Diamino-2-butyne (DABY) is a mechanism-based inhibitor of PSAO. The substrate saturation kinetic data for DABY and the pseudo-first order time-dependent irreversible inactivation of PSAO indicate that DABY is a mechanism-based inhibitor of this enzyme with a partition ratio of 17 and characteristic constants. The mechanism of the interaction involves an intermediate aminoallenic compound that forms covalently bound pyrrole in the reaction with an active site nucleophile (Peč and Frébort, 1992b; Frébort et al., 2000a).

1,5-Diamino-2-pentyne (DAPY) was found to be a weak substrate of GPAO and *Onobrychis viciifolia* AO (OVAO). Prolonged incubations, however, resulted in irreversible inhibition of both enzymes. For GPAO and OVAO, rates of inactivation of 0.1 to 0.3 min^{-1} were determined and the apparent K_i values were in the range of 10^{-5} *M*.

DAPY was found to be a mechanism-based inhibitor of the GPAO and OVAO. Its N^1-methyl and N^5-methyl analogues were tested with GPAO and were weaker inactivators (especially the N^5-methyl) than DAPY (Lamplot et al., 2004). The alkylamines 2-bromoethylamine and 2-chloroethylamine, and the shortest diamine 1,2-diaminoethane are irreversible inhibitors of several AOs. The compounds acted as poor substrates and irreversible inhibitors of LSAO. The irreversible inhibition is caused by the aldehyde product reacting with a highly reactive species of the TPQ-derived free radical catalytic intermediate that, after β-elimination, forms a stable six-membered ring (Medda et al., 1997a). 1,4-Diamino-2-chloro-2-butene, very similar to DABY in structure, is a mechanism-based inhibitor for PSAO, a 1.3 equivalent of which inhibits the enzyme by 94%. Considerable inactivation is seen for PSAO with 1,6-diamino-2,4-hexadiyne (Shepard et al. 2002). Tyramine appears to be a good substrate for LSAO. In the course of tyramine oxidation the enzyme gradually becomes inactivated with the concomitant appearance of a new absorption at 560 nm due to the formation of a stable adduct (Padiglia et al., 2004).

4.4 PHYSIOLOGICAL ROLE

Polyamines spermidine, spermine, and their diamine obligate precursor putrescine are ubiquitous in plant cells, where they are involved in many fundamental physiological processes such as proliferation, differentiation, and death (Kusano et al., 2007) and also in adaptive responses to environmental stresses (Bouchereau et al., 1999; Walters, 2003). The cellular levels of these important biomodulators are strictly regulated by biosynthetic and catabolic pathways. A physiological role has been claimed for CAO in the regulation of intracellular polyamine levels although strong relevance has been ascribed to the role played of aminoaldehydes and hydrogen peroxide (H_2O_2), reaction products of CAO-mediated polyamine oxidation during plant responses to biotic and abiotic stresses (Cona et al., 2006). Furthermore, aminoaldehydes are important intermediates in the biosynthesis of nicotine, tropane, and piperidine alkaloids (Heim et al., 2007).

By biochemical, histochemical, and immunohistochemical methods, CAO has been localized in the apoplasts of mature tissues undergoing lignification or extensive wall-stiffening events, e.g., xylem, xylem parenchyma, cortical parenchyma, sclerenchyma, and epidermis in species of the family *Fabaceae*, in which CAO is expressed at the highest level, especially by peas (*Pisum sativum*), lentils (*Lens culinaris*), and chickpeas (*Cicer arietinum*; Cona et al., 2006).

CAO has been hypothesized to play a key role as a delivery system for H_2O_2 in cell walls during cell growth and differentiation and during responses to pathogen attacks and wound healing after mechanical damage or herbivore feeding. Indeed, H_2O_2 is the co-substrate for the peroxidase-driven reactions occurring during cell wall maturation and stress-induced stiffening and behaves as a second messenger in signalling programmed cell death (PCD) in both defense mechanisms and development.

During plant growth and development, CAO has been demonstrated to possess species- and tissue-specific regulatory features and spatio-temporal expression patterns. Moreover, CAO activity is affected by endogenous and environmental stimuli such as hormones and light. A positive spatial correlation was found among CAO, peroxidase, and lignin levels in chickpea (Cona et al., 2006) and tobacco (Paschalidis and Roubelakis-Angelakis, 2005) stems, supporting the hypothesis of a functional correlation between the two enzymes.

Accordingly, CAO transcript level and activity are modulated during chickpea seedling development in parallel with cell maturation (Cona et al., 2006). Moreover, in chickpea epidermis, de-etiolation elicits an increase in CAO activity level, suggesting its involvement in growth processes (Cona et al., 2006). In barley developing grain, CAO activity is hypothesized as a source of H_2O_2, a co-substrate in peroxidase-catalyzed reactions occurring in vascular tissues and neighboring cells (Asthir et al., 2002). Interestingly, in *Arabidopsis thaliana*, the expression of a CAO encoding gene (*atao1*) in lateral root cap cells and differentiating xylem suggests CAO involvement in cell wall cross-linking and in developmental PCD, an event that both cell types are destined to undergo (Møller and McPherson, 1998). However, CAO-mediated H_2O_2 production has also been linked to cell expansion in the fast-growing tissues of soybean seedlings, inasmuch as the simultaneous apoplastic production of superoxide anion can lead to peroxidase-mediated H_2O_2 reduction to form the wall-loosening agent hydroxyl radical (Delis et al., 2006).

CAO contributes to plant defense responses and wound healing mainly via cell wall strengthening events triggered by the apoplastic production of H_2O_2. In chickpea seedlings, mechanical wounding induces a rapid increase of CAO transcript accumulation and enzyme activity. Moreover, plant treatment with specific CAO inhibitors aminoguanidine and 2-bromoethylamine decreases lignosuberized depositions and H_2O_2 accumulations along lesions (Cona et al., 2006).

The defense capacity of *Ascochyta rabiei*-resistant chickpea cv Sultano is strongly impaired by *in vivo* CAO inhibition (Cona et al., 2006). CAO expression is up-regulated in response to tobacco mosaic virus infection in resistant tobacco plants that develop hypersensitive responses but not in susceptible plants (Marini et al., 2001). In *Arabidopsis*, the expression of *atao1* increases in developing vascular tissues during interactions with nematode parasites to counteract the effects of mechanical pressure caused by infection (Møller et al., 1998).

The involvement of CAO in wound healing is further confirmed by evidence that CAO is a target enzyme in the signal transduction pathway leading to wound- or herbivore-induced systemic protection. Indeed, in chickpeas, CAO is strongly induced by treatment with jasmonic acid (Cona et al., 2006), a plant growth regulator known to activate several wound-responsive genes. Moreover, its methyl ester induces protection against powdery mildew infection in barley seedlings by altering polyamine catabolism through activation of both polyamine biosynthetic enzymes and CAO activity (Walters, 2003). CAO-mediated H_2O_2 production is also involved in symbiotic interactions in which the control of infection thread growth is ensured by peroxidase-mediated cell wall strengthening events that harden the intercellular matrix, counteracting bacteria penetration (Wisniewski et al., 2000).

4-Aminobutyraldehyde, the product of CAO-mediated oxidation of putrescine, is further oxidized by NAD-dependent aminoaldehyde dehydrogenase (AMADH, EC 1.2.1.19) to 4-aminobutyric acid (GABA), an important metabolite that participates in various physiological processes and is rapidly produced in response to biotic and abiotic stresses. In plants, the accumulation of GABA via AMADH is correlated to wound healing and salt stress resistance. In pea stems, mechanical injury elicits increases in CAO, peroxidase, and AMADH activities at wound sites, especially in cortical parenchyma and epidermal cells, in spatial correlation with intensive lignification (Petřivalský et al., 2007) while in soybean (*Glycine max*) roots, salt stress induces GABA accumulation via stimulation of CAO activity (Xing et al. 2007). Experimental data suggest that CAOs are developmentally regulated enzymes with a key role in wound healing and activation of defense responses after pathogen attack.

However, the increase of CAO activity in rapidly proliferating and growing tissues and its decrease during maturity indicate a correlation with physiological stages characterized by intense metabolism such as cell division or organ formation, suggesting a different CAO role in the regulation of cell cycle through the modulation of polyamine cellular content. Indeed, in *Fabaceae*, CAO activity is absent in dry seeds, appears early during germination, reaching a peak after a few days, and then decreases (Cona et al., 2006). Consistently, in topinambur (*Helianthus tuberosus*), CAO activity is high during the early cell division phase of tuber growth, decreases until flowering, and then again increases during the cell enlargement preceding entry into dormancy (Torrigiani et al., 1989).

In particular, CAO activity increases during G1 and early S phases of the tuber cell cycle and closely follows the peak of putrescine content (Torrigiani et al., 1989). The strict correlation between CAO activity and putrescine accumulation remains consistent with the hypothesis of a role for CAO in controlling polyamine cellular levels (Torrigiani et al., 1989). Furthermore, during the lag phase in the growth cycle of tobacco (*Nicotiana tabacum*) BY-2 cells, the content of putrescine, despite a marked induction of its biosynthetic enzyme ornithine decarboxylase activity, shows a transient decline. The simultaneous rise of CAO activity supports the view that catabolic enzymes play an important role in the regulation of polyamine levels in cells in the stage preceding the onset of cell division (Gemperlová et al., 2005). Moreover, CAO and polyamine oxidase (PAO) activities have been shown to be critical components in cell cycle–endocycle progression in vascular tissues of tobacco plants (Paschalidis and Roubelakis-Angelakis, 2005).

The dual role of CAO in cell division and cell wall stiffening is consistent with the occurrence of two alternative pathways for putrescine synthesis in plants, specifically via ornithine decarboxylase from ornithine and via arginine decarboxylase from arginine, involved, respectively, in the regulation of cell division in actively growing cell tissue and in stress responses or morphogenetic processes. However, it is still unclear whether cell cycle progression may be influenced by cell wall localized CAO activity through a signal transduction pathway triggered by apoplastic H_2O_2 production, analogous to the process reported for tobacco PAO in defense responses (Kusano et al., 2007) or alternatively by intracellular localized

CAO. Ten cDNAs encoding putative CAO can be retrieved by a database search of *Arabidopsis*; several showed predicted intracellular localization.

ACKNOWLEDGMENTS

This work was supported by funds from Grant MSM 6198959215 of the Ministry of Education, Youth and Sports of Czech Republic and by PRIN 2006 (Progetti di ricerca di interesse nazionale) of Italy.

5 Soluble Copper Amine Oxidases from Mammals

Paola Pietrangeli, Laura Morpurgo, Bruno Mondovì, Maria Luisa Di Paolo, and Adelio Rigo

CONTENTS

5.1 MOLECULAR AND GENERAL PROPERTIES

Soluble copper-containing amine oxidases (AOs) from mammals are dimeric proteins that react with a ping-pong mechanism consisting of enzyme reduction by primary amines and subsequent re-oxidation by molecular oxygen:

$$E_{ox} + R\text{-}CH_2\text{-}NH_3^+ + H_2O \rightarrow E_{red}\text{-}NH_3^+ + R\text{-}CHO$$

$$E_{red}\text{-}NH_3^+ + O_2 \rightarrow E_{ox} + NH_4^+ + H_2O_2$$

In addition to a metal, an AO contains an organic prosthetic group, trihydroxyphenylalanine quinone (Janes et al., 1990). Preliminary information about the structural organization of BSAO was obtained from a study of the complex thermal unfolding behavior of the protein by differential scanning calorimetry (Giartosio et al., 1988). The thermodynamic study revealed that the dimer is the fundamental cooperative unit. At least four different domains were recognized. Taking the transition enthalpy change as an indication of domain site, it was possible to define two sets of domains similar in pair. The two large domains differ in thermostability by only 1 to 2 degrees, while the melting temperatures of the two smaller domains differ by 20 degrees. Removal of copper displaces at least the two transitions with higher stability and produces a broad peak characterized by a greater increase in heat capacity than normal for globular proteins. The dramatic change in the endotherm after the removal of copper indicates a structural role of the metal cofactor.

The plot of BSAO activity as a function of temperature shows an inflection point at 40 to 43°C. This result associated with nonstrict Arrhenius curves and slightly different activation energies at various temperature intervals, suggests transitions between two different BSAO conformations, stable below 40°C and above 43°C, respectively (Befani et al., 1989). The cytotoxic effect of BSAO on tumor cells at elevated temperature may be related to these properties (Agostinelli et al., 1994b).

5.2 ACTIVE SITE

Depending on the carbonyl reagents, variable values from 1.0 to 2.0 TPQ per dimer were titrated in BSAO (De Biase et al., 1996), in PPAO (Lindstrom and Petterson, 1978; Falk et al., 1983; Collison et al., 1989), and in rhKDAO (Elmore et al., 2002). When more than one TPQ reacted in the two former enzymes, the first one always reacted far faster than the second (De Biase et al., 1996; Collison et al., 1989). The reactivity of a single subunit in BSAO correlates with the ability to be > 80% inactivated by the removal of a single Cu ion (Morpurgo et al., 1987). A fast reaction of one molecule per BSAO dimer, followed by the slower reaction of 0.2 to 0.3 more molecules was found with various hydrazide derivatives (De Biase et al., 1996). These properties suggested a half-of-the-site reactivity for BSAO. Flexible hydrazides with long aromatic tails were found to be highly specific inhibitors. This was taken to indicate the presence of an extended hydrophobic region near the active site (Artico et al., 1992; Morpurgo et al., 1992).

5.3 REDOX AND STRUCTURAL PROPERTIES OF COPPER SITE

The oxidation state of copper during the catalytic cycles of AOs has been investigated and debated for many years. Two EPR experiments on pig kidney enzyme produced conflicting results. In the first experiment, only a substrate-dependent conformational change was observed without reduction of Cu (Mondovì et al., 1967a). In the second, partial reduction of copper occurred after a few minutes' incubation (Mondovì et al., 1969). In agreement with the former result, no reduction of the Cu EPR signal was induced by the reaction of BSAO with benzylhydrazine, a slow substrate. In the presence of N,N-diethyldithiocarbamate (DDC), a copper chelating agent, the reaction was able to proceed, in air, with formation of the hydrazine adduct and liberation of benzaldehyde (Morpurgo et al., 1989).

Before the x-ray structures were elucidated (see Chapter 9), the issue of whether copper was within bonding distance of the organic cofactor was addressed in BSAO, which has been the target of many spectroscopic investigations in solution. A study of the reaction with phenylhydrazine and DDC (Morpurgo et al., 1987) showed that respective simultaneous reactions with organic cofactor and copper were possible. The EPR spectrum of the DDC complex and the band at 447 nm of the phenylhydrazine adduct were both affected to a limited extent by the presence of the other inhibitors. This behavior suggests that copper was not bound to the cofactor, but was probably close to it. Spectroscopic studies and semiempirical molecular orbital calculations on models for TPQ led to the assignment of TPQ vis-UV transitions and structural properties of some derivatives. The absence of any metal effect on both experimental and calculated TPQ optical properties strongly implied the lack of a direct bond between Cu^{2+} and the other cofactor (Bossa et al., 1994).

5.4 CATALYTIC MECHANISM

The catalytic mechanism described in Chapter 4 shows two possible reaction pathways, depending on the ability to form the Cu^{+}-semiquinolamine radicals of AOs from various sources. Under anaerobic conditions, the TPQ band is fully bleached in LSAO, with a fast release of aldehyde and formation of the Cu^{2+} quinolamine in rapid equilibrium with a radical intermediate, absorbing at 464 nm, 434 nm, and 348 (sh) nm. Under the same experimental conditions, BSAO forms the Cu^{2+} aminoquinol at slower rate than LSAO, without radical formation in the vis-UV spectrum. The signal of very low intensity detected in the EPR spectrum does not appear related to the catalytic reaction (Pietrangeli et al., 2000). In BSAO, the decay of Cu^{2+} aminoquinol on reaction with oxygen occurs with Michaelis–Menten kinetics. In LSAO, the radical decays to the oxidized native species with second order kinetics (Bellelli et al., 1991).

Based on these results, the reaction with oxygen is proposed to occur via different mechanisms in the two enzymes (Padiglia et al., 2001a; see Chapter 4) with formation of a labile oxygen complex during BSAO reoxidation (Bellelli et al., 2000). The failure to form the radical is supported by the catalytic activity of Co^{2+}-substituted BSAO (see Chapter 3). It was first reported by Suzuki et al. (1983), but the significance of the result was not immediately appreciated. The derivative was again characterized by Morpurgo et al. (1990) and its catalytic activity was taken to confirm that copper reduction does not occur in the catalytic cycle.

Scanning microcalorimetry measurements indicated a similar conformational role for both copper and cobalt (Morpurgo et al., 1994). These results were somewhat unusual because no other copper protein with similar properties was known. A very significant result was the 50-fold higher K_m value for oxygen in the Co^{2+} BSAO than in the native enzyme (De Matteis et al., 1999). Similar conclusions were reached by the Klinman group using Co(II)-substituted HPAO (Mills and Klinman, 2000; Mills et al., 2002).

The catalytic competence of Co^{2+} BSAO and the low rate of TPQ reduction by benzylamine in Cu^{2+}-depleted BSAO (Agostinelli et al., 1997) suggest a conformational role of Cu^{2+}. The metal acts as a Lewis acid, decreasing the pKa of the coordinated water molecule bridging Cu^{2+} to TPQ, thus facilitating the proton transfer to O_2. This makes it possible to identify the bridging water in the place of Asp 385 with the catalytic base with a pKa of about 8.0 controlling BSAO catalytic activity.

5.5 KEY ROLES OF CHEMICAL AND PHYSICAL FACTORS IN CONTROL OF CATALYTIC EFFICIENCY

The catalytic cycle of mammalian AOs involves the break of strong covalent bonds such as C–H, C–N, and N–H of the primary amino group; dissociation energies are about 100 Kcal/mol. These dissociation energies should be minimally dependent on the substituents of the 3C atom, that is, on the molecular structure of the amine.

According to these considerations, the experimental data obtained by measuring the reactivity of this class of enzymes toward various amines show that the k_{cat} value (which in mammalian AOs should mainly related to the rupture of the reported bonds) is minimally sensitive to the substrate structure. In particular, for BSAO, the k_{cat} values differ less than one order of magnitude while the k_{cat}/K_m values vary by more than five orders of magnitudes. This behavior shows the importance in catalysis of the moiety of the molecule far from the reactive NH_2 group. In fact, the characteristics of this moiety should match those of the amine active site and lighten the role of physical interactions to affect catalytic efficiency.

This section will examine what is known about the substrate specificity of most studied soluble mammalian AOs, focusing on structure–function studies carried out to explain the molecular characteristics that control physical interactions between substrates and active sites and the physio-pathological roles of these enzymes. Various types of substrates have been used as "probes" of the active sites. Based on the dependence of catalytic efficiency (k_{cat}/K_m) on pH, ionic strength, and isotopic effect, it was possible to obtain information about the forces controlling the enzyme–substrate interaction (van der Waals forces, hydrophobic effect, hydrogen bonding, and coulombic interactions) and on the nature of the enzyme residues playing a key role in this step.

Since most of the structure–function studies were carried out on the bovine serum enzyme (Palcic and Klinman, 1983; Farnum et al., 1986; Hartmann and Klinman, 1991; Stevanato et al., 1994; Di Paolo et al., 2003 and 2007) for which the resolved 3D crystal structure is available (Lunelli et al., 2005), BSAO will be treated in greater detail in the following section of the chapter.

5.5.1 Experiments with Various Types of Substrates

Table 5.1 reports some values of catalytic efficiency of BSAO toward primary amines characterized by different steric hindrances, chain lengths, charge distributions (monoamines and polyamines) (Palcic and Klinman, 1983; Hartmann and Klinman, 1991; Di Paolo et al., 2003). Figure 5.1 summarizes the effects of pH and ionic strength on kinetic parameters obtained using various substrates and Figure 5.2 shows a model of the BSAO active site resulting from these kinetic studies.

Primary amine with two or more amino groups and lacking large hydrophobic region — It appears from Table 5.1 that BSAO shows high k_{cat}/K_m values for polyamines such as spermine and spermidine which, at physiological pH, behave as highly positive charged "sticks." The positive amino group in position 5 of a polyamine hinders the docking as demonstrated by the higher catalytic efficiency for 1,8-diamineoctane with respect to that of spermidine and by the low k_{cat}/K_m for 1,3-diaminopropane or N^8-Ac-spermidine. This adverse effect suggests an electrostatic repulsion between the positive charge in position 5 of the substrate and a positive charge inside the active site (group IH in the active site model of Figure 5.2). The strong dependence of the plot log (k_{cat}/K_m) versus pH [Δlog (k_{cat}/K_m)/ΔpH] = (1.9 ± 0.2) in the pH range 6.5 to 7.5 in the case of spermine, spermidine, and N^8-Ac-spermidine (Figure 5.3), was explained by the presence of two groups with pKa ≈8.2 and ≈7.8 (GH and IH, respectively, in Figure 5.2) that, in their deprotonated forms favor catalytic efficiency.

TABLE 5.1
Kinetic Parameters of BSAO for Amines with Various Structure and Charge Distributions

Substrate	k_{cat}[a] (s^{-1})	K_d (M)	10–3 k_c/K_m ($M^{-1} \cdot s^{-1}$)	Relative k_{cat}/K_{mb}
Aliphatic amines[c]				
Spermine	2.1	N/A	105	175
Spermidine	1.8	N/A	8.00	13.3
N^8-Acetylspermidine	0.6	N/A	0.94	15.7
1,8-Diamine octane	0.9	N/A	2500	4167
1-Aminononane	0.4	N/A	46.0	77
1-Aminobutane	1.2	N/A	0.48	0.8
1,3-Diamine propane	0.3	N/A	0.02	0.033
Methylamine[d]	1.7	N/A	0.27(c)	0.45
Aromatic amines[e]				
Benzylamine	0.54	4.2×10^{-3}	0.32	1
Phenylethylamine	0.38	1.2×10^{-3}	0.45	1.406
p-Tyramine	0.39	1.7×10^{-2}	0.007	0.022
m-Myramine	0.51	1.6×10^{-4}	0.43	1.344
m-Methoxyphenylethylamine	1.2	4.1×10^{-5}	38	118.75
p-Methoxyphenylethylamine	1.2	9.5×10^{-4}	0.13	0.406
Para substituent of benzylamine[f]				
Br	1.40	5.4×10^{-4}	4.50	10.87
F	2.69	3.41×10^{-3}	3.77	9.11
CF_3	1.92	1.25×10^{-3}	6.49	15.68
CH_3	0.44	1.99×10^{-3}	0.415	1
OCH_3	0.971	9.06×10^{-4}	17.1	41.30
$CH(CH_3)_2$	0.279	2.32×10^{-4}	1.54	3.72
OH	0.944	2.02×10^{-3}	5.29	12.78
$COCH_3$	1.18	4.69×10^{-3}	7.64	18.45

Notes: [a] Per catalytic center; [b] Benzylamine taken as "reference" substrate to compare kinetic parameters obtained under various experimental conditions in different works. [c] Source: Di Paolo et al., 2003, in 25 *mM* HEPES, 150 *mM* NaCl, pH 7.20 at 37°C; [d] Unpublished data; [e] Source: Palcic and Klinman, 1983, in 0.1 *M* potassium phosphate, pH 7.20, at 25°C. [f] *Source:* Hartmann and Klinman, 1991, in 0.1 *M* potassium phosphate, pH 7.20, at 25°C; N/A = not available.

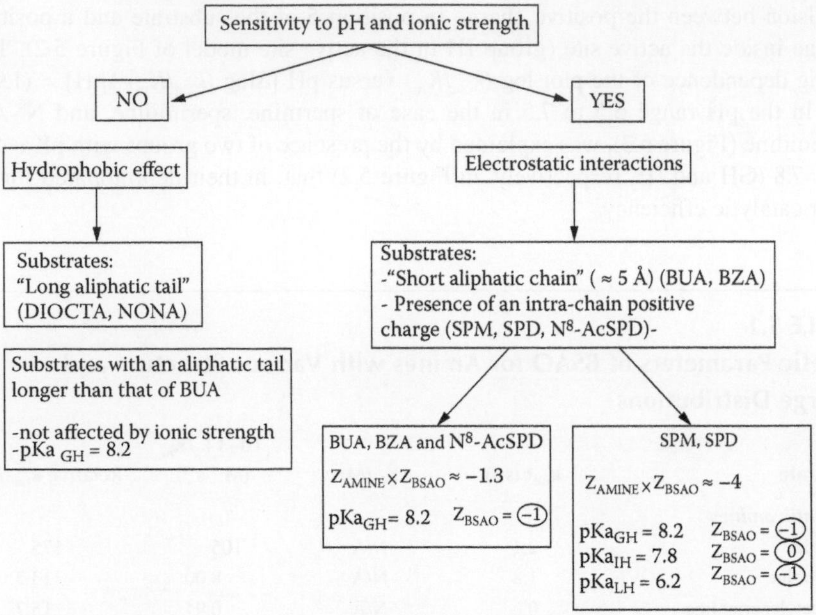

FIGURE 5.1 Summary of forces controlling the catalytic efficiencies of different types of substrate for the BSAO active site. This scheme is based on the effects of pH and ionic strength on catalytic efficiency. GH, IH, and LH are the groups reported in the active site model in Figure 5.2.

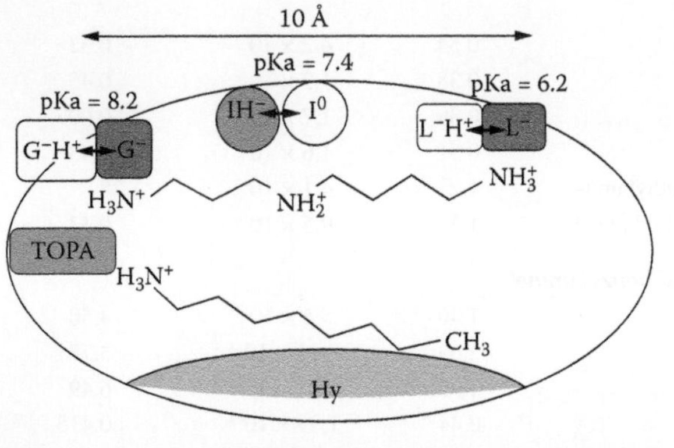

FIGURE 5.2 Model of BSAO active site–substrate interaction according to kinetic data. The active site model is based on the effects of pH and ionic strength using various substrates. Two different binding modes are shown. Apolar substrates such as nonylamine interact mainly with the hydrophobic region of the active site. Polyamines such as spermidine interact mainly by electrostatic interactions with the deprotonated form of G^-H^+, $I^{\circ}H^+$, and L^-H^+ titrated by the effect of pH on the catalytic efficiency toward polyamines. According to crystal structure analysis of the BSAO active site, the GH group can be identified with aspartic acid 385, IH with histidine 442, and LH with carboxylic groups of aspartic acid 445 and glutamic acid 417. (From Di Paolo et al. 2003. With permission.)

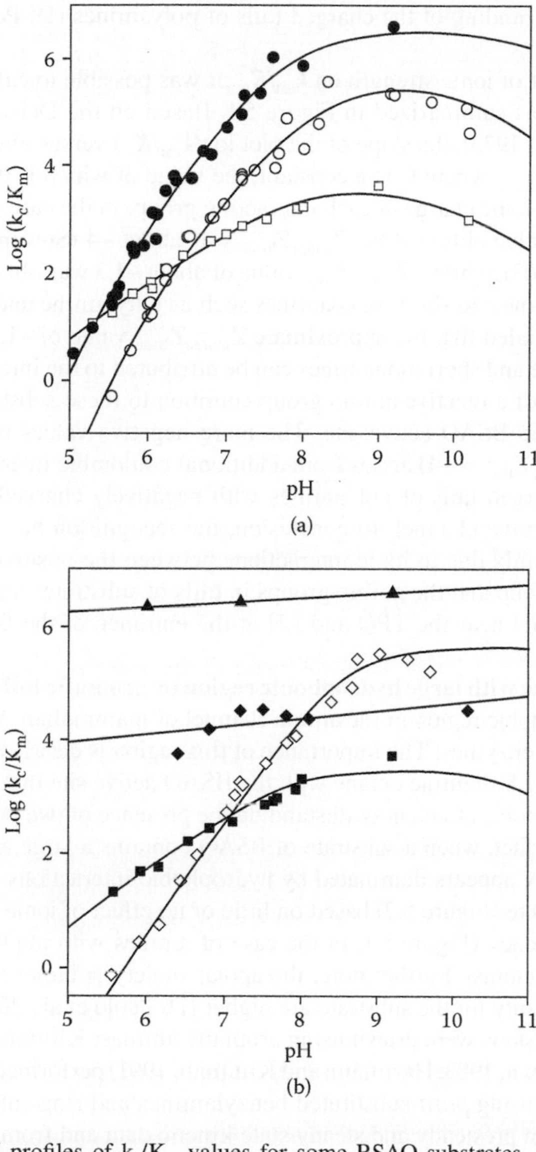

FIGURE 5.3 pH profiles of k_c/K_m values for some BSAO substrates. Plots of log k_c/K_m versus pH for (A) spermine (●), spermidine (○), benzylamine (□), (B) ^{N8}acetylspermidine (◇), butylamine (■), nonylamine (◆), 1,8-dimineoctane (▲). (From Di Paolo et al., 2003. With permission.)

The slope [$\Delta\log(k_{cat}/Km)/\Delta pH$] = (3.1 ± 0.1) in the pH range 5 to 6 in the case of spermine and spermidine suggested the involvement of another group with a pKa 6.2 ± 0.2 (group LH in Figure 5.2) involved in ionic interactions with the positive tails of polyamines. This site (LH) was found to bind some positive ions such as tetraphenyl phosphonium cations of group IA and II A with various affinities. These ions may compete with the binding of the charged tails of polyamines (Di Paolo et al., 2002 and 2004).

From the effect of ionic strength on k_{cat}/K_m, it was possible to calculate the signs of these residues as summarized in Figure 5.1. Based on the Debye–Huckel equation (Leidler et al., 1973), the slope of the plot $\log(k_{cat}/K_m)$ versus ionic strength ($I^{1/2}$) is $2 \times C \times Z_{amine}Z_{Bsao}$, where C is a constant, the value of which is 0.523 in water at 37°C, and Z is the ionic charge of each interacting group. In the case of spermine and spermidine, the value obtained for $Z_{amine}Z_{Bsao}$ was about –4 (Stevanato et al., 1994; Di Paolo et al., 2003), while a $Z_{amine}Z_{Bsao}$ value of about –1.3 was obtained in the case of N^8-Ac-spermidine and short monoamines such as butylamine and benzylamine.

It can be concluded that the approximate $Z_{amine}Z_{Bsao}$ value of –1.3 obtained with N^8-Ac-spermidine and short monamines can be attributed to the interaction between a positive charge of a reactive amino group common to these substrates and a negative charge of the BSAO active site. The more negative values of spermine and spermidine ($Z_{amine}Z_{Bsao} \approx -4$) arises from additional coulombic interactions between the positively charged tails of polyamines with negatively charged residues at the entrance of the amine channel. In conclusion, the recognition between BSAO and polyamines is mainly due to ionic interactions between the positive charges of the reactive amino group and the amino groups in tails of substrates and the negatively charged groups GH near the TPQ and LH at the entrance of the BSAO active site (Figure 5.2).

Primary amine with large hydrophobic region or aromatic tail — The presence of a large hydrophobic region in the amine channel of mammalian AO is a key characteristic of these enzymes. The importance of this region is clearly demonstrated by the interaction of 1,8-diamine octane with the BSAO active site that is largely dominated by a hydrophobic effect, notwithstanding the presence of two positively charged amino groups. In fact, when a substrate of BSAO contains a large apolar moiety, its catalytic efficiency appears dominated by hydrophobic interactions with a region of the BSAO active site (Figure 5.2) based on little or no effect of ionic strength and pH on catalytic efficiency (Figure 5.1, in the case of amines with aliphatic tails longer than four carbon atoms). Furthermore, the apolar moiety is larger and the catalytic efficiency and affinity for the substrate are higher (Di Paolo et al., 2004).

Similar conclusions were drawn using aromatic amines: Klinman and coworkers (Palcic and Klinman, 1983; Hartmann and Klinman, 1991) performed many structure reactivity studies using *para*-substituted benzylamines and ring-substituted phenylethylamines. From presteady and steady state kinetic data and from the dependence of deuterium isotopic effect on k_{cat} and k_{cat}/K_m values, they were able to calculate the rate constants of the C–H bond cleavage and the substrate dissociation constant values (K_d values, Table 5.1). By comparing ΔG values calculated from dissociation constant data, they found that the methoxy substituent binds tightly to the hydroxyl substituent and the binding is stronger when the substituent is in the *meta* position of phenylethylamine. This behavior was explained as an indication of a hydrophobic region in the active site (methoxy versus hydroxyl group) and formation of a hydrogen bond between a protein side chain and the *meta* substituent.

From a comparison of data obtained with *para*-substituted benzylamines and use of multiple linear regression analyses, Klinman and coworkers found that the unique determinant of relative substrate binding is substituent hydrophobicity, implying the presence of a lipophilic binding pocket in the BSAO active site. They also found that while hydrophobic interactions are favorable for binding of *p*-substituted benzylamines, they slow the following chemical steps. Similar effect was obtained using *n*-alkanamines. The presence of a large hydrophobic moiety in the substrate favors docking to the apolar region of the BSAO active site, but probably orients the substrate in a position less favorable to the following chemical steps and affects the microenvironment inside the active site (Di Paolo et al., 2007).

5.5.2 Amine Channel of Active Site

The crystallographic structure of BSAO recently obtained (Lunelli et al., 2005) shows that this enzyme consists of two structurally equivalent monomers, each containing a funnel-shaped amine channel allowing the access of amines to TPQ present at the bottom of the active site. A third entity is present between the two monomers and should permit the entrance of molecular oxygen to the Cu ion to oxidize the reduced TPQ. The amine channel of the active site is characterized by a mouth of ~25 Å × 9 Å, with a volume of ~4500 Å^3, and contains TPQ at the bottom, ~23 Å below the protein surface.

The opening of this cavity, present in each monomer, is partially occupied by the end of a hairpin pertaining to the other monomer. This hairpin ends with the aspartic residue D445 identified with the LH group and 14 Å above TPQ. This hairpin closes part of the wide mouth of the cavity which is ~9 Å wide between the side-chains of D445 and the facing residue L172. On the other side, the opening of the cavity entrance is delimited by a small loop of residues 85 through 87.

The bottom of the cavity is defined by the side chain of L468, ~6 Å from TPQ toward the protein surface. This residue along with Y238 and N211 reduces the lower part of the cavity to a small channel partly formed by the carbonyl 5 of TPQ when in the "productive" conformation. Close to the D445–TPQ line, ~7.5 Å from TPQ, lies histidine 442, a fully conserved residue in all AOs. On the opposite wall of this small channel is the side chain of D385, the active site base, whose carboxyl group is 5 Å away from the O5 of TPQ.

Presence of electrostatic gradient and hydrophobic region — Several positively and negatively charged residues are present on the surface of the enzyme around the entrance of the amine channel. Only negative residues are located inside the channel: D179, D445, and E417, near the entrance ~15 to 19 Å from TPQ; the D385 catalytic aspartate; and TPQ at the bottom of the cavity which appears the most negative region of the cavity. No positively charged residues are present on the wall of the cavity except for the H442 residue that may be positively charged at the working pH and plays an on–off role in the docking of biogenic polyamines.

These findings are consistent with electrostatic attraction of positively charged substrates into the cavity. This electrostatic gradient in the amine active site drives the positively charged amino group of the substrate toward TPQ in a correct orientation for catalytic reaction. With biogenic polyamines such as spermine and spermidine, the area between TPQ and the D385 residue (identified with the GH group of Figure 5.2) can accommodate the positively charged amino group of the head, while the protonated tail of the polyamine, according to its length, can interact electrostatically with D179 and E417 and/or with the mobile D445 (identified with the LH group of Figure 5.2).

The D445 aspartate residue is at the end of the highly movable hairpin of the other subunit. The flexibility of this arm is the origin of the high flexibility of BSAO in substrate docking, demonstrated by circular dichroism measurements (Di Paolo et al., 2007) and explaining the ability of this enzyme to accommodate substrates with different lengths and charge distributions. The presence of histidine 442 near the D445–TPQ line may explain the adverse effect on the catalytic efficiency toward short diamines and substrates containing an amino group ~6 Å from the reactive amino group. This histidine 442 may be the IH group of Figure 5.2.

The presence of an active site with the shape of a large funnel and the mobility of some constitutive elements of the channel serve as the basis of the broad substrate specificity, while the charge distribution is responsible for the strong differences in docking of physiological substrates and the low catalytic efficiency toward benzylamine and small polyamines. The presence of an electrostatic gradient inside of the active site is substantiated by the influence of ionic strength on the docking of charged amines (Figure 5.1). Finally, the array of negative charges at the entrance of the cavity makes this region a potential binding site for positively charged ions such as metal of groups IA and IIA and tetraphenyl phosphonium (Di Paolo et al., 2002 and 2004).

In addition to negatively charged groups, the amine channel is characterized by a hydrophobic region. Analysis of x-ray data shows the presence in the active site of a wall extending from the protein surface to the TPQ formed mainly by hydrophobic residues (F388, L468, M467, F393, A394, P396, L424, W412, F414, Y392). This wall may be where highly hydrophobic substrates such as aliphatic amines with long chains, i.e., 1-amino-nonane or 1,8 diamine octane, strongly bind by apolar interactions. It is worth noting that in the binding of this class of substrates only hydrophobic interactions appear to play a role; electrostatic interactions were excluded (Figure 5.1). Furthermore, x-ray data of the active site collected from BSAO crystals soaked with Xe revealed xenon peaks very near residue F393 of the hydrophobic wall of the surface cavity. This confirms the affinity of the active site region toward apolar molecules, since xenon binds preferentially to hydrophobic sites.

Presence of water in active site and dielectric constant — The large cavity of the amine active site is exposed to water molecules that can also reach the TPQ. Water is fundamental to the control of the dielectric constant of the enzyme microenvironment and to the formation of hydrophobic bonds. Water in the active site and a role in docking relate to the dependence of the kinetic parameter k_{cat}/K_m on temperature. As reported for natural substrates such as spermine and spermidine, the ratio k_{cat}/K_m appears strongly controlled by k_1 values, that is, the kinetic rate constant of the docking step. From the Arrhenius plot of k_{cat}/K_m, which is linear in the range 12 to 45°C, an activation entropy of 18.8 eu was calculated at 37°C (Stevanato et al., 1995). This high value of activation entropy of the binding step was attributed to the release of water molecules due to the docking of positive charged polyamines in the negative charged active site.

More information about the highly hydrophobic region was obtained from kinetic studies using *n*-alcanamines with increasing chain lengths (4 to 10 carbon atoms; Di Paolo et al., 2007). In a pH range of 6 to 8.5, ln (k_{cat}/K_m)/(number of carbon atoms in aliphatic tail) = +0.98 independent of pH, while the k_{cat} values were slightly sensitive to the number of C atoms, indicating the importance in the docking step of the number of C atoms characterizing the amine. The free energy of binding of one methylene group of the monoamine tail was calculated as –2.72 ± 0.29 kJ/mol at 37°C and pH 7.2.

This value is quantitatively similar to the difference in the free energy of partitioning of a methylene group between water and 1-octanol, which is characterized by a dielectric constant $\varepsilon_R = 10.3$, suggesting that the BSAO active site should have a similar ε_R. This finding was confirmed via a fluorescence probe, dansylethylenediamine immobilized by irreversible binding to TPQ. Since the dansyl group is very sensitive to solvent polarity, it was possible from the fluorescent emission and excitation spectra of the probe inside the active site to estimate the value of ε_R of the microenvironment surrounding the probe. An ε_R in the range of 7 to 12, depending on the value of the refraction index used, was calculated for the BSAO active site. Docking studies of dansylethylendiamine into the BSAO active site, showed that the dansyl ring of the probe docks in the hydrophobic pocket above the TPQ cofactor.

5.5.3 Substrate Specificities of Other Soluble Mammalian Amine Oxidases

Table 5.2 compares the affinities of some soluble mammalian AOs toward various substrates. PPAO has a broad substrate specificity and shows a binding affinity (K_m) in the range 140 to 560 μ*M* for tryptamine, spermidine, spermine, histamine, N-methylhistamine, and putrescine, but only low affinity for butylamine. The highest conversion rates were reported for polyamines (spermine and spermidine). Unlike the bovine enzyme, PPAO is also efficient toward histamine and N-methylhistamine (Buffoni and Blaschko, 1964; Feurle and Schwelberger, 2007).

Table 5.3 compares the steady state kinetic parameters of BSAO, measured on a series of common substrates, with those for LCAO and PSAO (Pietrangeli et al., 2007). The two latter enzymes show almost identical parameters, while BSAO is much less reactive, especially with aliphatic diamines and histamine. The k_{cat} for putrescine is four orders of magnitude lower. Differences are smaller with some aromatic monoamines. RhKDAO is also less reactive than plant AOs; the activities of spermine and benzylamine are barely noticeable (Elmore et al., 2002). No absolute values were reported for PKAO, but the values decreased from putrescine to benzylamine (Schwelberger and Bodner, 1997) as in plant AOs.

The strong pH dependence of BSAO kinetic parameters on the substrate nature (see Section 5) is explained by the several charged groups present in the large BSAO channel that interact with the substrate polar groups (Di Paolo et al., 2003; Lunelli et al., 2005). A limited pH dependence of kinetic parameters was observed in plant enzymes. The plot of k_{cat}/K_m versus hydrogen ion concentration produced approximately parallel bell-shaped curves with similar pK_a couples, in agreement with the assignment as free enzyme and free substrate pK_a, suggesting a predominance of

TABLE 5.2
Kinetic Parameters for Diamine Oxidases and Plasma Amine Oxidases

Substrate	hKDAOa K_m (μM)	PKAOb K_m (μM)	PPAO c K_m (μM)	BSAO K_m (μM)
Histamine	2.8 [2.50]	300 [0.57]	318	2400[d]
1-Methylhistamine	3.4 [1.25]		484	N/A
Putrescine	20 [1.00]	500 [1.00]	560	1300[d]
Cadaverine	30 [0.62]	1200 [0.49]	N/A	N/A
1-3-Diaminopropane	130 [0.16]	500 [0.37]		17000[e]
1-6-Diaminehexane	150 [0.08]	1300[0.29]		N/A
Spermidine	1100 [0.01]		329	225[e]
Spermine	Too low		272	20[e]
BZA	Too low		267	1500[e]
Butylamine	N/A		1711	2500[e]
Tryptamine			143	

Notes: [a] Relative catalytic efficiency with respect to putrescine. $k_{cat}/K_m = 24\ \mu M^{-1} min^{-1}$ for putrescine. Adapted from Elmore et al. 2002; [b] relative catalytic efficiency with respect to putrescine. $V_{max}/K_m = 2.16\ \mu l\ M^{-1} min^{-1}$). Adapted from Bardsley et al., 1971; [c] From Feurle and Schwelberger, 2007; [d] From Pietrangeli et al., 2007; [e] From Di Paolo et al., 2003.

TABLE 5.3
Steady State Kinetic Parameters of BSAO, LCAO, and PSAO for Oxidative Deamination of Primary Amines at 25°C in Potassium Phosphate Buffer, pH 7.2

Substrate	BSAO[a]		LCAO[b]		PSAO[b]	
	k_{cat} (M^{-1})	K_m (s^{-1})	k_{cat} (M^{-1})	K_m (s^{-1})	k_{cat} (M^{-1})	K_m (s^{-1})
Spermine	2.04	7.3×10^{-5}	28.3	6.3×10^{-4}	N/A	
Spermidine	1.87	1.7×10^{-4}	100	2.1×10^{-3}	115	2.9×10^{-3}
Benzylamine	1.23	1.7×10^{-3}	3.7	2.9×10^{-4}	2.8	2.8×10^{-4}
4-Aminomethylpyridine	1.21	6.7×10^{-6}	3.0	1.0×10^{-4}	3.9	9.6×10^{-5}
Phenylethylamine	0.37	8.4×10^{-4}	15.8	1.2×10^{-3}	10.6	8.2×10^{-4}
2-Aminoethylpyridine	0.34	1.4×10^{-4}	N/A		N/A	
Tyramine	0.085	6.5×10^{-4}	32.9	3.1×10^{-3}	27.7	2.3×10^{-3}
Histamine	0.024	2.4×10^{-3}	10.3	7.9×10^{-4}	N/A	
Putrescine	0.017	1.3×10^{-2}	262	2.7×10^{-4}	281	4.3×10^{-4}
Cadaverine	N/A		159	1.0×10^{-4}	N/A	
Agmatine	N/A		45.9	4.9×10^{-4}	N/A	

Notes: [a] From De Matteis et al., 1999 ; [b] From Pietrangeli et al., 2007.

hydrophobic interactions (Pietrangeli et al., 2007) that appear restricted to the immediate proximity of the TPQ region.

With phenylethylamine in which a polar group other than the amino group is absent, K_m is independent of pH and the reaction seems to be controlled only by the hydrophobic interaction (stacking) with Phe 298, the "gate" residue at the active site. Thus, the pH dependence of putrescine K_m should be related to an interaction of the second amino group. Highest k_{cat} values were produced by putrescine, cadaverine, and spermidine (Pietrangeli et al., 2007); they all give rise to cyclic oxidation products Δ^1-pyrroline, Δ^1-piperidine, and 1-(3-aminopropyl)-pyrroline, respectively (Saysell et al., 2002). The cyclization is considered a spontaneous reaction (Ameyama et al., 1984) because it does not involve other enzymes. It may facilitate the turnover since the k_{cat} of spermidine involving the $(CH_2)_4NH_2$ moiety is larger than that of spermine, the oxidation of which involves the $(CH_2)_3NH_2$ terminal moieties with release of acrolein. The k_{cat} of spermine is comparable to those of some aromatic amines (Pietrangeli et al., 2007).

Rabbit plasma amine oxidase has a good affinity for benzylamine, substituted phenylethylamine (dopamine and tyramine), and long chain diamines (such as 1,8-diamine octane), while low activity was measured toward polyamines (spermine and spermidine) (McEwen et al., 1966).

Equine plasma amine oxidase (Carter et al., 1994) displays a markedly different substrate specificity respect to the bovine and porcine enzymes: it has a higher affinity for benzylamine ($K_m = 0.13$ *mM* versus $K_m = 1.5$ *mM* for the bovine enzyme). Unlike other plasma enzymes, it shows a high reactivity toward norepinephrine. In particular, equine plasma amine oxidase shows a better reactivity toward dopamine, histamine, and serotonin than to polyamines (Carter et al., 1994).

The substrate specificity of human plasma SSAO/VAP-1 should be that of its membrane-bound form, which shows a different substrate preference than the above reported plasma AOs. The known physiological substrates of human SSAO/VAP-1 are methylamine and aminoacetone, but it shows good affinity for benzylamine (K_m ~0.2 *mM*), it is active on phenethylamine (K_m ~3.8 *mM*; Holt et al. 2007), but under physiological conditions it is inactive with polyamines and putrescine. In addition to benzylamine, high catalytic efficiencies of human SSAO/VAP-1 are reported for 4-fluorobenzylamine, 4-phenylbutylamine, and 3-fluoro-4-methylbenzylamine (Yraola et al., 2006). The human SSAO/VAP-1 is treated in greater detail in Chapter 6.

PKAO shows the highest catalytic efficiency for putrescine and cadaverine, but also toward histamine (Buffoni, 1966; Bardsley et al., 1970 and 1971). For the human enzyme (hKDAO), the preferred substrates are, in decreasing order, histamine, 1-methylhistamine, putrescine, and cadaverine; low activity was found toward benzylamine and polyamines (Elmore et al., 2002). On the basis of specificity toward various substrates, Bardsley et al. (1970) proposed a model for the active site of PKAO. A negatively charged substrate binding group should anchor the positive tails of putrescine and cadaverine; it should be situated at ~6 to 9 Å from the carbonyl cofactor and they should be separated by a hydrophobic or methylene binding site to which the aliphatic chain of the substrate may dock. A group on the free enzyme attributed to the catalytic aspartate with an apparent pK_a of 8.0 controls the K_m for putrescine of human kidney amine oxidase (Elmore et al., 2002).

All these studies indicate that the physiological substrates of these mammalian AOs are attracted to the active site by electrostatic interaction: the catalytic aspartate, conserved in all the AOs near the TPQ, form a negative area at the bottom of the active site and guide the entry of the positive head of amine that deprotonates before the reaction with the cofactor. The presence of a negatively charged site, probably due to Asp and

Glu residues should favor the anchoring of a positive group of the tail of substrate, if present. The distance of this second negative site from the TPQ depends on the source of the enzyme. In AOs with substrate specificity for diamines (cadaverine, putrescine), the distance should be ~6 to 8 Å. In enzymes with high specificity for polyamines (spermine, spermidine), the distance should be higher (10 Å or more). Furthermore, a hydrophobic region appears to be present in all mammalian AOs.

5.5.4 Inhibitors without NH_2-Reactive Groups

Many studies have been conducted on mammalian AO inhibitors for therapeutic and pharmacological purposes and to find highly selective inhibitors for the various SSAOs to clarify their physiological role in tissues. Since Chapter 14 will discuss this issue in detail, this section will focus on inhibitors that reversibly compete with amine substrates for binding sites. Usually, these types of inhibitors rapidly bind to the enzyme by hydrogen bonds, Van der Waal's forces, electrostatic attraction, and hydrophobic interactions without covalent bond formation, so modifying the kinetic characteristics of AOs. Certain studies revealed how some inhibitory effects may be substrate-dependent, that is, they may depend on the structural properties of both the enzyme and the tested substrate.

Several reversible BSAO inhibitors behave as modulators in a substrate-specific manner. Alkali metal ions with large ionic radii, such as Cs^+ and Ca^{2+} may be complexed (K_d values of 23 and 27 *mM*, respectively) by the anionic site (LH in Figure 5.2) at the entrance of the BSAO channel which is also involved in anchoring the polyamine tail. These ions behave as competitive inhibitors with these substrates (Di Paolo et al., 2002) but also attenuate the substrate inhibition effect observed at high substrate concentration (Holt et al., 2007).

Unlike BSAO, only small cations (Li^+ and Na^+) were found to behave as irreversible inhibitors for porcine and bovine kidney diamine oxidase (Padiglia et al., 2001b). Kinetic and spectroscopic studies indicated that the small cations in these diamine oxidases may reach the carboxylic groups of aspartate or glutamate residues in the pocket near the TPQ, inducing a rearrangement of the ligands of copper and a rotation of TPQ in a nonproductive "on-copper" configuration.

Larger cations such as tetraphenylphosphonium and other phosphonium compounds bind to BSAO with high affinity (K_d ~0.15 to 0.4 μ*M*; Di Paolo et al., 2004; Holt et al., 2008). In fact, phosphonium compounds interact with both negative LH sites (through the positive moiety of the inhibitor) and the hydrophobic area of the active site (through their hydrophobic moiety) preventing the docking of substrates and attenuating substrate inhibition.

The presence of a hydrophobic region inside the BSAO active site and its role in the enzyme–inhibitor complex formation was suggested also by the finding that increasing the lengths of the aliphatic chains of *N*-substituted analogues of 1-methyl-4-phenyl-1,2,3,6-tetrahydropyridine increases the inhibition of the enzyme (Bhatthi et al., 1988). Some imidazoline receptor ligands (clonidine, cirazoline, 2-benzofuranyl-2-imidazoline, etc.) were found to modulate some SSAOs in a substrate-dependent manner, with K_i values dependent on the source of the enzyme (Holt et al., 2004). In particular, guanfacine, guanabenz, and 2-BFI were found to inhibit both PKAO (K_i

values = 33, 47, and 168 μM, respectively, using putrescine substrate) and BSAO (K_i = 2.5, 4.8, and 41 μM, respectively, with benzylamine substrate).

Cirazoline, clonidine, and oxymetazoline were found to act as weak inhibitors in the case of PKAO (K_i ~0.8 to 0.9 *mM* for clonidine; Federico et al., 1997; Holt et al., 2004) while they modulate BSAO kinetic depending on the substrate structure. In this case, a mixed or biphasic inhibition was found using spermidine, methylamine, *p*-tyramine, or phenylethylamine as a substrate at a cirazoline concentration that increased the benzylamine oxidation rate. Kinetic, spectroscopic, radioligand binding experiments and structure studies indicate that this activation of BSAO by imidazoline ligands results from disinhibition of the enzyme by competition with the substrate for the reduced form of the enzyme, leading to an apparent reduction in K_m values (Holt et al., 2008). Multiple binding sites may explain the behavior of these imidazoline compounds with BSAO.

Unlike the phosphonium compounds that are too large to enter the active site cavity of BSAO, the imidazoline compounds should bind both to the tail of the spermidine binding site (LH in Figure 5.2) and inside the BSAO active site cavity. From the resolved crystal structure of the BSAO–clonidine complex, clonidine was found to bind near the TPQ, at the bottom of the active site funnel.

Binding of clonidine causes a displacement of TPQ, inducing a rotation of the TPQ side chain to the "on-copper" nonproductive orientation. The phenyl ring of clonidine lies almost perpendicular to the imidazoline moiety of the molecule and is trapped by stacking interaction in a sandwich-like orientation within the aromatic side chains of TPQ and Tyr 472 residues. A noncovalent electrostatic interaction between the positively charged inhibitor (amiloride, clonidine, or gabexate mesylate) and the negatively charged TPQ–Asp diad was also proposed for PKAO on the basis of the three-dimensional structures of homologous copper amine oxidases from *E.coli* and *Pisum sativum* (Federico et al., 1997). The increase of the affinities of amiloride compounds for human placenta diamine oxidase found increasing the apolar moiety of the inhibitor (from K_i = 5.1 and K_i = 1.75 μM for amiloride and ethylpropylamiloride, respectively, to K_i 47 *nM* for phenamil; Novtony et al., 1994) confirms the presence and the role of a hydrophobic region also in mammalian diamine oxidases. The specific behavior of each inhibitor, depending on the source of the enzyme, underlines the critical role of physical interactions among mammalian amine oxidases and substrates or modulator molecules.

5.6 MODULATION OF AMINE OXIDASE ACTIVITY BY H_2O_2

Recent research suggests a broad spectrum of possible functions ranging from cell adhesion, cell growth regulation, and immunological processes to clinical applications to treat cancers and allergies. Their action may be related, at least in part, to the control of H_2O_2 levels as they are reported to be inactivated by the same H_2O_2 they produce upon prolonged exposure to amine in excess of oxygen. This was demonstrated in pea seedlings (Mann, 1955), pig kidney amine oxidase (Mondovì et al., 1967b), and BSAO (Pietrangeli et al., 2000). For BSAO, inactivation by aldehydes, assisted by H_2O_2, was also reported (Lee et al., 2001a).

One condition for inactivation is that the TPQ cofactor of BSAO must be in reduced form. The reduced form is stabilized by inactivation, since a band appears at 310 nm in the UV-vis difference spectrum with respect to native BSAO, while the reactivity with carbonyl reagents decreases (Pietrangeli et al., 2000). The Cu^{2+} EPR signal is not affected by inactivation. As reported above, a radical of low intensity forms at g = 2.001. The radical may derive from a conserved residue in proximity of the active site, such as the Tyr at hydrogen-bonding distance of TPQ ionized hydroxyl, which becomes ionized upon TPQ reduction. In agreement with this proposal, the mutation of the corresponding tyrosine caused a decrease of activity in *Hansenula polymorpha* AO (Hevel et al., 1999) and in *Escherichia coli* AO (Murray et al., 2001).

The inactivation by H_2O_2 may be part of an autoregulatory process *in vivo*, possibly relevant to cell adhesion and redox signaling. The VAP-1 adhesion protein was demonstrated to be an AO, with high homology with BSAO, but very low AO activity (Salminen et al., 1998). H_2O_2, which was formerly considered an undesirable toxic product of oxygen reduction, is now believed to be involved in the regulation of cellular functions and membrane receptor signalling (Rhee, 1999).

5.7 PHYSIOLOGICAL FUNCTIONS

The source of soluble AOs in mammals is reported in Chapter 7. A plethora of physiological functions, sometimes in contrast with each other, are described for mammal AOs and their physiopathology. One of the main roles of these enzymes is their ability to oxidize many biologically active, biogenic amines producing amino aldehydes, ammonia, and hydrogen peroxide as catabolic products.

Various cellular processes depend on the actions of AO substrates and reaction products. Animal AOs oxidize both monoamines and polyamines, including diamines. In addition, BSAO can oxidize free amino groups of nonconventional substrates such as polylysine up to 50 Kda and some proteins such as lysozyme and ribonuclease (Wang et al., 1996a).

Histamine is oxidized by DAO and not by a specific enzyme (Mondovì et al., 1964). Many functions imputable to histamine and possible therapeutic applications are described in Chapters 11 and 15. The enhancement of histamine content in humans may be attributed to many factors. Histaminase-blocked animals exhibited several clinical symptoms (hypotension, flushing, vomiting) and elevation of plasma histamine levels (Sattler et al., 1988). The histamine content in food is variable and high in some allergenic extracts. The largest amounts of histamine and tyramine are found in fermented foods.

The mechanisms of pseudoallergic reactions of histamine caused by food seem to be mainly ascribed to intestinal hyperpermeability caused by irritant species or changes of enzymatic activities likely occurring in patients with atopic dermatitis (Ionescu and Kiehl, 1989). High histamine content in foods and beverages results from microbial contamination (Bodmer et al., 1999).

Inhibition of DAO activity is a risk in specific immunotherapy (Wantke et al., 1993). Some antidepressants and antipsychotics with antihistamine activities stimulate food intake and increase body weight in agreement with an anorexic action of histamine (Morimoto et al., 2001). See Chapter 11 discussion of the role of AOs in intestinal diseases.

DAO activity in the sera of pregnant women in placenta indicates a possible protecting role against the release of polyamines from placental units (Bardsley et al., 1974a). During pregnancy, plasma SSAO is elevated (Boomsma et al., 2003), unrelated to markers of endothelial cell activation in pre-eclampsia (Sikkema et al., 2002). A decrease of placental AO activities in premature births was found (Kirkel et al., 1983). DAO activity increases up to 1000-fold in the blood of pregnant women (Holinka and Gurpide, 1984) to protect the fetus from excess biogenic amines in the placental unit (Bardsley et al., 1974a).

The scavenging ability of the plasma enzyme against oxygen free radicals and their byproducts (Mateescu et al., 1997) partially explains the protective effect of AO, purified from plants, in heart ischemia (see Chapter 15). AOs are considered biological regulators, especially in cell growth and differentiation. Their primary involvement in the inhibition or progression of cancer is probably attributable to a balance between the actions of AOs and antioxidant enzymes like copper–zinc superoxide dismutase (Cu,ZnSOD) on the products of biogenic amine oxidation, H_2O_2, and ROS. Oxidation products may be involved in cataract formation, as shown by the direct proportionality between AO content and lens opacity (Befani et al., 1999).

In Cu,ZnSOD-deficient mice, animals surviving 21 months developed hepatocellular carcinoma. Microinjection of AO in chicken embryo fibroblasts induced by the use of Rous sarcoma virus provoked death. In nontransformed cells, the damage was reversible (Bachrach et al., 1987). The lethal effect on tumors cells may have been imputable to the higher concentration of polyamines in tumor cells.

A correlation between tumor malignancy and AO activity level was demonstrated in human astrocytomas; the activity was proportional to the degree of malignancy (Mondovì et al., 1983). However, in some patients, even those affected by metastatic tumors, the level of circulating DAO activity was unchanged (Baylin and Luk, 1985). A possible use of AOs for tumor therapy (Mondovì et al., 1994) was suggested by the inhibition of Ehrlich ascites tumor growth in mice treated with pig kidney DAO (Mondovì et al., 1982). Both hydrogen peroxide and aldehydes produced by polyamine oxidation of AO contributed to cytotoxicity (Averill-Bates et al., 1994).

BSAO, added to an incubation mixture of cultured hamster fibroblasts in the presence of added polyamines showed cytotoxic effects. Cytotoxicity was inhibited by catalase during the first 20 min of incubation. During longer incubation times, aldehyde dehydrogenase protected the cells. Complete protection was observed in the presence of catalase and aldehyde dehydrogenase (Averill-Bates et al., 1994). The cytotoxicity induced by BSAO was enhanced at 42°C and occurred much more rapidly than at 37°C. The observed enhancement of the cytotoxic effect may be attributed to an increased reaction rate with formation of toxic products leading to cell death. Considering the aminoaldehyde interaction with DNA, the involvement of biogenic amine oxidation products in the hyperthermic sensitivity of tumor cells may be postulated (Mondovì et al., 1994). For details see Chapters 10 through 14.

In summary, it should be suggested that BSAO in particular and AOs in general may be considered key enzymes in cell growth and differentiation processes and thus in tumor development and growth regulation (Pietrangeli and Mondovì, 2004; Toninello et al., 2006). The double activity and contradicting effects of mammal AO are apparent also in other pathologies such as damage to organs by H_2O_2 and/or aldehydes and protection against oxygen free radicals (ROS). Acrolein produced by the oxidation of spermine and spermidine by AOs (Alarcon, 1970; Sakata et al., 2003) appears to be both carcinogenic (Cohen et al., 1992) and cytotoxic (Averill-Bates, 1994) and is recovered in quantities equivalent to H_2O_2 recovery (Gahl and Pitot, 1982). This compound is considered a component of a universal cell growth regulatory system (Alarcon, 1972). It may act as a mediator of cell transformation under oxidative stress, when cells are pretreated with benzo[a]pyrene, a major carcinogen found in smoke (Takabe et al., 2001). Acrolein and its glutathione adduct, glutathionylpropionaldehyde, induce ROS formation (Adams and Klaidman, 1993) and are involved in cell transformation from oxidative stress (Takabe et al., 2001). For information about the possible involvement of mammalian AOs in other pathologies and therapeutic applications, see Chapters 10 through 15.

6 Membrane-Bound Copper Amine Oxidases*

Andrew Holt

CONTENTS

* This chapter is dedicated to Dr. Brian A. Callingham on the occasion of his 75th birthday.

6.1 INTRODUCTION

6.1.1 Background

Copper amine oxidase enzymes were first identified in mammalian tissues by Rucker and O'Dell (1971) who observed inhibition of a nonmitochondrial amine oxidase in bovine aorta by semicarbazide and other carbonyl reagents. A similar activity was observed in rabbit aorta shortly thereafter (Rucker and Goettlich-Riemann, 1972), and then in membrane fractions from mesenteric and femoral arteries (Coquil et al., 1973) and hearts (Lyles and Callingham, 1975) of rats. A lack of sensitivity toward clorgyline clearly distinguished these novel activities from classical mitochondrial monoamine oxidase (MAO). The similarities between properties of these enzymes and of soluble plasma amine oxidases were noted (Coquil et al., 1973; Rucker and O'Dell, 1971), and an early suggestion was that the source of plasma amine oxidase in some mammals may be the vascular wall.

6.1.2 Historical Nomenclature of Membrane-Bound CAO Enzymes in Mammals

The sensitivity of membrane-bound CAOs to inhibition by semicarbazide led quickly to the use of the term *semicarbazide-sensitive amine oxidases* (SSAOs) to distinguish these enzymes from MAOs. However, soluble copper-containing diamine oxidases (DAOs), enzymes closely related to membrane-bound SSAOs, are also inhibited by semicarbazide (Bieganski et al., 1980), as is copper-containing lysyl oxidase (Kagan et al., 1979), leading to a lack of clarity with this nomenclature approach. Adding to the confusion, some researchers chose terms such as benzylamine oxidase (Buffoni et al., 1968) and clorgyline-resistant amine oxidase (Lyles and Callingham, 1981) to refer both to membrane-bound SSAO enzymes and to the soluble form found in the blood.

More recent data demonstrating a role for membrane-bound SSAO in the vascular endothelium as a lymphocyte adhesion molecule (Smith et al., 1998) created a divergence in nomenclature such that some researchers now refer to the enzyme protein as vascular adhesion protein-1 (VAP-1). Consequently, almost 40 years after membrane-bound SSAO enzymes were first described, no satisfactory universal nomenclature system has been agreed upon for these ubiquitous enzymes that are somewhat anomalously grouped under a single Enzyme Commission (EC) identification number as EC 1.4.3.6. The development of a logical system should be seen as a priority on the part of researchers in the field.

In this chapter, only membrane-bound CAO enzymes will be referred to by the SSAO acronym. The species and tissue and/or cellular source of the enzyme will be included for clarity. Where mentioned, soluble CAO in the blood will be referred to as plasma SSAO, although the enzyme present in ruminant blood and secreted by the liver (Hogdall et al., 1998) will be cited as plasma amine oxidase. Plasma amine oxidases that can oxidize the terminal primary amine moieties of polyamines will not be referred to as PAOs because that acronym is often used for polyamine oxidase, an enzyme unrelated to plasma amine oxidases.

6.2 DISTRIBUTION OF MEMBRANE-BOUND SSAO IN MAMMALS

6.2.1 Gene Products

Five gene families coding for mammalian CAO enzymes have been identified. They are described in Chapter 7. The *AOC3* gene encodes the predominant membrane-bound SSAO found in numerous tissues throughout the body (Bono et al., 1998a; Heniquez et al., 2003). This chapter will focus on the AOC3 enzyme, referred to hereafter as SSAO, a member of the EC 1.4.3.6 sub-subclass. Most of the literature discussing membrane-bound SSAO enzymes refers to the product of the *AOC3* gene.

6.2.2 Tissue and Organ Distribution of SSAO

The occurrence of SSAO in tissues and organs is usually expressed in the literature in terms of enzyme activity versus a substrate at a single concentration (Andres et al., 2001; Boomsma et al., 2000a; Lewinsohn, 1984; Lewinsohn et al., 1978). Significant variation exists in estimated activities because they are influenced by the substrate used, its concentration, and the chosen assay method. Nevertheless, it is apparent that while SSAO can be detected to a degree in most tissues of mammalian species studied (Andres et al., 2001; Boomsma et al., 2000a; Lewinsohn, 1984; Lewinsohn et al., 1978), a few tissues show much higher SSAO activities. Arterial tissue, lung (vasculature), and gastrointestinal smooth muscle appear to possess high SSAO activity in most species, while activity in adipose tissue is high in several species including rats (Ochiai et al., 2005) and mice (Moldes et al., 1999) and moderate in humans.

Activity levels in human heart, venous tissue, and adrenal glands are modest and are usually lower in other species. Liver, kidney, brain, and most other major organs of most species exhibit low or negligible activity, although hepatic sinusoidal endothelium stains positively for SSAO and expresses high activity (Lalor et al., 2007). While the existence of an endogenous SSAO inhibitor in the brain has been proposed (Obata and Yamanaka, 2000a), it is generally accepted that measures of SSAO activity in most tissues provide acceptable comparative indications of relative enzyme protein levels in different tissues.

6.2.3 Subcellular Distribution

Most membrane-bound SSAO exists as an ectoenzyme (type II protein) anchored to the plasma membranes of specific cell types within tissues (Holt and Callingham, 1994). In the vasculature, SSAO is located primarily on the outer membranes of vascular smooth muscle cells (Lyles and Singh, 1985; Precious and Lyles, 1988). Inflammatory stimuli result in expression of SSAO at the surfaces of vascular endothelial cells in many small venules (Smith et al., 1998). However, it seems possible that prior to trafficking to the endothelial surface, intracellular SSAO detected in cytoplasmic granules of high endothelial venules in lymphatic organs (Salmi and Jalkanen, 1995; Salmi et al., 1993) may also possess oxidase activity. SSAO has also been detected via antibody studies on the surfaces of other smooth muscle cell types in the intestine, bladder, and uterus, but not in skeletal or cardiac muscle (Jaakkola et al., 1999). In fat, SSAO is located at the plasma membrane in both brown (Barrand et al., 1984) and white (Carpéné et al., 2003; Morris et al., 1997; Raimondi et al., 1991) adipocytes.

A strong association of SSAO with caveolae in vascular smooth muscle (Jaakkola et al., 1999) and in adipocytes (Souto et al., 2003) suggests that SSAO may interact in a functional manner, directly or via regulation of substrate or product levels, with other caveolar proteins. Furthermore, in adipocytes, SSAO accounts for almost 2.5% of total cell surface proteins, with approximately 7 million SSAO dimers expressed on a single adipocyte (Morris et al., 1997). This remarkable degree of expression and caveolar localization of the enzyme suggests an important physiological role. Intracellular punctate staining for SSAO in human vascular endothelium suggests a vesicular location (Salmi and Jalkanen, 1992). Lower levels of SSAO activity are detectable in low density microsomes isolated from rat adipocytes, compared with activity at fat cell membranes (Morris et al., 1997), consistent with the view that no selective endocytosis of SSAO from plasma membranes takes place.

In all cell types, SSAO appears to be anchored to cell membranes by a short, hydrophobic single membrane-spanning domain at the N-terminal of the protein, with a cytosolic N-terminal consisting of as few as four residues (Smith et al., 1998). Since the enzyme usually exists as a homodimer, each dimer is thus anchored by two membrane-spanning regions. It has been suggested that proteolytic cleavage of these anchor sequences releases SSAO into the blood (Abella et al., 2004; García-Vicente et al., 2005), resulting in low levels of circulating AOC3 protein (Lyles et al., 1990) that is distinct from the AOC4 amine oxidase in ruminant plasma.

6.3 MAJOR STRUCTURAL PROPERTIES OF HUMAN MEMBRANE-BOUND SSAO

Primary sequence data are available for membrane-bound SSAO from several species, including humans, rats, mice, cattle, and orang-utans. Positions of conserved or critical residues determined from sequence analyses and alignments are listed in Table 6.1. A remarkable degree of sequence similarity exists between human SSAO and enzymes from other species; this is particularly true when several critical conserved amino acids are compared.

To generate crystals of human SSAO, the full-length human enzyme has been overexpressed in CHO (Airenne et al., 2005) and HEK293 (Jakobsson et al., 2005) cells prior to purification. The solubilized full-length proteins generated possess a hydrophobic N-terminal tail that would normally tether the protein to the plasma membrane. However, in the crystallized enzyme, the position of this tail is undefined, although data indicate a possible interaction between Cys 748 near the C-terminal of the enzyme adjacent to the plasma membrane and an unidentified cysteine residue that may be Cys 41 (Jakobsson et al., 2005). Accordingly, it is likely that the hydrophobic N-terminal tail of the solubilized full-length enzyme adopts a low energy compact conformation and interacts with hydrophobic residues near the base of the enzyme. The possibility that the three-dimensional structure and catalytic behavior of the enzyme are influenced by the unnatural position of the anchor tail should not be discounted.

TABLE 6.1
Primary Sequence Comparisons of Mammalian Membrane-Bound SSAO Enzymes

	Species				
Parameter	**Human**	**Rat**	**Mouse**	**Cattle**	**Orang-utan**
Accession code	Q16853	O08590	O70423	Q9TTK6	Q5R9I0
Residues	763	763	765	762	763
Subunit (protein) mass (Da)	84,622	84,981	84,534	84,500	84,602
TPQ	Tyr 471	Tyr 471	Tyr 471	Tyr 470	Tyr 471
Catalytic base	Asp 386	Asp 386	Asp 386	Asp 385	Asp 386
Histidines ligating copper	His 520, 522, 684	His 520, 522, 684[a]	His 520, 522, 684	His 519, 521, 683	His 520, 522, 684
Transmembrane spanning residues	6 to 26	7 to 27	7 to 27	7 to 27	6 to 26
Identity with human SSAO	(100%)	82.5%	83.2%	84.3%	97.4%
Reference	Airenne et al., 2005	Ochai et al., 2005	Moldes et al., 1999	Høgdall et al., 1998	[b]

Notes: [a] His 443, 520, and 522 initially proposed as copper ligands; [b] Predicted sequence submitted to UniProtKB database by the German cDNA Consortium; last modified 2008.

Human SSAO, like related copper-containing amine oxidases, is a homodimeric enzyme. Each monomer has an amino acid mass of 84.6 kDa and contains one active site reached by an entrance channel that faces toward the cell membrane. Six potential N glycosylation sites and four O glycosylation sites are identifiable on each monomer; extensive glycosylation of native SSAO increases the subunit molecular weight by ~5 to 9 kDa to ~90 kDa. The presence of N- and O-linked carbohydrates is claimed to attenuate SSAO enzymatic activity to a degree (Maula et al., 2005), perhaps through steric hindrance of substrate access or product departure. However, we have observed no significant change in kinetic constants for methylamine oxidation following deglycosylation of human SSAO (our laboratory, unpublished data). A comprehensive review of SSAO structural features may be found in Chapter 9.

6.4 CATALYTIC MECHANISM OF MAMMALIAN MEMBRANE-BOUND SSAO

As with all CAO enzymes, the overall reaction catalyzed by SSAO is comprised of reductive (i) and oxidative (ii) half-reactions usually written as:

(i) $E_{ox} + RCH_2NH_3^+ + H_2O \leftrightarrow E_{ox}\text{-}NH^+CH_2R \rightarrow E_{red}\text{-}NH_2 + RCHO + H_2O$

(ii) $E_{red}\text{-}NH_2 + O_2 \rightarrow E_{ox} + H_2O_2 + NH_4^+$

In this ping-pong reaction, an aldehyde product is released anaerobically, with subsequent addition of oxygen reoxidizing enzyme and releasing hydrogen peroxide and ammonium. The K_M for oxygen with ox lung SSAO may be estimated as ~20 μM, based on steady state experiments with microsomes (Lizcano et al., 2000b), around 12-fold lower than the dissolved oxygen concentration in air-saturated buffer (Hubálek et al., 2005) and in a low micromolar range typical of other CAO enzymes (Johnson et al., 2007). While the K_M for oxygen with human SSAO is not known, it is reasonable to assume that it is also below the concentration of dissolved oxygen in air-saturated buffer. Kinetic experiments in which varying concentrations of amine substrate are examined may be expected to yield hyperbolic plots of initial velocity versus amine. However, reduced initial rates at elevated substrate concentrations have been observed with bovine (Lizcano et al., 2000b), human (Holt et al., 2008), and rat (Lyles and Callingham, 1975) SSAO enzymes, suggesting that the reaction mechanism, like that of other mammalian (Ignesti, 2003; McEwen, 1965b) and bacterial (Shepard and Dooley, 2006) CAO enzymes, is more complex than the accepted reaction scheme suggests.

6.4.1 Steady State Enzyme Behavior

Careful consideration of the CAO half-reactions suggests that even in the presence of saturating dioxygen, initial reaction velocity may not be related hyperbolically to amine concentration, since amine can bind both to TPQ_{ox} and to TPQ_{img}. Substrate binding presumably occurs with different affinities to the distinct enzyme forms, while the hydrolytic pathway, *via* E_{ox}, probably occurs with a different rate constant than does transimination to yield TPQ_{ssb} directly. Furthermore, the possibility that amine may bind to one or more forms of the enzyme in a nonproductive complex is raised by the observation that clonidine, an active site-directed reversible inhibitor of bovine plasma amine oxidase (BPAO) (Holt et al., 2004; Holt et al., 2008), crystallizes within the enzyme active site with TPQ_{ox} in an inactive "on-copper" conformation. These possibilities are illustrated in Figure 6.1. Unless the affinities of substrates for forms of the enzyme other than E_{ox} are at least two orders of magnitude higher than that for E_{ox}, this may be observed as nonhyperbolic substrate behavior at concentrations of amine typically present in kinetic assays. Indeed, examples of nonhyperbolic data may be found for bovine (Lizcano et al., 2000b), rat (Lyles and Callingham, 1975), and human (Holt et al., 2008) membrane-bound SSAO, and for several soluble CAO enzymes (Bardsley et al., 1980; Ignesti, 2003; McEwen, 1965b; Shepard and Dooley, 2006). Typically observed as reduced initial rates at higher substrate concentrations, the important consequence of this behavior is that the Michaelis–Menten equation is inappropriate for analyzing kinetic data obtained with many SSAO enzymes, including membrane-bound SSAO, and more complex computational approaches are necessary (Holt et al., 2008). The value of many published kinetic constants (K_M, V_{max}) for SSAO obtained from Michaelis–Menten analyses including linear Lineweaver–Burk plots may be called into question. In addition, neither the mechanism nor the potency of novel SSAO inhibitors can be determined appropriately if an incorrect enzyme–substrate interaction model is used.

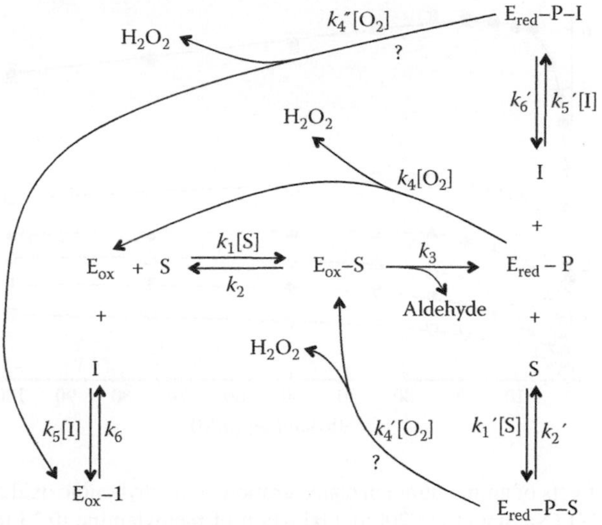

FIGURE 6.1 Proposed (simplified) reaction scheme for human SSAO with substrates and inhibitors (Holt et al., 2008). Substrates and inhibitors interact with oxidized enzyme and with at least one form of reduced enzyme following release of aldehyde but prior to release of ammonium (indicated as E_{red}–P). Partial inhibition observed in the presence of high concentrations of substrate suggest that reoxidation of E_{red}–P may occur while the substrate or inhibitor is bound to the reduced enzyme, although these pathways (indicated by question marks) have not been observed directly.

The kinetic behavior of SSAO enzymes is influenced markedly by the ionic environment—both concentrations and identities of buffer salts—in which assays are conducted (Di Paolo et al., 1995a; Di Paolo et al., 2002; Holt et al., 2007). Cations may compete with substrate amine for binding to the enzyme active site, exhibiting different affinities when different states of the TPQ cofactor are present (Holt et al., 2007). With human SSAO, small cations appear to compete more effectively with amine binding to the enzyme form through which substrate inhibition is elicited (Holt et al., 2007), with the result that kinetic curves may show clear substrate inhibition at low cation concentrations, but may tend toward a hyperbolic appearance at higher cation concentrations (Figure 6.2). It has been suggested that metal cations and ammonium may bind close to the oxyanion at C4 of TPQ_{img} in *Hansenula polymorpha* CAO (Plastino et al., 1999), supporting the view that substrate inhibition may occur as a result of amine binding at or near an on-copper TPQ_{img} species.

The practical consequence of such observations is that the choice of buffer for a kinetic assay of SSAO activity can exert profound effects on the results obtained, and a case may be made for use of a buffer that to some degree mimics the extracellular ionic environment in which SSAO enzymes usually operate (Holt and Palcic, 2006).

The ability of ammonium to interact directly with CAO enzymes and influence activity (Narang et al., 2008) is interesting from the point of view that ammonium is a product of SSAO-mediated amine oxidation, and SSAO inhibition by ammonium may thus be of physiological significance. Similarly, H_2O_2 is reported to cause a

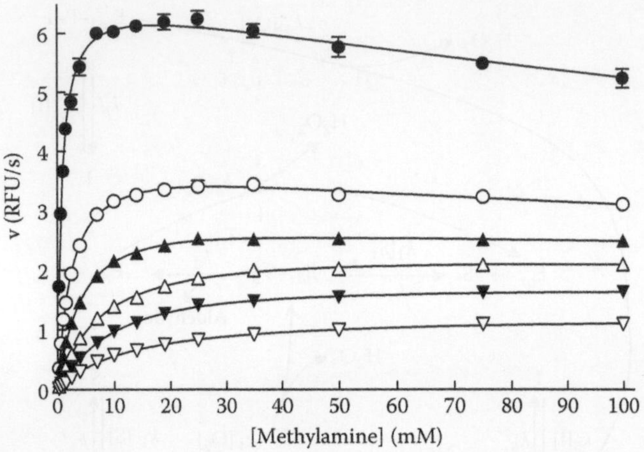

FIGURE 6.2 Effects of ammonium ion concentration on methylamine oxidation by human recombinant SSAO (Narang et al., 2008). Oxidation of methylamine (0.25 to 100 *mM*) by human recombinant SSAO (1 *nM*) was determined by following peroxidase-dependent generation of resorufin from Amplex Red in HEPES buffer (50 *mM*, pH 7.4) in the presence of ammonium chloride at 22 (●), 82 (○), 172 (▲), 372 (△), 722 (▼) and 1272 (▽) *mM*. As ammonium concentration increases, a clear reduction in turnover of methylamine and a loss of substrate inhibition at higher methylamine concentrations occur.

progressive, time-dependent, and substantial inhibition of BPAO when it is generated from substrate benzylamine at 1.5 × K_M (Lee et al., 2001a; Pietrangeli et al., 2004). Inhibition was maximal and activity leveled off at 20 to 25% of that in a control preparation following 4 hours of amine turnover (Lee et al., 2001a). However, when H_2O_2 was added directly to the enzyme and preincubated for 4 hours prior to the addition of benzylamine, more than 80% of control activity remained, indicating that the site of H_2O_2 formation is important. In contrast, we have seen no evidence of time-dependent inhibition of human recombinant SSAO when incubated with benzylamine for 4 hours at 3 × K_M, compared with a sample containing catalase (unpublished data). Activity in the sample containing catalase was 5 to 10% higher than in the sample lacking catalase, with the rate of product generation linear in both samples.

6.5 SUBSTRATES

Substrate selectivities of a variety of mammalian SSAO enzymes have been reviewed extensively by Lyles (1995 and 1996). Common substrates for human, rat, and bovine SSAO are listed in Table 6.2, along with comparative K_M values obtained from the literature.

6.5.1 Endogenous Substrates

6.5.1.1 Unique Substrates

Only two endogenous substrates for membrane-bound SSAO are not also oxidized by EC 1.4.3.4 MAO enzymes. Methylamine was identified as a substrate by Lyles and

TABLE 6.2
K_M Values (*μM*) for Oxidation of Amine Substrates by Mammalian SSAO (AOC3) Enzymes

	Human	Rat	Cattle
Endogenous Unique Substrates			
Methylamine	576[a] (Holt et al., 2008), 832[b] (Lyles et al., 1990), 22 to 56[b] (Castillo et al., 1998)	247[b] (Yu, 1990), 182[b] (Lyles et al., 1990)	235[c], 340[b] (Lizcano et al., 1994; Lizcano et al., 2000a), 210[c] (Lizcano et al., 1998), 8 to 14[b] (Castillo et al., 1998)
Aminoacetone	92[b] (Lyles and Chalmers, 1992), 126[b] (Deng and Yu, 1999)	19[b] (Lyles and Chalmers, 1995)	94[b] (Lizcano et al., 1994)
Endogenous Shared Substrates			
Tryptamine	1320[a,d,e]	54[b] (Andree and Clarke, 1982b), 67[b] (Lyles and Taneja, 1987)	ND[c] (Lizcano et al., 1998)
2-Phenylethylamine	3,780[a] (Holt et al., 2008), 7900[b] (Precious and Lyles, 1988)	16[b] (Yu, 1990), 15[b] (Andree and Clarke, 1982b)	312[b] (Lizcano et al., 1994), 1187[c] (Lizcano et al., 1998)
p-Tyramine	10,500[b] (Precious and Lyles, 1988)	40[b] (Barrand and Callingham, 1982), 52[b] (Guffroy et al., 1985)	ND[b,c] (Lizcano et al., 1990; Lizcano et al., 1998)
Histamine	400[a,d,e]	583[b] (Yu, 1990)	1110[c] (Lizcano et al., 1998)
Dopamine	Not available	270[b] (Andree and Clarke, 1982b)	ND[c] (Lizcano et al., 1998)
Exogenous Substrates (Excluding Pharmaceuticals)			
Benzylamine	194[a] (Holt et al., 2008), 222[b] (Lyles et al., 1990), 155[b] (Yu et al., 1994), 161[b] (Precious and Lyles, 1988), 8 to 14[b] (Castillo et al., 1998)	4-6[b] (Clarke et al., 1982), 6[b] (Yu, 1990), 3[b] (Andree and Clarke, 1982b)	40[b] (Lizcano et al., 1994), 50[c] (Lizcano et al., 1998), 1 to 3[b] (Castillo et al., 1998)
Allylamine	Not available	145[b] (Yu, 1990), 20[b] (Holt, 1990)	Not available

Notes: ND = not detectable; [a] Recombinant enzyme; [b] Homogenate or cells; [c] Purified enzyme; [d] Unpublished; [e] Very low activity at V_{max}.

colleagues (Precious et al., 1988) after observation of anomalously high methylamine oxidation in human plasma (McEwen, 1965a). While iproniazid, a carbonyl MAO inhibitor, enhanced urinary excretion of methylamine in rats (Dar et al., 1985), this enhancement was not due to MAO inhibition, and the involvement of SSAO was confirmed following administration of semicarbazide (Lyles and McDougall, 1988).

Methylamine is found endogenously as a metabolite of adrenaline (Schayer et al., 1952), sarcosine, or creatinine (Davis and de Ropp, 1961), and has exogenous sources such as the products of degradation of ingested choline, lecithin, and creatinine by intestinal bacteria (Jones and Brunett, 1975; Lowis et al., 1985; Zeisel et al., 1983). However, no regulated mechanisms appear to exist for the synthesis and release of methylamine, particularly in tissues rich in SSAO activity. Furthermore, the aldehyde product of methylamine oxidation, formaldehyde, is more toxic than the parent amine (Gubisne-Haberle et al., 2004; Yu et al., 1997; Yu and Zuo, 1996). Methylamine possesses several functional capabilities such as activation of α-2 macroglobulin (Suda et al., 1997) and regulation of voltage-activated potassium channels (Pirisino et al., 2004). However, while SSAO certainly oxidizes methylamine *in vivo*, it is difficult to argue that methylamine represents a physiological substrate for the enzyme since the consequences of methylamine metabolism are largely pathological.

Aminoacetone was initially identified as a substrate for plasma amine oxidases in cattle (Winter et al., 1973) and goats (Ray and Ray, 1983) and was later confirmed as a substrate for membrane-bound SSAO from human (Lyles and Chalmers, 1992) and rat (Lyles and Chalmers, 1995) blood vessels. Aminoacetone can be synthesized from glycine in the presence of acetyl-CoA or via catabolism of L-threonine (Elliott, 1960), with the rate of aminoacetone synthesis partially determined by the ratio of CoA to acetyl-CoA (Callingham et al., 1995; Fubara et al., 1986; Tressel et al., 1986). The product of aminoacetone oxidation, methylglyoxal, is, like formaldehyde, toxic (Bechara et al., 2007; Callingham et al., 1995).

Aminoacetone does not have any known direct physiological effects, and may simply be a metabolic intermediate. Nevertheless, the availability of this amine appears to be more closely regulated than that of methylamine, a more favorable property when potential physiological substrates are considered. While regulation of methylamine and aminoacetone concentrations by SSAO would not appear to offer physiological benefits of great consequence, it is tempting to suggest that generation of hydrogen peroxide is the ultimate role of membrane-bound SSAO. Whether either of these amines represents a physiological source of SSAO-generated hydrogen peroxide is presently unknown, largely because no exhaustive search has sought other potential endogenous substrates unique to SSAO.

Both methylamine and aminoacetone are metabolized efficiently by SSAO from several mammalian species (Table 6.2). In general, the affinity of mammalian SSAO for aminoacetone is higher than for methylamine. Tissue and plasma levels of these amines in mice lie below K_M values and appear unrelated to the level of SSAO in the organs and tissues examined (Xiao and Yu, 2008).

6.5.1.2 Shared Substrates

Several endogenous amines are substrates both for SSAO and for MAO enzymes (Table 6.2). These include several trace amines (Lindemann and Hoener, 2005), some of which have been proposed to act as neuromodulators (Berry, 2004; Xie and Miller, 2008). While it is widely believed that MAO is exclusively responsible for trace amine metabolism, inhibition of both MAO and SSAO results in a more substantial elevation in trace amines than does inhibition of MAO alone (Dewar et al., 1988), suggesting a role for SSAO in trace amine clearance. Dopamine and histamine, substrates for

MAO and DAO respectively, are also metabolized by SSAO in several rat tissues (Buffoni et al., 1994; Lizcano et al., 1991; Raimondi et al., 1993).

6.5.2 Exogenous Substrates

Benzylamine is a naturally occurring amine that may be isolated from the roots and bark of *Moringa pterygosperma* (Chakravarti, 1955). It is an excellent substrate for membrane-bound SSAO in most species (Table 6.2), to the extent that mammalian SSAO enzymes were initially called benzylamine oxidases (Buffoni et al., 1968; Lewinsohn et al., 1978). Human SSAO is somewhat anomalous in that it has a high K_M for benzylamine, and turnover of methylamine is more rapid at a K_M concentration, compared with benzylamine (Lyles et al., 1990; Precious et al., 1988). Benzylamine is also a substrate for MAO-B, with the consequence that with benzylamine as substrate, examination of SSAO in tissue samples containing MAO-B requires prior inactivation of MAO-B, usually with an acetylenic irreversible inhibitor such as pargyline or selegiline (deprenyl) (Lizcano et al., 1991). However, direct or indirect compensatory alterations in SSAO activity have been observed (Fitzgerald and Tipton, 2002; Yu et al., 1992), and such experiments must be carefully controlled.

Allylamine is an extremely reactive alkylamine used industrially to manufacture pharmaceuticals, rubber, and polymer plastics. Highly selective cardiovascular toxicity has been observed with allylamine in a number of animal models (Boor and Ferrans, 1985; Boor and Hysmith, 1987), including myocardial necrosis, intimal proliferation in the aorta and coronary arteries, and ventricular aneurysm formation. SSAO inhibition prevents allylamine-induced cardiovascular toxicity (Boor and Nelson, 1980), and the toxic metabolite has been identified as the aldehyde acrolein generated by SSAO-mediated allylamine oxidation in the vasculature (Boor et al., 1990). Although kinetic data are not available for oxidation of allylamine by human SSAO, acrolein generation from allylamine in human blood vessels rich in SSAO has been demonstrated (Conklin et al., 2006).

Although literature coverage is rather sparse, it is likely that SSAO enzymes have a role to play in the plasma clearance of some pharmaceuticals. In particular, it has been suggested that amlodipine clearance may be due, at least in part, to SSAO (Strolin-Benedetti, 2001), and this area merits greater consideration than it has already been afforded (Strolin-Benedetti et al., 2007).

6.6 INHIBITORS

The current situation with respect to inhibitor availability has been reviewed extensively in Chapter 14, and comments here are limited to mentions of additional examples relating to inhibition of membrane-bound SSAO. Inhibitors mentioned are classified based upon selectivity.

6.6.1 SSAO Inhibitors That Also Inhibit Monoamine Oxidase

Most available SSAO inhibitors also affect MAO activity to some extent. This is perhaps not surprising, given the degree to which substrates are shared between these enzymes, and thus the presumed similarities between the ionic and hydrophobic environments within the active site cavities of SSAO and MAO.

6.6.1.1 Nonhydrazine Substrate Analogues

Many of these compounds are α-methylated primary amines; for example, α-methylation of substrate phenylethylamine yields amphetamine—a competitive inhibitor of MAO that displays some selectivity for MAO-A (Mantle et al., 1976). In contrast, noncompetitive inhibition of SSAO has been described (Clarke et al., 1982), although our data for human SSAO suggest that inhibition does indeed occur at the active site, and reversibly, but with a preference for binding to the reduced form of the enzyme (Holt et al., 2008). The anti-arrhythmic drug mexiletine, the MAO-A inhibitor amiflamine, and several compounds related to amiflamine are modest inhibitors of rat SSAO, but mexiletine at least is less potent against the human enzyme (Clarke et al., 1982; Eriksson and Fowler, 1984). Although inhibition by α-substituted amines is presumed due to an inability of the enzyme to initiate catalysis upon these substrate analogues, an irreversible component has been attributed to amiflamine (Fowler et al., 1984). This observation is surprising, and may raise questions with regard to the purity of the inhibitor.

6.6.1.2 Hydrazine-Based Inhibitors

Based originally upon the antitubercular drug isoniazid, hydrazine compounds are substrate analogues that often display reasonable selectivity for SSAO over MAO but are otherwise less than ideal as pharmaceuticals due to hepatotoxicity and a lack of selectivity with respect to other carbonyl-containing enzymes. Inhibition of SSAO is usually potent and irreversible or slowly reversible. Phenelzine is a very potent SSAO inhibitor displaying an IC_{50} versus bovine lung SSAO of 20 *nM* (Lizcano et al., 1996) and versus rat aorta SSAO of 1.6 *nM* (Lyles and Callingham, 1982a). Inhibition of the bovine enzyme was slowly reversible, and kinetic plots revealed a complex mechanism of interaction suggestive of multiple binding sites for the inhibitor (Lizcano et al., 1996).

Our recent experiences with reversible inhibitors indicate that multisite inhibitor kinetics are suggested in cases where an inhibitor binds to the same site on the enzyme in different oxidation states (Holt et al., 2008), although the presence of two distinct sites for phenelzine cannot be dismissed. Alternatively, SSAO-mediated oxidation of phenelzine has been reported (Andree and Clarke, 1982a), with the oxidized product causing slowly reversible inhibition of the rat enzyme. This may have contributed to the apparent complexity of inhibition kinetics for phenelzine.

Both phenylhydrazine and hydralazine show potent, time-dependent inhibition of bovine aorta SSAO that is not reversible (Lizcano et al., 1996). Hydralazine is a potent inhibitor of rat vascular SSAO, exhibiting an IC_{50} of 25 *nM* (Lyles and Callingham, 1982a). Potency against human SSAO is less impressive than against the rat enzyme, displaying an IC_{50} of ~1.6 μM with human recombinant SSAO (unpublished data). The ability of hydralazine to inhibit vascular SSAO has led to claims that SSAO inhibition by hydralazine may be responsible for its vasodilator properties (Vidrio, 2003; Vidrio et al., 2000). While this property of hydralazine may contribute to effects on the vasculature, other properties of hydralazine must also be considered in this regard (Bang et al., 1998).

The lack of selectivity of hydrazine-based compounds (Binda et al., 2008; Lyles, 1984) is further illustrated by benserazide and carbidopa. They are relatively effective SSAO inhibitors (Andree and Clarke, 1982a; Lewinsohn et al., 1978; Lyles and Callingham, 1982b), but are used clinically as inhibitors of peripheral dopa decarboxylase for administration to Parkinsonian patients. Dopa decarboxylase is a carbonyl-containing enzyme, with pyridoxal 5′-phosphate as a cofactor. Such enzymes are susceptible to inhibition by hydrazines.

6.6.1.3 (Halo)allylamine Derivatives

A series of halogenated aromatic allylamine derivatives were synthesized initially as suicide substrates for MAO enzymes (McDonald et al., 1985). Several were later found to possess good potency as inhibitors of SSAO (Lyles et al., 1987; Palfreyman et al., 1986). Indeed, one of these (MDL-72274) was far more potent *in vitro* compared to rat SSAO (IC_{50} 8 *nM*) than rat liver MAO-A (IC_{50} 520 *μM*) or MAO-B (IC_{50} 11.7 *μM*). Nevertheless, while potent and selective for SSAO *in vitro*, we found that following chronic interperitoneal administration of MDL-72274 to rats at a 2 mg/kg dose daily for 14 days—a dose causing around 70% inhibition of SSAO—selectivity over MAO-B was entirely lost (Holt et al., 2003). This may have been a consequence of phase I metabolism or the result of a lack of purity of the compound (McDonald et al., 1985).

6.6.1.4 Miscellaneous Compounds

Tranylcypromine, an antidepressant that inactivates MAO irreversibly, has been shown to act as a competitive reversible and rather weak inhibitor of human recombinant SSAO (K_i of ~600 *μM*), displaying significant selectivity for the oxidized form of the enzyme (Holt et al., 2008). Tranylcypromine inhibits other soluble CAO enzymes with modest affinity (Knowles et al., 2007; Shepard et al., 2003) and it has been suggested that administration of tranylcypromine as an antidepressant may result in side effects due to SSAO inhibition (Wilmot et al., 2004). This may also be true following administration of other antidepressants (Obata and Yamanaka, 2000b), although SSAO inhibition, even following chronic antidepressant administration, is likely to be modest at best, given the relatively low potency of these drugs as SSAO inhibitors.

Guanabenz is an imidazoline binding site ligand shown previously to inhibit rat liver MAO-A and MAO-B with IC_{50} values of 4 and 40 *μM*, respectively (Ozaita et al., 1997) and pig plasma amine oxidase with an IC_{50} of 5.1 *μM* (Banchelli et al., 1986). We demonstrated reversible competitive inhibition of the oxidized form of human recombinant SSAO by guanabenz (K_i 11 to 14 *μM*), although appreciable binding to the active site of reduced SSAO (K_i 26 to 65 *μM*) also occurs at concentrations of guanabenz effective against MAO-B (Holt et al., 2008). Other imidazoline ligands such as guanfacine and oxymetazoline also inhibit human recombinant SSAO at mid-micromolar concentrations (unpublished data). Oxymetazoline is a very weak MAO inhibitor (Ozaita et al., 1997) while the effects of guanfacine on MAO have not been assessed. We observed inhibition of human recombinant SSAO by aminoguanidine (unpublished data), although with an IC_{50} (~30 *μM*) more than 1000-fold less potent than that versus porcine kidney diamine oxidase (Holt and Baker, 1995).

6.6.2 SSAO Inhibitors Showing Selectivity *in Vitro*

Only a relatively few selective inhibitors of membrane-bound SSAO have been identified. Selectivity is defined against MAO enzymes, although it is possible and indeed likely that many of these compounds exert other effects elsewhere.

6.6.2.1 Hydrazine-Based Inhibitors

Semicarbazide is the archetypal inhibitor upon which the nomenclature of SSAO enzymes is based. Originally defined as enzymes that may be inhibited by carbonyl reagents, Zeller (1959) argued that the term *carbonyl reagent* should be replaced by *semicarbazide* since some carbonyl reagents may also inhibit MAO enzymes. Concentrations of semicarbazide between 0.1 and 1 *mM* are usually sufficient to inhibit SSAO completely, with negligible effects on MAO (Lyles, 1996; Lyles and Callingham, 1975).

The carcinostatic agent procarbazine is a remarkably effective SSAO inhibitor *in vitro* (Holt et al., 1992b), displaying an IC_{50} versus rat SSAO of 210 *nM*, compared with values in excess of 1 *mM* against MAO enzymes. This selectivity was maintained *in vivo* (Holt and Callingham, 1995), with ID_{50} values versus SSAO lower than 10 mg/kg and negligible effects on MAO-A and MAO-B at intraperitoneal doses up to 100 mg/kg. The proximal metabolite of procarbazine, azoprocarbazine, was more potent against rat SSAO *in vitro* (Holt et al., 1992a), while maintaining selectivity over MAO enzymes. A distal metabolite, monomethylhydrazine, was an extremely potent SSAO inhibitor *in vitro* with an IC_{50} of 7 *nM* compared with 63 and 98 μM against hepatic MAO-A and MAO-B, respectively (Holt et al., 1992a). This potency was maintained *in vivo*; 1 mg/kg monomethylhydrazine administered intraperitoneally was sufficient to cause complete SSAO inhibition in most tissues, with no inhibition of MAO (Holt and Callingham, 1995). However, in brown adipose tissue, both procarbazine and monomethylhydrazine caused an intriguing dose-dependent increase in MAO-A, with 1 μg/kg monomethylhydrazine causing a fourfold increase in MAO-A within 2 hours of administration. These effects, which did not occur *in vitro*, are consistent with responses of rat brown adipose tissue MAO-A to benserazide reported by Lyles and Callingham (1982b), but have yet to be explained.

Although many hydrazine-based SSAO inhibitors display excellent potency and selectivity, the likelihood of toxicity, including hepatotoxicity, with prolonged administration means that development of alternative safer compounds remains a priority.

6.6.2.2 Nonhydrazine Inhibitors

Hydroxylamine is a potent inhibitor of rat SSAO, with complete inhibition observed at 1 μM, a concentration ineffective versus MAO (Lyles, 1996; Lyles and Singh, 1985). Human recombinant SSAO is also inhibited by hydroxylamine with an IC_{50} of 71 *nM* (our laboratory, unpublished data).

Several pyridine analogues show modest potency as SSAO inhibitors, with inhibition perhaps due to their actions as slow substrates for the enzyme. Pyridoxamine showed time-dependent inhibition of rabbit SSAO from lung and heart with an IC_{50} of ~10 μM (Buffoni et al., 1989), while reversibility with dialysis was very slow. Data did not reveal whether pyridoxamine was simply a slow-binding inhibitor or whether catalytic activation to a reactive transition state intermediate first occurred.

An analogue of pyridoxamine, B24, was more potent, showing time-dependent inhibition with an IC_{50} versus rat SSAO of 0.3 μM and greater than 1000-fold selectivity over MAO (Banchelli et al., 1990; Buffoni et al., 1990). While also effective *in vivo* (Buffoni et al., 1990), the enzyme recovered rapidly following cessation of drug administration, suggesting that inhibition was not irreversible.

A series of novel allylamine derivates structurally related to compounds described in Section 6.1.3 were recently synthesized; one, LJP 1586 (Z-3-fluoro-2-(4-methoxybenzyl)allylamine), is a potent and selective inhibitor displaying IC_{50} values versus SSAO ranging from 4 *nM* (mouse) through 9 *nM* (rat) to 27 and 43 *nM* (human umbilical cord SSAO and recombinant enzyme, respectively). The compound showed 80-fold selectivity over human MAO-B and more than 3000-fold selectivity over MAO-A. Potency and selectivity were maintained *in vivo* and pharmacokinetic parameters in the rat were acceptable, although oral bioavailability (15%) was modest and the volume of distribution at steady state (5l/kg) suggests extensive partitioning into the tissue compartment. Nevertheless, this compound should prove highly useful as a research tool in experiments designed to determine physiological roles for SSAO.

6.7 ASSAYING SSAO ACTIVITY

Research focusing on SSAO function and development of novel inhibitors requires the ability to determine SSAO activity in a quantitative, reproducible manner with a sensitive assay system. Furthermore, the capability to examine large numbers of samples in a single assay is critical in order to generate kinetic curves with sufficient numbers of points across a sufficiently wide substrate concentration range, and with an appropriate number of replicate determinations, to optimize the likelihood of success of nonlinear regression analyses.

The most popular approaches to SSAO quantification may be grouped under four main headings: chromatography approaches, techniques measuring disappearance of oxygen, radioisotopic approaches, and spectrophotometric and/or spectrofluorimetric approaches. Each general approach presents advantages and disadvantages, with availability of appropriate instrumentation perhaps representing the single most important factor in the choice of assay.

Techniques based upon quantification following HPLC separation, while perfectly valid, do not lend themselves to extensive quantitative analyses of kinetic behavior. Therefore, beyond indicating literature examples of such approaches (Li et al., 2004; van Dijk et al., 1995; Yu et al., 2003a), no further discussion will ensue. A more extreme example of a low throughput approach is use of an oxygen electrode (DuBois and Klinman, 2005), a technique highly suited to specialized applications but not ideal for day-to-day SSAO quantification.

6.7.1 Radiochemical Assays

Radiochemical analyses of SSAO activities have typically examined oxidation of [^{14}C]benzylamine followed by organic solvent extraction of nonpolar metabolites (Lyles and Callingham, 1982b) or of [^{14}C]methylamine followed by column chromatographic separation of nonpolar metabolites (Lyles et al., 1990). With benzylamine, a single user can complete around 300 individual assays in 8 hours, whereas with methylamine, column preparation and separation reduces that number by 50%.

One major advantage of a radiochemical approach is the ability to use whole cells, tissue homogenates, or perfused vascular preparations as the SSAO source (Elliott et al., 1989). Furthermore, the sensitivity of the technique facilitates measurement of extremely low enzyme activity. However, due to the discontinuous nature of the assay, experiments must be carefully controlled for linearity of product formation (Holt, 2007). Also, the presence of other enzymes such as MAO-B that contribute to benzylamine metabolism must be addressed by use of an irreversible inhibitor. Experiments are also relatively expensive and generate radioactive waste.

A problem with radioisotopic procedures that has not generally been perceived as such relates to the limited availability of substrates in radiolabeled form, with the label in a position where it will not be lost during catalysis or undergo isotopic exchange with bulk solvent (Farnum et al., 1986). Recent data from our laboratory indicate that reversible SSAO inhibitors affect enzyme activity in a substrate-dependent manner (Holt et al., 2008). Therefore, it is inappropriate to examine the behavior of SSAO with, for example, [^{14}C]benzylamine, or the effects of an SSAO inhibitor versus [^{14}C]benzylamine oxidation and then presume that results with other substrates would be similar or identical. For this reason, techniques facilitating the use of a variety of substrates are preferable.

6.7.2 Spectroscopic Assays

Several spectroscopic techniques are suitable for quantifying oxidation of a specific substrate. For example, generation of benzaldehyde from benzylamine may be monitored directly at 254 nm (Tabor et al., 1954), while formaldehyde dehydrogenase-mediated formation of NADH may be monitored at 340 nm in a coupled assay as an indirect measure of methylamine oxidation (Lizcano et al., 2000a).

Other coupled assays for SSAO use H_2O_2 generated by the enzyme to drive a horseradish peroxidase-dependent stoichiometric generation of a colored or fluorescent species. Indeed, such is the adaptability of peroxidase-coupled assays that luminescent species may also be generated and quantified (Schwelberger and Feurle, 2007). Nevertheless, approaches relying upon absorbance or fluorescence detection are used widely and have become more popular with increasing availability of multiwell plate readers. With access to a suitable plate reader and no requirement for automated plate preparation, a single user can complete up to 1000 individual assays in 8 hours.

Several absorbance-based assays make use of a variety of chromogenic peroxidase substrates. Perhaps the most common is an assay developed by Sharman that uses 4-aminoantipyrene and either 2,4-dichlorophenol (Elliott et al., 1991) or vanillic acid (Holt and Palcic, 2006; Holt et al., 1997) as chromogens. Color formation is monitored at around 500 nm and may thus be followed in inexpensive polystyrene microplates. The sensitivity of the assay may be enhanced (albeit modestly) through use of a fluorogenic peroxidase substrate such as Amplex Red (Zhou and Panchuk-Voloshina, 1997).

Several substrates including dopamine can act as hydrogen donors in the peroxidase reaction (Nelson and Huggins, 1974) and are thus unsuitable for use in a peroxidase-coupled SSAO assay procedure. Furthermore, we have found that several substrates and a number of other test compounds including some inhibitors interfere with the Amplex Red assay through, for example, fluorescence quenching. It is prudent to screen compounds to be included in these assays for the potential to cause problems. The problems likely to be encountered and the methodologies designed to identify and circumvent them are outlined elsewhere (Holt and Palcic, 2006).

Regardless of the assay method chosen, the choice of buffer is very important if a realistic enzymatic behavior is to be mapped (Holt et al., 2007; Holt and Palcic, 2006). Furthermore, because it is now clear that SSAO enzymes do not follow simple Michaelis–Menten kinetics (Holt et al., 2008), appropriate nonlinear regression analyses should be applied to data. Linear regression of transformed data, such as Lineweaver–Burk or Hanes–Woolf curve-fitting, may generate misleading axes intercepts, potentially resulting in a misunderstanding of the effects that inhibitors or site-directed mutagenesis may exert upon enzyme behavior.

6.8 WHAT DOES SSAO DO?

This chapter has outlined characteristics and behaviors typical of membrane-bound SSAO enzymes. Such an outline is important because of the significant differences among these enzymes and soluble CAO and mitochondrial MAO enzymes that share some substrates and inhibitors with SSAO. What have not been discussed are the potential physiological and pathophysiological roles played by SSAO; other chapters in this book will discuss these in depth. However, some final remarks on potential roles of SSAO, as they relate to the enzyme's characteristics, are nevertheless pertinent.

The association of membrane-bound SSAO with vascular smooth muscle and adipocyte membranes in particular, and the remarkably high degree of expression of the enzyme on the surfaces of these cells, strongly suggests that SSAO plays an important role in the vascular wall. Protection against deleterious circulating amines would be better achieved by an enzyme associated with vascular endothelium. Furthermore, both Yu (Gubisne-Haberle et al., 2004) and Boor (Conklin et al., 2001) have shown clearly that generation of toxic aldehydes within vessel walls has undesirable consequences. Thus, it seems unlikely that vascular SSAO functions simply as a scavenger enzyme, and a more subtle role seems likely. It is also possible that under conditions of elevated SSAO substrate levels (Yu and Zuo, 1993), cytoprotective mechanisms may be overwhelmed and vascular damage that would not occur when SSAO is operating at a physiological level becomes evident.

The importance of H_2O_2 in eliciting SSAO-mediated responses in adipocytes (Enrique-Tarancón et al., 2000; Enrique-Tarancón et al., 1998; Marti et al., 1998; Visentin et al., 2003; Zorzano et al., 2003), coupled with the role of H_2O_2 as a vasoactive signaling molecule (Ardanaz and Pagano, 2006) leads to the obvious conclusion that the physiological function of membrane-bound SSAO may be as a source of H_2O_2, with physiological responses to SSAO thus dependent upon the tissue and the availability of substrate. Some regulation of substrate availability would then be important, and neither methylamine nor aminoacetone is an ideal substrate in this regard. Perhaps the physiological substrate for SSAO has yet to be identified.

The changes of SSAO activity under several disease conditions (Lewinsohn, 1984)—and the presence of marked changes in differentiating cells (Bour et al., 2007a; El Hadri et al., 2002; Moldes et al., 1999)—may offer clues as to physiological roles for SSAO and also to the transduction pathways within which H_2O_2 may be active. Alternatively, an involvement of H_2O_2 need not be required if SSAO plays a lysyl oxidase-like role, recognizing amine groups on cell surface proteins (Olivieri et al., 2007; Wang et al., 1996a) or sugars (O'Sullivan et al., 2003a) and catalyzing reactions that lead to structural changes within connective or elastic tissue or changes in cell–cell interactions. The effects of SSAO inhibition (Langford et al., 1999; Langford et al., 2002), knock-out (Mercier et al., 2006), and over-expression (Göktürk et al., 2003a) on vascular architecture and observations of such effects in aortic aneurysm (Sibon et al., 2008) lend credibility to this view.

Progress in understanding the implications of SSAO-mediated amine oxidation has increased our understanding of what these enzymes *can* do. However, most consequences are pathological rather than physiological, and what SSAO *does* do is far from clear. We cannot be confident of the safety of chronic administration of SSAO inhibitors to address pathological consequences of oxidation until we develop a better understanding of the reason for high SSAO activity in smooth muscles and adipocytes. Addressing this point must be made a priority if SSAO is to become a validated therapeutic target.

ACKNOWLEDGMENTS

Research in the author's laboratory is supported by the Canadian Institutes of Health Research.

7 Copper Amine Oxidase Genes

Ivo Frébort and Hubert G. Schwelberger

CONTENTS

7.1 INTRODUCTION

Copper/quinone containing amine oxidases (EC 1.4.3.6, amine: O_2 oxidoreductase [deaminating, copper containing]) have been intensively studied since the 1950s. They appear to be widespread enzymes and are found in bacteria, yeast and fungi, plants, fish, birds, and mammals. The first gene encoding a copper amine oxidase from the yeast *Hansenula polymorpha* (now *Pichia angusta*) was cloned and sequenced in 1989 (Bruinenberg et al., 1989) followed by genes from lentil seedlings, *Lens esculenta* (Rossi et al., 1992), bacteria, *Klebsiella aerogenes* (Yamashita et al., 1993) and *Arthrobacter* strain P1 (Zhang et al., 1993), rat colon (Lingueglia et al., 1993), human kidney (Novotny et al., 1994), *Escherichia coli* K12 (Azakami et al., 1994), bovine serum (Mu et al., 1994), *Arthrobacter globiformis* (Tanizawa et al., 1994; Choi et al., 1995), pea seedlings (Tipping and McPherson, 1995), and several others.

Further information about amine oxidase genes from various organisms was generally obtained by a similarity search against newly sequenced genes and genomes and such genes were annotated as putative copper amine oxidase. Due to the strictly conserved active site amino acid consensus sequence NYD/E that contains the tyrosine precursor (Y) of the organic cofactor TPQ, three absolutely conserved histidine residues binding the copper ion, an aspartic acid residue functioning as a catalytic base, and other structurally important residues identified by structural studies (see Chapters 2 and 5), the prediction of copper amine oxidase genes from genomic sequence assemblies is highly reliable. The first copper amine oxidase identified by sequence similarity was the human kidney amiloride-binding protein cloned in 1990 and identified as amine oxidase four years later (Mu et al. 1994).

This chapter focuses on the distribution of copper amine oxidase genes in diverse organisms from prokaryotes to eukaryotes, trying to address their evolutionary aspects and, in eukaryotes, their functional characteristics such as expression pattern, organ distribution, and subcellular localization. Data for this study were obtained from publicly accessible databases on the Internet and processed with widely available search and prediction software.

7.2 GENES IN PROKARYOTES

7.2.1 Gram-Negative Bacteria: *Escherichia coli* and *Klebsiella aerogenes*

Periplasmic tyramine oxidase from *Escherichia coli* K12 is a product of the *maoA* gene (Azakami et al., 1994). A BLAST search (tblastn 2.2.18; Altschul et al., 1997) of bacterial genomes with the *E. coli* tyramine oxidase as a protein query sequence (NP_415904) revealed over 80 hits indicating the presence of copper amine oxidase genes in other bacteria such as *Acinetobacter baumannii*, *Legionella pneumophila*, *Mycobacterium* sp., *Rhodococcus* sp., *Arthrobacter* sp., *Bradyrhizobium* sp., etc., but certainly not in all bacterial species.

Klebsiella aerogenes tyramine oxidase (Sugino et al., 1991) is closely related to that of *E. coli*. Both bacteria share the same regulatory mechanism to control the copper amine oxidase gene related to the synthesis of arylsulfatase (Murooka et al., 1996).

7.2.2 Gram-Positive Bacteria: Arthrobacter Species

Arthrobacter globiformis can produce two different amine oxidases when induced by either phenethylamine or histamine. A gene coding for phenethylamine oxidase was cloned first (Tanizawa et al., 1994). Histamine oxidase from *A. globiformis* has different substrate specificity; its cloned gene yielded a sequence different from that of phenethylamine oxidase (Choi et al., 1995). The phenethylamine oxidase gene was used for producing recombinant apoenzyme in *E. coli* and solving TPQ biosynthesis (Kim et al., 2002; Moore et al., 2007).

Arthrobacter strain P1 produces a methylamine oxidase that is distinct from the above enzymes (Zhang et al., 1993).

7.3 GENES IN LOWER EUKARYOTES AND PLANTS

7.3.1 Copper Amine Oxidase Genes in Fungi and Yeast

Studies of copper amine oxidases focused mainly on the *Pichia angusta* yeast and the *Aspergillus niger* fungus. *P. angusta* (formerly *Hansenula polymorpha*) is known to contain two genes that can produce different amine oxidases, depending on the inducing amine, a peroxisomal methylamine oxidase (Bruinenberg et al., 1989, Cai and Klinman, 1994) or a benzylamine oxidase (Mu et al., 1992).

Amine oxidase ANAO-I (EC 1.4.3.6) from *A. niger* AKU 3302 that is induced by adding *n*-butylamine into a growth medium of the fungus has been studied for several decades (Frébort and Adachi, 1995; Frébort et al., 1996). The effort finally resulted in cloning the respective structural gene (GenBank Accession No. AF362473), understanding its regulation, deducing the amino acid sequence of the enzyme (GenBank Protein Accession No. AAK51081), and modeling the 3-D structure (Frébort et al., 2003). Another amine oxidase that is more specific to methylamine was purified and characterized from the same strain of *A. niger* (Frébort et al. 1999). *A. niger* is an organism of growing biotechnological importance, currently as a host for the production of heterologous proteins. A BLAST search (tblastn 2.2.18) with ANAO-I as a protein query sequence on NCBI Map Viewer against the recently published genome sequence of *A. niger* strain CBS 513.88 (Pel et al., 2007) detected six map elements bearing genes coding for copper amine oxidase, including An09g01550 that encodes a homologue to ANAO-I from *A. niger* AKU 3302 and An17g00010 that probably represents the methylamine oxidase gene (Frébort et al. 1999).

Localization of ANAO-I via electron microscopy clearly shows that the protein is secreted (Frébort et al., 2000b), although it lacks a cleavable N-terminal signal peptide. A recombinant ANAO-I was secreted by a *Saccharomyces cerevisiae* host (Kolaříková et al., 2008). The functions of the other enzymes and their cellular localizations are still unknown. Target P1.1 (Emanuelsson et al., 2007) indicated the product of An13g00710 as a possible mitochondrial protein. The other amine oxidases were also shown as nonsecreted by the Secretome P1.0b (Dyrløv-Bendtsen et al., 2004) analysis for nonclassical and leaderless secretion of proteins.

A. niger has a high capacity secretory system, but its secretory pathway is still poorly characterized compared to yeast (Conesa et al., 2001; Sims et al., 2005). Recent results show *A. niger* homologues to *S. cerevisiae* proteins Srp54p and Kar2p/BipA, suggesting that both signal recognition particle-dependent and -independent routes of yeasts are also present in filamentous fungi. Apart from copper amine oxidases, *A. niger* was also shown to express a FAD monoamine oxidase MAO-N (Schilling and Lerch, 1995). BLAST identified the gene coding for this enzyme as the entry NT_166527/An12g03290 and found two other genes for FAD amine oxidases with less similarity (NT_166518/An01g01840 and NT_166519/An02g08370). In its natural habitat, *A. niger* secretes large amounts of a wide variety of enzymes needed to release nutrients from biopolymers, mainly coming from decaying plant material. Its genome contains six and three genes coding for copper amine oxidase and FAD amine oxidase, respectively. It is probable that different substrate specificities of these enzymes direct the highly developed metabolic apparatus of the fungus to support its saprophytic lifestyle.

A BLAST search (tblastn) over the entire database of nucleotide sequences of fungi with ANAO-I protein sequences as a query produced over 200 hits indicating that copper amine oxidase genes are abundant, especially in filamentous fungi. In accordance with previous results based on electrophoretic screening for amine oxidase (Frébort et al., 1997), copper amine oxidases were detected in Aspergillus, Penicillium, Fusarium, Trichoderma, and Gibberella genera and also in other yeast and fungi such as Candida, Coccidioides, etc. Of 17 completed fungal genomes to date, 13 contain two or more copper amine oxidase genes as shown in Table 7.1. No genes for copper amine oxidase were found in baker's yeast *Saccharomyces cerevisiae* S288c, a genetically closely related yeast *Candida glabrata* CBS138 (cause of human candidiasis), a soil-borne fungus *Cryptococcus neoformans* JEC21 (opportunistic pathogen, causes cryptococcosis), or an intracellular parasite *Encephalitozoon cuniculi* GB-M1 that belongs to the microsporidia (infects mammals, causes a variety

TABLE 7.1
Copper Amine Oxidase Genes in Completed Fungal Genomes

Organism	Copper Amine Oxidase Genes*	Genome Map Element/Gene Locus Tag	Number of Introns	Protein ID
Aspergillus clavatus NRRL1	(5)	NW_001517097/ ACLA_042070	3	XP_001271411
		NW_001517103/ ACLA_063240	3	XP_001273783
		NW_001517104/ ACLA_010630	2	XP_001274063[p]
		NW_001517102/ ACLA_005880	2	XP_001273258
		NW_001517110/ ACLA_003780	0	XP_001276391[t]
Aspergillus fumigatus Af293	4(2)	NC_007194/AFUA_1G13440	4	XP_752714
		NC_007196/AFUA_3G14590	3	XP_754167
		NC_007198/AFUA_5G01470	0	XP_748195[p]
		NC_007200/AFUA_7G04180	2	XP_749090
		NZ_AAHF01000003		
		NZ_AAHF01000010		
Aspergillus niger CBS 513.88	6	NT_166519/An02g10920	2	XP_001400221
		NT_166520/An03g00730	3	XP_001389976
		NT_166523/An07g06400	3	XP_001391717
		NT_166525/An09g01550	3	XP_001393472
		NT_166528/An13g00710	3	XP_001396211
		NT_166532/An17g00010	2	XP_001398232
Candida glabrata CBS138	0			
Cryptococcus neoformans JEC21	0			
Debaryomyces hansenii CBS767	2	NC_06047/DEHA0E03355g	0	XP_459444
		NC_06049/DEHA0G06336g	0	XP_461794
Encephalitozoon cuniculi GB-M1	0			
Eremothecium gossypii ATCC 10895	1	NC_005788/AGOS_AGL073W	0	NP_986593

(Continued)

TABLE 7.1 (*Continued*)

Gibberella zeae PH-1	5(3)	NT_086521/not annotated		
		NT_086559/not annotated		
		NT_086532/not annotated		
		NT_086560/not annotated		
		NT_086558/not annotated		
		NZ_AACM02000091		
		NZ_AACM02000070		
		NZ_AACM02000377		
Kluyveromyces lactis NRRL Y-1140	1	NC_006041/KLLA0E18436g	0	XP_454779
Magnaporthe grisea 70-15	5	NW_001798725/MGG_10751	6	XP_360439
		NW_001798722/MGG_05808	2	XP_369656
		NW_001798736/MGG_03332	2	XP_360789
		NW_001798739/MGG_02681	1	XP_366605
		NW_001798743/MGG_09602	1	XP_364757
Neurospora crassa OR74A	2	NW_001849808/NCU05518	3	XP_960480[p]
		NW_001849755/not annotated		
Pichia stipitis CBS 6054	2	NC_009042/PICST_83878	0	XP_001382852
		NC_009045/PICST_55523	0	XP_001384892[p]
Saccharomyces cerevisiae S288c	0			
Schizosaccharomyces pombe 972h-	2	NC_003423/SPAC2E1P3.04	0	NP_596841
		NC_003424/SPBC1289.16c	0	NP_593985
Ustilago maydis 521	2	NW_101080/not annotated		
		NW_101082/not annotated		
Yarrowia lipolytica CLIB122	2	NC_006067/YALI0A16445g	0	XP_500130
		NC_006069/YALI0C18315g	0	XP_501971

Notes: * Numbers indicate genes found by BLAST search (tblastn) on NCBI Map Viewer using *Aspergillus niger* ANAO-I protein sequence as a query (AAK51081). Numbers in parentheses show additional hits not mapped to chromosomal locations; [p] = Annotated as peroxisomal protein; [t] = Truncated protein sequence, possible pseudogene.

of conditions of the nervous system and respiratory and digestive tracts). Of other fully sequenced aspergilla, *A. fumigatus* Af293 shows four mapped copper amine oxidase genes including homologues to *A. niger* ANAO-I and methylamine oxidase, AFU_3G14590 and AFUA_5G01470 (annotated as a peroxisomal enzyme), respectively. The genome also shows two other genes mapped to chromosomal locations. *A. clavatus* NRRL 1 contains five unmapped genes. Exon/intron analysis showed that typical yeast genes usually do not contain introns (*Debaryomyces hansenii*, *Kluyveromyces lactis*, *Pichia stipis*, *Schizosaccharomyces pombe*, *Yarrowia lipolytica*), whereas aspergilla genes contain up to four introns. Compositions of exons and introns of fungal genes are shown in Figure 7.1.

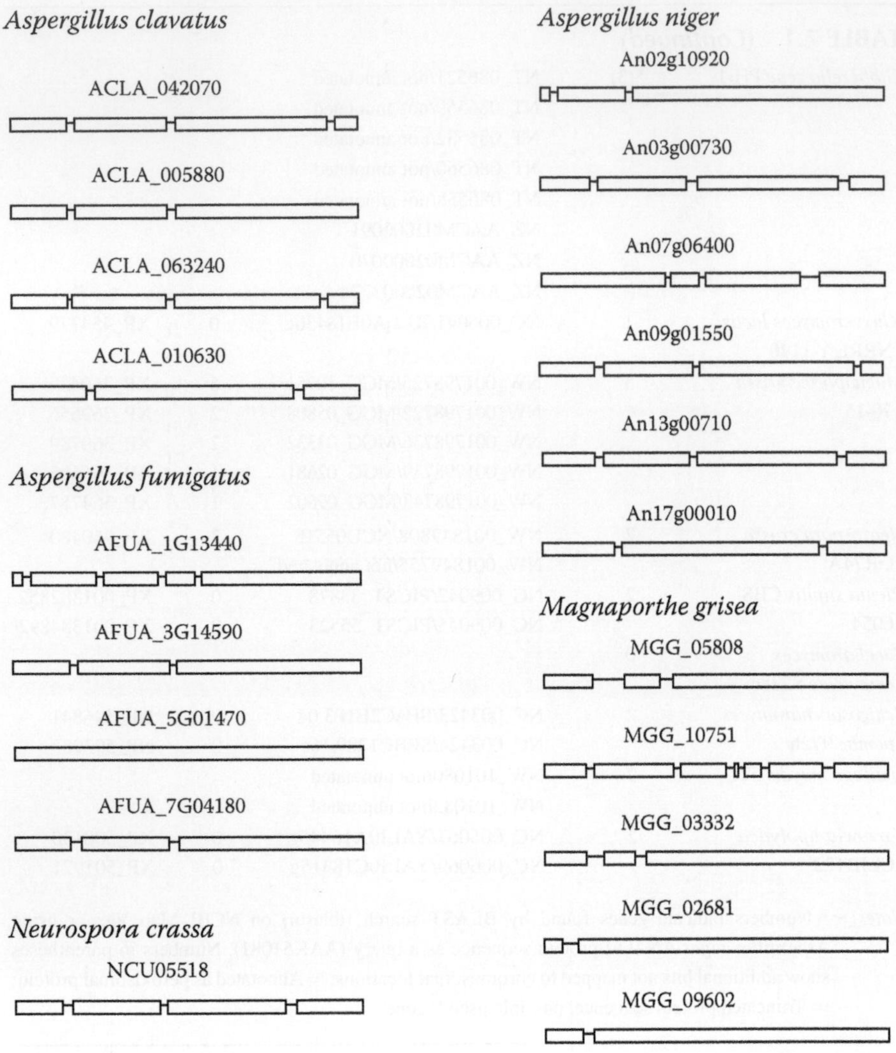

FIGURE 7.1 Exon/intron composition of copper amine oxidase genes from complete fungal genomes.

Promoter analysis of the 5′-upstream region of the *A. niger anao-1* gene revealed 14 potential binding sites for NIT2 (Frébort et al., 2003), a global transcription factor already found in other fungi such as *Neurospora crassa* and *Aspergillus nidulans*. NIT2 positively regulates the expression of up to 100 genes related to nitrogen metabolism expression including those of nitrite and nitrate reductases (Fu and Marzluf, 1990). This observation agrees with the general view that amine oxidase allows an organism to use amines as carbon and nitrogen sources for growth.

7.3.2 Copper Amine Oxidase Genes in Plants

Genes that were cloned and sequenced as cDNA include those coding for copper amine oxidase from the seedlings of *Lens culinaris* (Rossi et al., 1992, corrected by Tipping and McPherson, 1995; S78994), *Pisum sativum* (Tipping and McPherson, 1995), PEACAO, (Koyanagi et al., 2000), *Lathyrus sativus* (Kopečný et al., unpublished, AJ786401), *Cicer arientium* (Rea et al., 1998; AJ009825), *Euphorbia characias* latex (Padiglia et al., 1998a, AF171698), *Arabidopsis thaliana* (Møller and McPherson, 1998), *Glycine max* (Delis et al., 2006; AJ583529), *Lycopersicon esculentum* (Katinakis, unpublished, AJ871578), *Canavalia lineate* (Song et al., unpublished, AF172681), *Brassica juncea* (Jiao and Pua, unpublished, AF449459), and *N*-methylputrescine oxidase from *Nicotiana tobacco* (Katoh et al., 2007, AB289457; Heim et al., 2007, DQ873385).

A BLAST search (tblastn) of the entire database of plant nucleotide sequences with *Pisum sativum* amine oxidase protein sequence (AAA62490) as a query produced over 200 hits indicating that copper amine oxidase genes are abundant in plants. Corresponding gene sequences were found also in the genomic sequences of *Vitis vinifera* (AM438619), *Medicago truncatula* (CU104689), *Solanum lycopersicum* (AP009268), and the moss *Physcomitrella patens* (XM_001757151).

Further search of mRNA clones detected copper amine oxidase genes in species of cereals, where they were less expected due to the prevalence of FAD-containing polyamine oxidases (Šebela et al., 2001). Sequences bearing TPQ and copper histidyl ligand signatures were found in the mRNA clone BT018375 from endosperm of *Zea mays* and BG607371 from early reproductive apex of *Triticum monococcum*.

BLAST on NCBI Map Viewer using the *Pisum sativum* amine oxidase sequence AAA62490 as a query against fully sequenced genomes of *Arabidopsis thaliana*, *Oryza sativa*, and *Populus trichocarpa* revealed the presence of three to five genes in each species as shown in Table 7.2, including the already cloned gene *atao1* (Møller and McPherson, 1998), NC_003075/AT4G12290. Exon/intron analysis (Figure 7.2) revealed up to 4 introns in *Oryza sativa* and up to 11 in *Arabidopsis thaliana*.

7.3.3 Copper Amine Oxidase Genes in Other Organisms

Copper amine oxidase genes were found in complete genomes of cyanobacteria *Synechococcus* sp. (CP000240), *Trichodesmium erythraeum* (CP000393), and *Nostoc* sp. (BA000019), in green algae *Chlamydomonas reinhardtii* (XM_001700279) and red algae *Cyanidioschyzon merolae* (AP006491), but not in brown algae. Interestingly, copper amine oxidases genes were found in Trypanosomas, especially in plant pathogens *Phytophthora sojae* and *Phytophthora ramorum*, but also in *Toxoplasma gondii* and *Cryptosporidium hominis* pathogens affecting humans. No such genes were found in insects.

7.3.4 Phylogenetic and Expression Analysis of Fungal and Plant Copper Amine Oxidases

Phylogenetic analysis performed on the copper amine oxidase proteins from completed yeast and plant genomes revealed basic classes of the enzymes that correspond

TABLE 7.2
Copper Amine Oxidase Genes in Completed Plant Genomes

Organism	Copper Amine Oxidase Genes*	Genome Map Element/ Gene Locus Tag	Protein ID	Number of Introns
Arabidopsis thaliana	5	NC_003070/AT1G62810	NP_176469	6
		NC_003070/AT1G31710	NP_174452	4
		NC_003071/AT2G42490	NP_181777	11
		NC_003074/AT3G43670	NP_189953	6
		NC_003075/AT4G12290	NP_192966	3
Oryza sativa	3	NC_08397/ Os04g0269600	NP_001052338	4
		NC_08399/ Os06g0338700	NP_001057560	3
		NC_08400/ Os07g0572100	NP_001060056[t]	2
Populus trichocarpa	4	NC_008474/not annotated		
		NC_008476/not annotated		
		NC_008478/not annotated		
		NC_008481/not annotated		

Notes: * Numbers indicate genes found by BLAST search (tblastn) on NCBI Map Viewer using *Pisum sativum* amine oxidase protein sequence as a query (AAA62490); [t] = Truncated protein sequence, gene splice sites unclear.

well to substrate specificity and subcellular localization. Multiple protein sequence alignment using ClustalW (BioEdit 7.0.9.0; Hall, 1999) indicated that the predicted protein encoded by the single gene from *Eremothecius gossypii* is truncated at position 469, shortly after the histidinyl ligand motif HQH, thus missing the C-terminal structural part.

Conversely, the product of the gene SPAC2E1P3.04 from *Schizosaccharomyces pombe* contains an additional stretch of unaligned sequence (including unexpected 8xHis) between the amino acid residues 658 and 765 that may come from an incorrectly predicted splice site. Similarly, *Aspergillus fumigatus* XP_749090 may contain an intron between the amino acids 610 and 629. These residues were deleted when preparing the tree. An unrooted tree calculated by the Protein Maximum Likehood program is shown in Figure 7.3. The amine oxidases form three basic groups that may be classified as peroxisomal methylamine oxidases related to *Hansenula polymorpha* P12807, fungal monoamine oxidases with *Aspergillus niger* ANAO-I (AF362473) as a typical representative, and plant diamine oxidases such as the *Pisum sativum* AAA62490.

Recent data from GENEVESTIGATOR (Zimmermann et al., 2004) show that the expression pattern of Arabidopsis copper amine oxidases is in agreement with previous data on *atao1* (Møller and McPherson, 1998). This enzyme preferably oxidizes putrescine, serving as a producer of hydrogen peroxide for lignifications. The

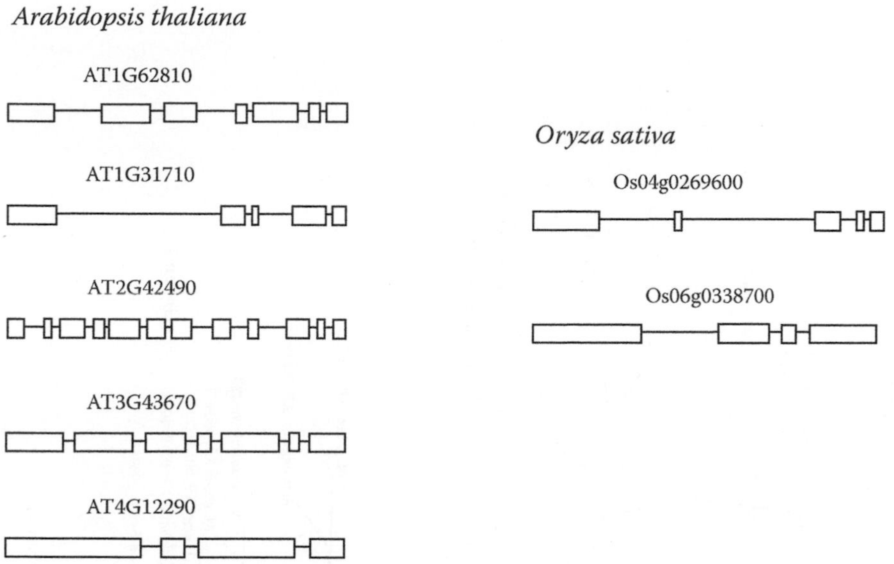

FIGURE 7.2 Exon/intron composition of copper amine oxidase genes from completed plant genomes.

corresponding gene AT4G12290 shows enhanced expression in hypocotyl and vascular tissue and in senescent leaves. The gene is induced by various stimuli, such as infection by *Pseudomonas syringae*, application of abscisic acid, and osmotic stress. Other genes do not show distinct patterns of expression, except for AT2G42490, the closest homologue of *Nicotiana tabacum* methylputrescine oxidase (ABI93948, Figure 7.3) that is preferentially expressed in roots and AT1G31710 that expresses in hypocotyls.

7.4 MAMMALIAN COPPER AMINE OXIDASE GENES

We presently know five different members of the class of mammalian copper-containing amine oxidases defined by their respective genes: diamine oxidase (DAO) encoded by the *AOC1* or *ABP1* gene, retina-specific amine oxidase (RAO) encoded by the *AOC2* gene, vascular adhesion protein-1 (VAP-1) encoded by the *AOC3* gene, a soluble paralog of VAP-1 called serum or plasma amine oxidase (SAO) encoded by the *AOC4* gene, and lysyl oxidase (LOX) and its homologs called lysyl oxidase-like proteins (LOXLs) encoded by the *LOX* and *LOXL1* through *LOXL4* genes, respectively (Schwelberger, 2007; Kagan and Li, 2003). The major facts about these proteins are summarized in Figure 7.4.

We know several members of the class of FAD-containing amine oxidases that are mentioned here only for clarity and will not be discussed further in this chapter: two isoforms of monoamine oxidase (MAO-A and MAO-B) encoded by the *MAOA* and *MAOB* genes, respectively (Shih, 2004), three isoforms of polyamine oxidase (PAO) all encoded by the *PAOX* gene (Vujcic et al., 2003; Seiler, 2004), and four isoforms of spermine oxidase (SMO) all encoded by the *SMOX* gene (Vujcic et al., 2002).

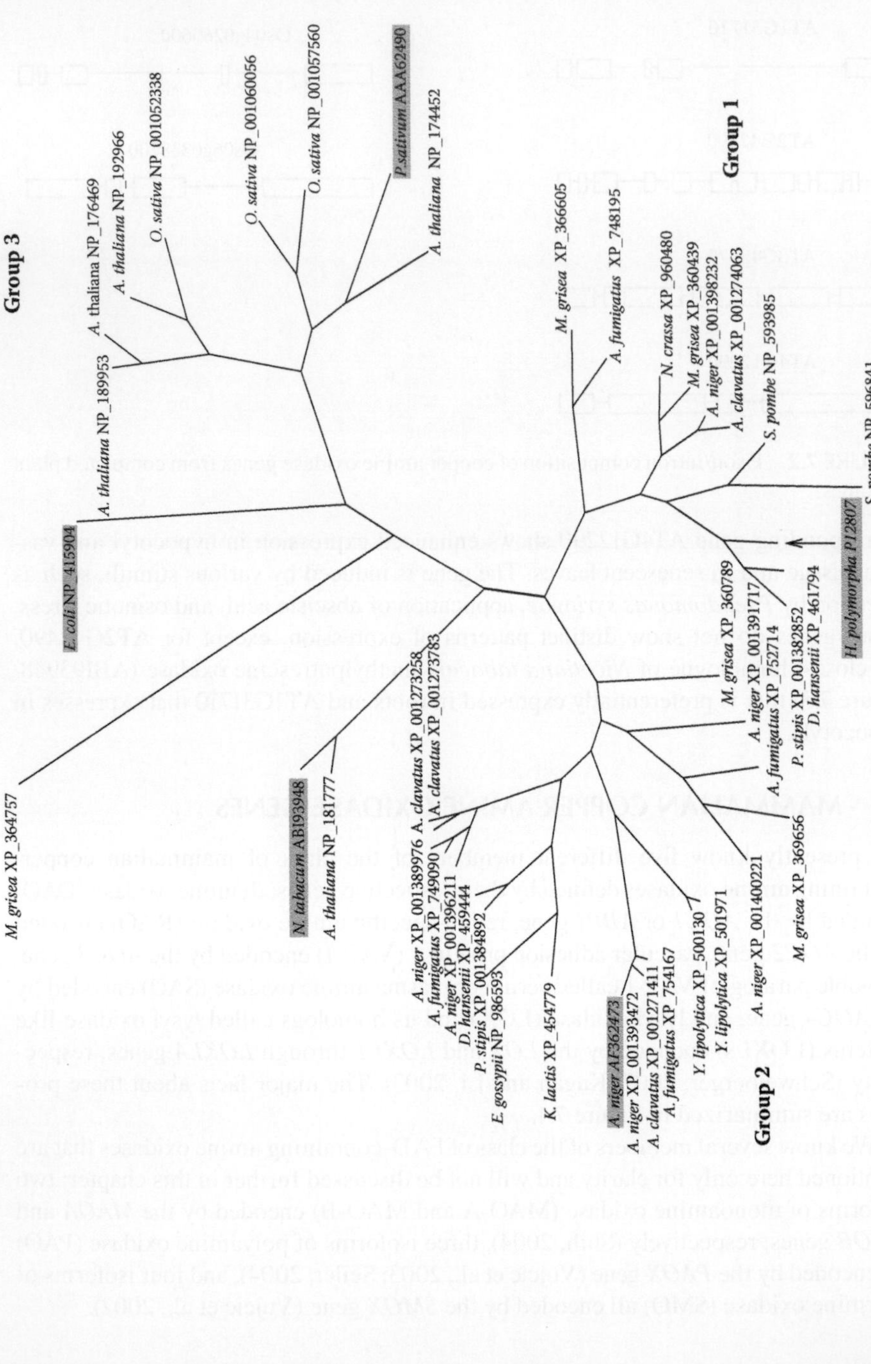
Group 3
Group 1
Group 2
M. grisea XP_364757
E. coli NP_415904
A. thaliana NP_189953
A. thaliana NP_176469
A. thaliana NP_192966
O. sativa NP_001052338
O. sativa NP_001060056
O. sativa NP_001057560
P.sativum AAA62490
A. thaliana NP_174452
M. grisea XP_366605
A. fumigatus XP_748195
N. crassa XP_960480
M. grisea XP_360439
A. niger XP_001398232
A. clavatus XP_001274063
S. pombe NP_593985
S. pombe NP_596841
H. polymorpha P12807
D. hansenii XP_461794
P. stipis XP_001382852
A. fumigatus XP_752714
A. niger XP_001391717
M. grisea XP_360789
M. grisea XP_369656
A. niger XP_001400221
Y. lipolytica XP_501971
Y. lipolytica XP_500130
A. fumigatus XP_754167
A. clavatus XP_001271411
A. niger XP_001393472
A. niger AF362473
K. lactis XP_454779
E. gossypii NP_986593
P. stipis XP_001384892
D. hansenii XP_459444
A. niger XP_001396211
A. fumigatus XP_749090
A. niger XP_001389976
A. clavatus XP_001273258
A. clavatus XP_001273783
N. tabacum AB193948
A. thaliana NP_181777

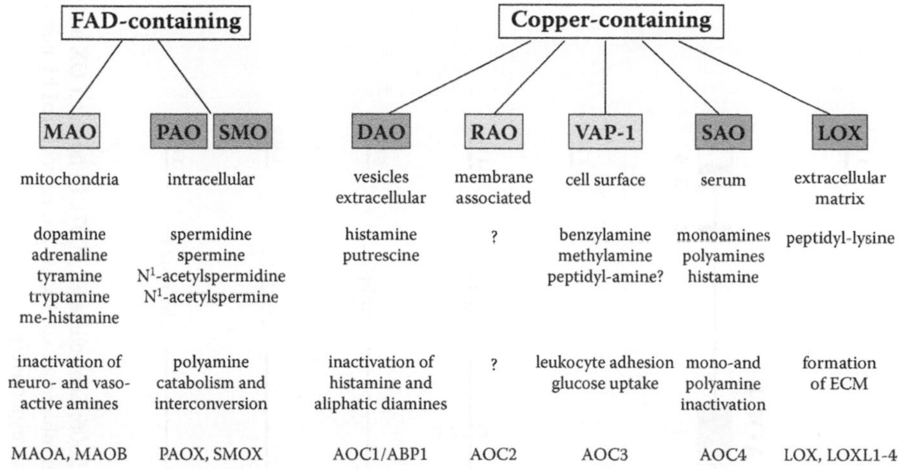

FIGURE 7.4 Classification and properties of mammalian FAD- and copper-containing amine oxidases. For each enzyme, the subcellular localizations, known substrates, putative functions, and gene symbols are shown below the enzyme abbreviation; membrane-associated proteins are shaded in lighter grey and soluble proteins in darker grey.

7.4.1 MAMMALIAN *AOC* GENES

The genes encoding DAO, RAO, VAP-1, and SAO (Figure 7.5A) are structurally related and possess considerable sequence similarities that indicate their common evolutionary origin. The encoded proteins of ca. 750 amino acid residues are similar to the copper amine oxidases found in microorganisms and plants (Tipping and McPherson, 1995) and carry the characteristic sequence motifs of this class of enzymes.

DAO is a soluble enzyme that converts histamine and other diamines and is encoded by the *AOC1* or *ABP1* gene located on chromosome 7q34-q36 in the human genome. *AOC1* has five exons and is not linked with other AOC genes (Chassande et al., 1994; Schwelberger, 2004). DAO was the first mammalian copper amine oxidase to be cloned although originally as the protein binding the diuretic drug amiloride (Barbry et al., 1990; Novotny et al., 1994). The DAO genes of different species have highly conserved structures and encode polypeptides with highly conserved sequences (Chassande et al., 1994; Schwelberger, 1999; Schwelberger et al., 2000; Schwelberger, 2004).

DAO expression is restricted to specific tissues; the major sites of expression are intestine, kidney, and placenta. DAO is a secretory protein with an N-terminal signal peptide sequence stored in membrane-associated vesicular structures in epithelial cells and secreted into the circulation upon stimulation (Schwelberger et al., 1998; Schwelberger, 2004). Despite studies of transcription factor binding in the promoter

---

FIGURE 7.3 (Opposite) Unrooted tree of copper amine oxidase proteins calculated by Protein Maximum Likehood program (ProML; Bioedit 7.0.9.0) and drawn by TreeView 1.6.6 (Page, 1996). Group 1 = methylamine oxidases (peroxisomal). Group 2 = monoamine oxidases. Group 3 = diamine oxidases. Typical representatives of each group are shaded.

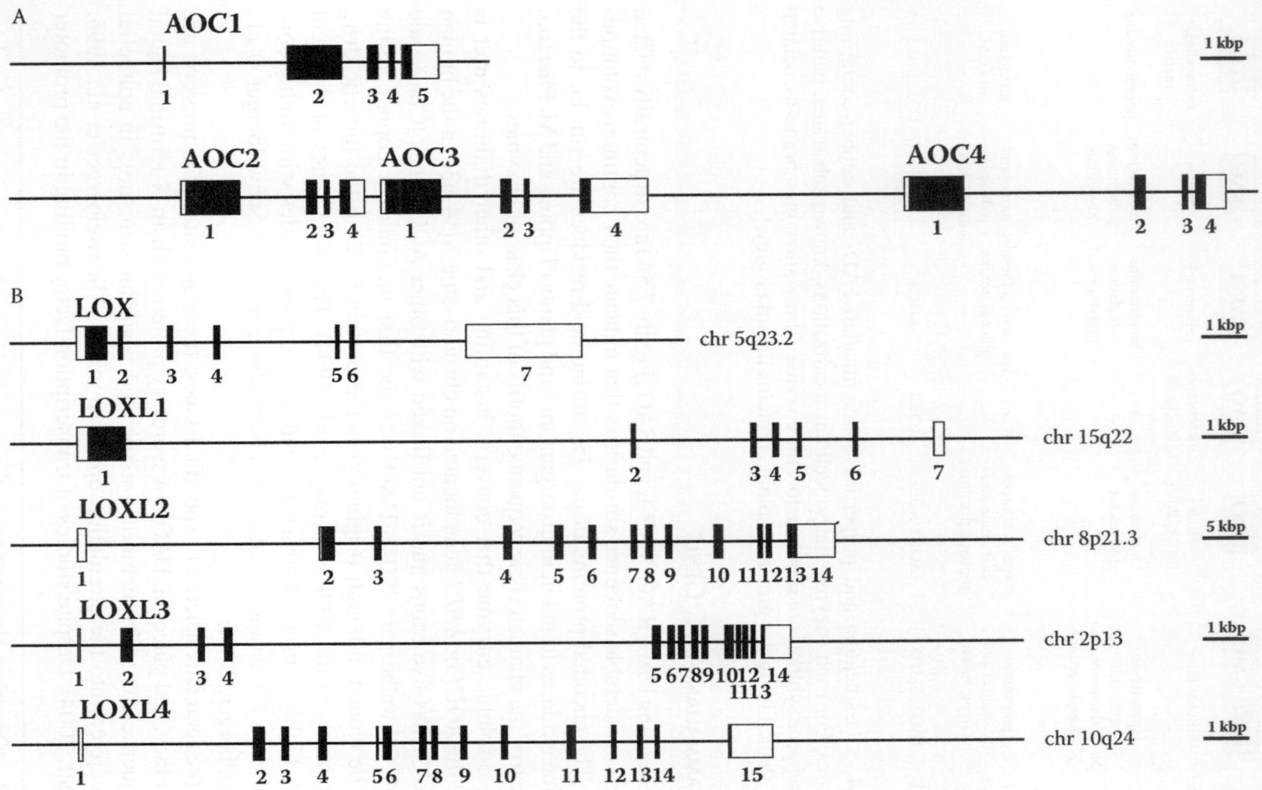

FIGURE 7.5 Structures of mammalian AOC and LOX genes. A: porcine AOC genes (Schwelberger, 2006; Schwelberger, 2007). B: human LOX and LOXL genes (from most recent human genome reference assembly: www.ncbi.nlm.nih.gov/sites/entrez?db=genome). Exons are represented by numbered boxes. Open boxes represent noncoding regions and black boxes represent protein coding regions, respectively.

regions of DAO genes of different mammals and expression analyses in cell models (Chassande et al., 1994), how the tissue-specific expression of DAO is regulated remains unclear. One proposal is that DAO serves as a barrier and scavenger that inactivates all sorts of diamines to prevent their transport into and accumulation in the circulation. However, the main function of DAO appears to be the inactivation of histamine after its release into the extracellular space, thus providing temporal and local control of the action of this inflammatory mediator (Schwelberger, 2004).

RAO is a membrane protein with a short N-terminal transmembrane domain and is encoded by the *AOC2* gene (Imamura et al., 1997; Imamura et al., 1998). The *AOC2* gene is tandemly arranged with the *AOC3* and *AOC4* genes on the same chromosome (17q21 in the human genome). All three genes have the same structural organization of four exons that correspond to exons 2 through 5 of the *AOC1* gene (Schwelberger, 2006 and 2007). Originally cloned in a search for genes involved in hereditary retinal disease, RAO has been deduced from its gene and cDNA sequence. Nothing is actually known about the protein and its enzymatic properties. RAO appears to be exclusively expressed in the ganglion cell layer of the retina but analysis of the promoter region of the *AOC2* gene provided no clear explanation for this restricted expression pattern and its regulation (Imamura et al., 1998). The dearth of studies of the RAO protein mean that its function remains a matter of speculation.

VAP-1 is a type II peripheral plasma membrane protein with a short N-terminal transmembrane domain that mediates the binding of leukocytes to endothelial cells (Salmi and Jalkanen, 1992 and 1996). Cloning of mouse and human VAP-1 showed that it is a CAO exhibiting enzymatic activity and highest conversion rates for monoamine substrates (Bono et al., 1998b; Smith et al., 1998). VAP-1 is encoded by the *AOC3* gene located downstream of the *AOC2* gene on human chromosome 17q21. VAP-1 is expressed in endothelial cells, smooth muscle cells, and adipocytes, ubiquitous cell types found in most tissues; thus its expression may be detected in many tissues (Bono et al., 1998b; Smith et al., 1998). Analysis of the mouse *AOC3* gene promoter region identified multiple transcription initiation sites and several putative transcription factor binding sites (Bono et al., 1998a). VAP-1 expression was significantly up-regulated at sites of inflammation where it facilitates the binding and extravasation of leukocytes; both the surface structure and oxidase activity of the protein are required in this process (Salmi and Jalkanen, 2001).

SAO is a soluble copper amine oxidase found in serum or plasma and prefers monoamine substrates (Lyles, 1996). Although a cDNA encoding a bovine SAO protein was cloned earlier (Mu et al., 1994) it was not clear whether SAO is encoded by a separate gene. A detailed study of the porcine copper amine oxidase genes and their products revealed an additional gene designated *AOC4* in addition to *AOC1*, *AOC2*, and *AOC3*. *AOC4* encodes a protein that is highly similar to VAP-1 but has a signal peptide sequence instead of a transmembrane domain at its N-terminus, hence the initial designation as soluble or secretory VAP-1 (Schwelberger, 2006 and 2007).

Subsequent analyses revealed that the *AOC4* gene is located immediately downstream of the *AOC2* and *AOC3* genes, is expressed mainly in the liver, and encodes the major fraction of the amine oxidase present in porcine blood plasma (Schwelberger, 2007). Complete *AOC4* genes in a cluster with *AOC2* and *AOC3* were found in the genomes of cows, dogs, macaques, and humans. However, in human *AOC4*, a single base change converts a codon for a conserved tryptophan at position 225 to a stop codon, thus leading to a truncated and nonfunctional protein. The genomes of mouse and rat contain only small fragments of an *AOC4* gene. Based on available genomic sequence data, functional *AOC4* genes are present in all species that have high SAO activities, confirming the assumption that *AOC4* does indeed encode the major fraction of SAO in most mammals.

Humans and rodents that have defective or absent *AOC4* genes, respectively, exert very low SAO activity, probably derived from partial proteolytic release of the large extracellular part of the membrane-associated *AOC3* gene product VAP-1 (Stolen, 2004b; Abella, 2004). The regulation of *AOC4* gene expression has not yet been studied in detail. Its protein appears to be constitutively produced and secreted by the liver, giving rise to relatively constant SAO activity in the blood. The broad substrate specificity of SAO with a preference for monoamines led to the notion that the enzyme functions as a general scavenger of biogenic and xenobiotic amines in the circulation (Lyles, 1996).

Within each species, the DAO protein is ca. 40% identical in sequence to the paralogous RAO, VAP-1, and SAO proteins, RAO is ca. 65% identical to VAP-1 and SAO, and VAP-1 and SAO are usually more than 90% identical. Overall interspecies protein sequence identity is ca. 80% for each of the orthologues. Although the promoter regions of the AOC genes have been studied and conserved transcription factor binding sites have been identified (Chassande et al., 1994; Imamura et al., 1998; Bono et al., 1998a), how the tissue-specific expression of these genes is regulated in different species remains unknown. Whether and how genetic polymorphisms of the AOC genes and the resulting enzyme variants contribute to the development of human diseases are also unknown.

7.4.2 Mammalian *LOX* and *LOXL* Genes

Lysyl oxidase (EC 1.4.3.13) and the lysyl oxidase-like proteins are members of the CAO class. They possess quinone cofactors but are structurally distinct from the enzymes discussed above. LOX is a ca. 30-kDa protein of the extracellular matrix that catalyzes the oxidative deamination of lysine residues in elastin and collagens, thereby initiating the formation of inter- and intrachain cross links in these proteins (Pinnell and Martin, 1968; Kagan, 1986). Mature LOX is produced from a ca. 50-kDa precursor by proteolytic removal of an N-terminal propeptide (Trackman et al., 1992) and contains a conserved copper binding site and the active-site cofactor lysyl tyrosyl quinone (LTQ) formed by oxidation of a conserved tyrosine residue and subsequent intramolecular cross linking with the ε-amino group of a conserved lysine (Wang et al., 1996b).

Human *LOX* is encoded by a gene with seven exons located on chromosome 5q23.2 (Figure 7.5B). This gene structure is highly conserved in all mammals. *LOX* appears to be expressed in most tissues reflecting the general requirement of forming and remodeling the extracellular matrix. Because this process is highly important for tumor growth and cell evasion, *LOX* and its homologues have recently attracted considerable attention for their possible role as tumor suppressors and in cancer metastasis (Erler and Giaccia, 2006; Payne et al., 2007).

LOX and *LOXL1* through *LOXL4* share a conserved C-terminal catalytic domain of ca. 250 amino acid residues with approximately 50% sequence identity between any two proteins but have distinct N-terminal domains. *LOXL1* resembles *LOX* in the exon organization of its gene and the sequence similarity of its N-terminal domain. *LOXL2*, *LOXL3*, and *LOXL4* possess similar N-terminal sequences that have no similarity to *LOX* and *LOXL1*.

Moreover, the *LOXL2* through *LOXL4* genes have conserved structures with 14 or 15 exons compared with 7 exons of *LOX* and *LOXL1* (Figure 7.5B). While we have information about their tissue-specific expression and cellular localization, little is known about the enzymatic activity, regulation, and function of the LOXL proteins (Kagan and Li, 2003; Molnar et al., 2003). Future studies of this enzyme family would be aided by determinations of the three-dimensional structures of mammalian LOX proteins that should reveal whether their catalytic domains are structurally related to the catalytic domains of the AOC proteins despite the lack of sequence conservation.

ACKNOWLEDGMENTS

This work was supported in part by Grant MSM 6198959216 of the Ministry of Education, Youth and Sports of the Czech Republic (IF) and by grants from the Austrian Science Fund (HGS). The authors thank Petr Galuszka for help with Arabidopsis databases.

8 Mechanism of TPQ Biogenesis in Prokaryotic Copper Amine Oxidase

Toshihide Okajima and Katsuyuki Tanizawa

CONTENTS

8.1 PREFACE

Copper amine oxidases (CAOs; EC 1.4.3.6) contain a redox-active quinone cofactor covalently linked to their polypeptide chains: 2,4,5-trihydroxyphenylalanine quinone, abbreviated as TOPA quinone or TPQ. Because TPQ occurs as an integral amino acid residue of a polypeptide chain and is encoded as a Tyr residue in all genes coding for CAOs, a specific mechanism of post-translational modification must occur to generate TPQ from the precursor Tyr residue in the nascent polypeptide chain.

The mechanism of TPQ generation has been studied extensively with recombinant CAOs from a bacterium (*Arthrobacter globiformis*) and yeast (*Hansenula polymorpha*), both of which are expressed heterologously in an inactive precursor form without bound cupric ion and TPQ. These studies indicate that TPQ is generated by a cupric ion-dependent self-catalyzed reaction of the enzyme protein. Chemically, the reaction of TPQ biogenesis is defined as oxygenation and/or oxidation of the phenol ring of a Tyr residue in the protein to TPQ, consuming 2 mol of dioxygen and producing 1 mol of hydrogen peroxide (Equation 8.1).

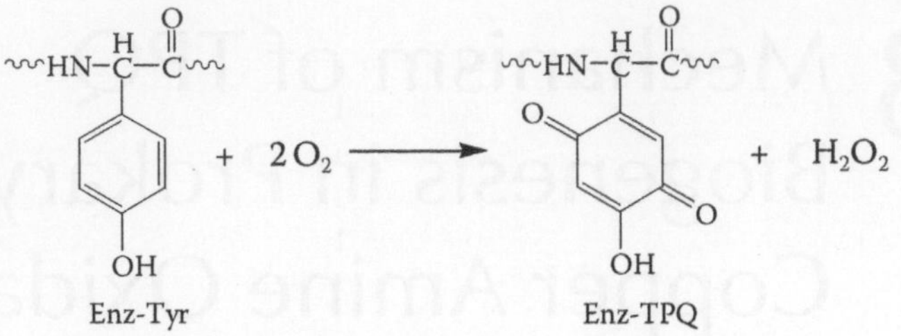

(8.1)

Although the TPQ biogenesis reaction is apparently simple, it must be controlled in a very sophisticated manner so that only the precursor Tyr residue in the active site among many others contained in the whole protein is specifically converted to TPQ. In addition, the phenol ring of Tyr residues in proteins is generally resistant to air oxidation unless otherwise activated. The bound cupric ion essential for catalytic activities of CAOs has also been shown to be essential for TPQ biogenesis, most likely playing an important role in initially activating the phenol ring of the precursor Tyr residue.

Unlike free TPQ and other synthetic analogs, most of which are unstable in aqueous solution, the cofactor TPQ thus formed in the active site of CAOs is rather stable via its accommodation inside the protein. It is only active toward substrate amines and analogous inhibitors that enter through the substrate channel, catalyzing oxidative deamination of primary amines to an aldehyde, hydrogen peroxide, and ammonia (Equation 8.2):

$$RCH_2NH_2 + O_2 + H_2O \rightarrow RCHO + H_2O_2 + NH_3 \quad (8.2)$$

Thus, CAOs are enzymes with dual functions, catalyzing the single turnover reaction of the cofactor generation and the multi-turnover catalytic reaction. This chapter will focus on TPQ biogenesis and describe the detailed molecular mechanism of the self-catalyzed reaction of TPQ generation, particularly in a bacterial CAO.

8.2 EVIDENCE FOR SELF-CATALYZED GENERATION OF TPQ

TPQ was first identified as a covalently bound organic cofactor in bovine plasma CAO (Janes et al., 1990). A few years later, TPQ was shown to be the common cofactor in both eukaryotic and prokaryotic enzymes. Comparison of the quinone-containing peptide sequences with the gene-encoded sequences (see Chapter 7) demonstrated that the precursor to the cofactor is a specific Tyr residue occurring in a highly conserved sequence: Asn–Tyr (TPQ)–Asp/Glu–Tyr (Janes et al., 1992).

These findings immediately led to the prediction that CAOs would have a co- or posttranslational modification system that acted on the precursor Tyr residue to generate TPQ. However, the mechanism of TPQ biogenesis remained unsolved for a few more years until an inactive precursor form of the protein without TPQ became available and the TPQ generation was demonstrated *in vitro* with the isolated protein. In 1994, the inactive precursor protein of CAO was obtained by expressing the recombinant enzyme from *Arthrobacter globiformis* (phenylethylamine

amine oxidase, AGAO) in *Escherichia coli* cells grown in a Cu^{2+}-limited medium (Tanizawa et al., 1994) that could be purified in the inactive form and activated after purification (Matsuzaki et al., 1994). This heterologous expression system developed by our group in Osaka afforded an important clue to elucidation of the mechanism of TPQ biogenesis (Tanizawa, 1995). Around the same time, Klinman's group in Berkeley also provided a system for studying the mechanisms of TPQ biogenesis using a eukaryotic CAO, *Hansenula polymorpha* amine oxidase (HPAO) (Cai and Klinman, 1994b).

The TPQ-less inactive protein without the bound Cu^{2+} ion was purified from *E. coli* cells carrying the expression plasmid for AGAO cultured in a Cu^{2+}-depleted medium (Matsuzaki et al., 1994). The Cu^{2+}-deficient AGAO showed negligible catalytic activity and no UV-vis absorption peak other than the 280-nm peak of a simple protein. Remarkably however, when the purified enzyme was incubated with excess Cu^{2+} ions ($CuSO_4$), the catalytic activity increased to a level comparable to that of the enzyme purified from the original bacteria (Table 8.1). After removal of the unbound Cu^{2+} ion by rigorous dialysis against a buffer containing EDTA, about one mol atom of copper was detected per mol of subunit of the fully activated enzyme (Table 8.1).

Concomitant with the Cu^{2+}-dependent appearance of enzymatic activity, the enzyme solution gradually colored brownish pink with an absorption maximum at 470 to 480 nm (Figure 8.1) ascribable to the formation of a quinone compound. In contrast, no change was observed in the absorption spectrum under anaerobic conditions, indicating that the formation of the chromophore was strictly dependent on the presence of dissolved dioxygen. The Cu^{2+}-deficient recombinant AGAO purified to homogeneity was thus competent for *in vitro* cofactor generation. The quinone formation was demonstrated by *in situ* incubation with Cu^{2+} ion of the precursor enzyme immobilized on a polyvinylidene difluoride membrane by electroblotting. Hence, we concluded that quinone formation is a self-catalytic process and requires no external enzymatic systems.

The sequence of a quinone-containing peptide isolated after the Cu^{2+}-activated AGAO was treated with *p*-nitrophenylhydrazine and digested with a protease was Ile–Gly–Asn–X–Asp–Tyr–Gly (X was an unidentifiable residue; Matsuzaki et al., 1994). The X residue was located at the position corresponding to Tyr 382 in the amino acid sequence deduced from the nucleotide sequence of the gene, and the sequence surrounding X coincided with the consensus Asn–Tyr (TPQ)–Asp/Glu–Tyr sequence, highly conserved in CAOs. Thus, the quinone compound generated by aerobic incubation with Cu^{2+} ion of the Cu^{2+}-deficient inactive enzyme was assumed to be TPQ, derived from Tyr 382.

Resonance Raman spectra of the *p*-nitrophenylhydrazones of the Cu^{2+}-activated enzyme, the isolated peptide, and also the TOPA quinone hydantoin model compound were virtually identical, showing unequivocally that the Cu^{2+} ion-generated quinone compound was TPQ. Moreover, the Y382F mutant enzyme in which the precursor Tyr was mutated to Phe showed little activity and no absorption band characteristic of TPQ around 480 nm, even after incubation with excess Cu^{2+} ions, confirming that the precursor to TPQ must be a Tyr residue, not a Phe residue.

TABLE 8.1
Characteristics of Cu- and Metal-Substituted AGAOs

Enzymes	Method of Preparation	Specific Activities (U/mg)	TPQ/Subunit	Metal Content (mol atom/subunit)[d]	K_m for 2-PEA[e] (mM)	k_{cat} (s^{-1})
Cu-AGAO[a]		67.5 (100)[c]	0.68 ± 0.01	0.88 (Cu) ± 0.0015	2.5 ± 0.03	75.7 ± 0.8
Co-AGAO	Biogenesis	0.35 (0.52)	0.51 ± 0.0002	0.0082 (Cu) ± 0.00056 1.12 (Co) ± 0.0032	2.5 ± 0.3	0.92 ± 0.02
	Substitution[b]	1.51 (2.2)	0.67 ± 0.02	0.0063 (Cu) ± 0.0004 1.06 (Co) ± 0.0011	1.9 ± 0.001	1.77 ± 0.001
Ni-AGAO	Biogenesis	0.23 (0.34)	0.58 ± 0.0008	0.0065 (Cu) ± 0.00062 0.93 (Ni) ± 0.011	3.4 ± 0.4	0.63 ± 0.02
	Substitution[b]	0.58 (0.90)	0.78 ± 0.02	0.0088 (Cu) ± 0.00017 0.800 (Ni) ± 0.024	1.6 ± 0.005	0.68 ± 0.002

Notes: [a] Recombinant enzyme produced by Cu^{2+}-assisted TPQ biogenesis; [b] Kishishita et al., 2003; [c] Relative activities; [d] Values for metals indicated in parentheses; [e] 2-Phenylethylamine.

Source: Okajima et al., 2005. With permission.

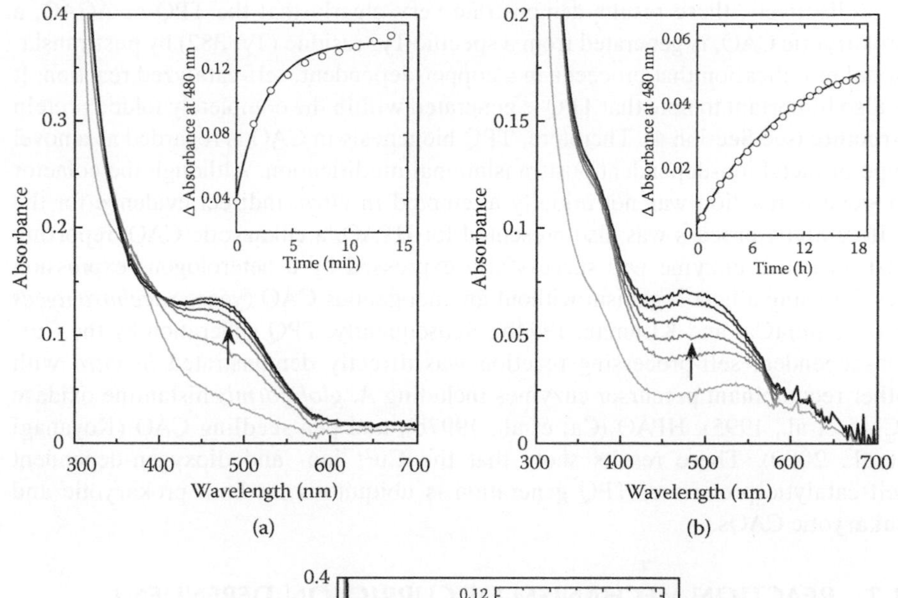

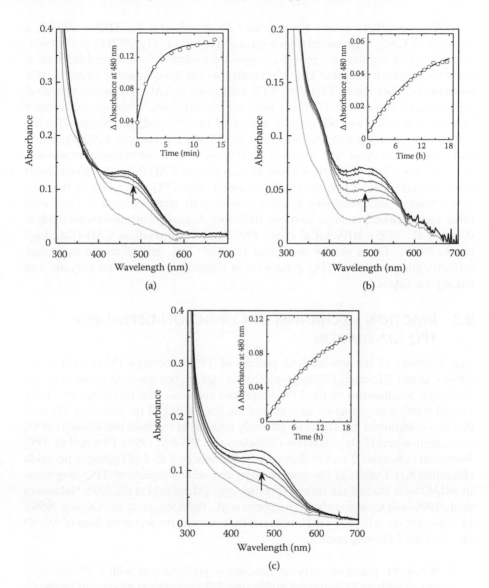

FIGURE 8.1 UV-vis spectral changes during incubation with Cu^{2+}, Co^{2+}, and Ni^{2+} ions under O_2-saturated conditions. Apo AGAO (0.1 *mM* subunit) was incubated with 0.5 *mM* Cu^{2+} (a), Co^{2+} (b), or Ni^{2+} (c) under O_2-saturating conditions (~1.2 *mM*). Absorption spectra were recorded at 0, 0.5, 1.0, 2.0, and 2.5 min for Cu^{2+} and 0, 3, 6, 9, 12, and 15 hr for Co^{2+} and Ni^{2+}; arrows indicate directions of spectral changes. Insets: Absorbance changes at 480 nm were plotted against incubation time and fitted to pseudo-first order kinetics by the least squares method. (*Source:* Okajima et al., 2005. Reproduced with permission.)

Collectively, these results demonstrate very clearly that the TPQ of AGAO, a prokaryotic CAO, is generated from a specific Tyr residue (Tyr 382) by posttranslational modification that proceeds in a copper-dependent, self-catalyzed reaction. It is also important to note that TPQ is generated within the completely folded protein structure (see Section 4). Therefore, TPQ biogenesis in CAO is regarded as a novel type of metal ion-dependent posttranslational modification. Although the cofactor biogenesis reaction was not initially attempted in *vitro*, indirect evidence for the self-catalytic process was also presented for HPAO, a eukaryotic CAO, reporting that an active enzyme was successfully expressed by a heterologous expression system using a host organism without an endogenous CAO (yeast *Saccharomyces cerevisiae*) (Cai and Klinman, 1994b). Subsequently, TPQ generation by the Cu^{2+} ion-dependent self-processing reaction was directly demonstrated *in vitro* with other recombinant precursor enzymes including *A. globiformis* histamine oxidase (Choi et al., 1995), HPAO (Cai et al., 1997b), and pea seedling CAO (Koyanagi et al., 2000). These results show that the Cu^{2+} ion- and dioxygen-dependent self-catalytic process of TPQ generation is ubiquitous for both prokaryotic and eukaryotic CAOs.

8.3 REACTION MECHANISM OF CUPRIC ION-DEPENDENT TPQ GENERATION

The discovery of the self-catalytic process of TPQ generation (Matsuzaki et al., 1994, Cai and Klinman, 1994b) promoted various studies intended to elucidate the molecular mechanisms of the TPQ biogenesis reaction. The stoichiometry of the overall reaction requiring $6e^-$ oxidation of the phenol ring of the precursor Tyr residue was determined for AGAO by precisely measuring the rates and amounts of O_2 consumption and H_2O_2 production (Ruggiero and Dooley, 1999). One mol of TPQ formation consumes 2 mol of dioxygen and produces 1 mol of hydrogen peroxide (Equation 8.1). Details of the reaction steps and related aspects of TPQ biogenesis in AGAO were studied via various spectroscopic (Matsuzaki et al., 1995; Nakamura et al., 1996) and kinetic analyses (Ruggiero et al., 1997; Ruggiero and Dooley, 1999). In summary, these *in vitro* TPQ biogenesis studies using the precursor form of AGAO provided the following results:

1. When the precursor enzyme adequately preincubated with Cu^{2+} ion is mixed with an O_2-saturated buffer, the TPQ generation monitored by the increase of absorbance at 480 nm follows first order kinetics (k_{obs} = 1.5 min^{-1}) regardless of the amount of Cu^{2+} initially added (Table 8.2; Ruggiero et al., 1997; Okajima et al., 2005).
2. No spectroscopically distinct intermediate is accumulated within a measurable time scale ($2 < t < 10$ sec) when the reaction is monitored by conventional UV-vis spectroscopy (Matsuzaki et al., 1994).
3. The rate of TPQ generation increases linearly with the increase of dioxygen concentration up to about 1.2 *mM* (saturating concentration of dissolved oxygen) (unpublished data), suggesting that AGAO has a binding site for dioxygen with very weak affinity or that dioxygen may react directly with a

TABLE 8.2
Metal Ion Specificities for TPQ Biogenesis

Metal Ion	Biogenesis Rate (min^{-1})[a]	pK_a Value of Metal-Bound Water[b]	Coordination Geometry in AGAO
Cu^{2+}	1.50	7.5	Square pyramidal[c] Tetrahedral[d]
Co^{2+}	1.32×10^{-4}	9.6	Tetrahedral[a] Octahedral[e]
Ni^{2+}	1.25×10^{-4}	9.4	Octahedral[e]
Zn^{2+}	Inert	9.6	Tetrahedral[f]

Notes: [a] Okajima et al., 2005; [b] Huheey et al., 1972; [c] Kim et al., 2002; [d] Wilce et al., 1997; [e] Kishishita et al., 2003; [f] Unpublished.

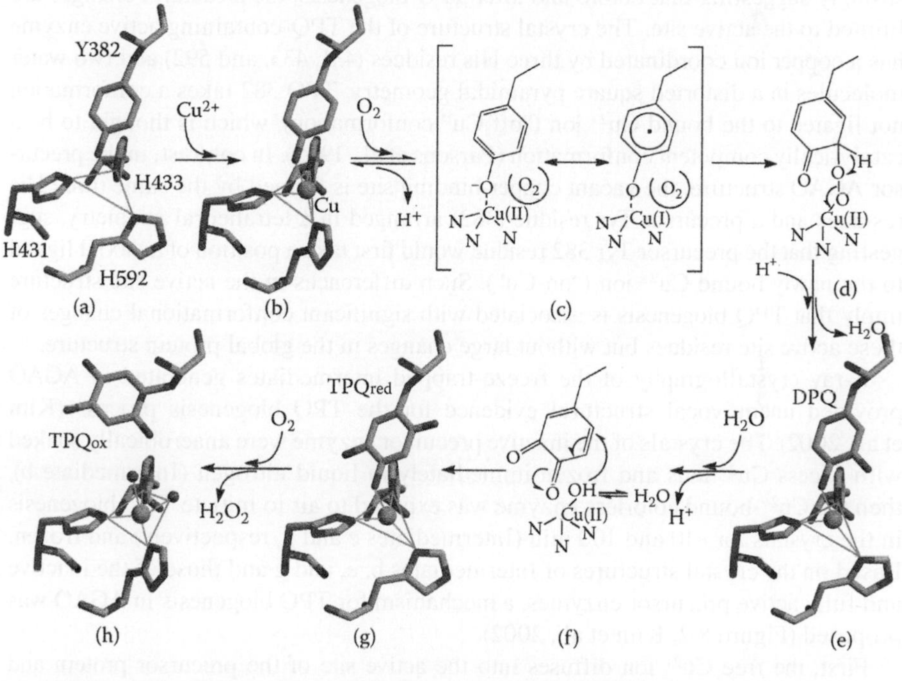

FIGURE 8.2 (See color insert following page 202.) Molecular mechanism of TPQ biogenesis. Incorporated structures of the metal-unbound precursor form (a), Intermediates b, e, and g and the TPQ-containing active form (h) were determined by x-ray crystallography. (*Source:* Kim et al., 2002. Reproduced with permission.)

tyrosinate intermediate (4-O⁻-dissociated Tyr axially ligated to the bound Cu^{2+} ion; see Figure 8.2) by bimolecular collision.

4. Based on resonance Raman spectroscopy of TPQ formed in ^{18}O-labeled water (Nakamura et al., 1996), the C2 oxygen atom of TPQ originates from H_2O rather than O_2.

5. A low-temperature electronic paramagnetic resonance (EPR) spectrum of an anaerobic complex of the precursor AGAO and Cu^{2+} ions shows super-hyperfine splitting ($A_{N\perp} = 16 \pm 1$ G; $g_{\perp} = 2.062$) consisting of seven lines, suggesting three equivalent nitrogen nuclei bound to Cu^{2+} ion (Kim et al., 2002).

As a specific observation for HPAO, a Cu^{2+}-tyrosinate intermediate exhibiting a ligand-to-metal charge transfer (LMCT) band at 350 nm is formed (Dove et al., 2000). The TPQ biogenesis reaction has also been studied using synthetic model compounds (Ling and Sayre, 2005).

Observations of the active site structure and its changes during TPQ biogenesis are also available for AGAO, which is the only enzyme that x-ray crystal structures of both the inactive precursor and the active TPQ-containing mature forms of CAO have been determined (Wilce et al., 1997). The two structures are almost superimposable, strongly suggesting that before and after TPQ biogenesis, the structural changes are limited to the active site. The crystal structure of the TPQ-containing active enzyme has a copper ion coordinated by three His residues (431, 433, and 592) and two water molecules in a distorted square pyramidal geometry. TPQ 382 takes a conformation not ligated to the bound Cu^{2+} ion ("off-Cu" conformation), which is thought to be a catalytically competent conformation (Parsons et al., 1995). In contrast, in the precursor AGAO structure, the vacant copper-binding site is formed by the same three His residues and a precursor Tyr residue (382) arranged in a tetrahedral geometry, suggesting that the precursor Tyr 382 residue would first take a position of an axial ligand to the newly bound Cu^{2+} ion ("on-Cu"). Such differences in the active site structure imply that TPQ biogenesis is associated with significant conformational changes of these active site residues but without large changes in the global protein structure.

X-ray crystallography of the freeze-trapped intermediates generated in AGAO provided unequivocal structural evidence for the TPQ biogenesis process (Kim et al., 2002). The crystals of the inactive precursor enzyme were anaerobically soaked with excess Cu^{2+} ions and frozen immediately in liquid nitrogen (Intermediate b), then the Cu^{2+}-bound colorless enzyme was exposed to air to initiate TPQ biogenesis in the crystals for ~10 and 100 min (Intermediates e and g, respectively) and frozen. Based on the crystal structures of Intermediates b, e, and g and those of the inactive and fully active precursor enzymes, a mechanism for TPQ biogenesis in AGAO was proposed (Figure 8.2; Kim et al., 2002).

First, the free Cu^{2+} ion diffuses into the active site of the precursor protein and is bound to the vacant copper-binding site (Figure 8.2, step a→b). In this state (Intermediate b), the bound Cu^{2+} ion is coordinated by the three His residues and Tyr 382 (precursor to TPQ) arranged in a trigonal pyramid geometry. The structure of Intermediate b clearly shows that the unmodified Tyr 382 is a direct axial ligand to Cu^{2+} at a distance of 2.5 Å. This distance is not short enough to trigger a typical LMCT interaction but may be sufficient for partially withdrawing electrons of the phenol ring of Tyr 382 by the Lewis acidity of Cu^{2+} (see below). A dioxygen molecule can then attack the ring C3 position of Tyr 382 with a decreased electron density (step c→d).

The formation of a Tyr radical (Tyr•) by full $1e^-$ transfer to Cu^{2+} may allow the facile reaction of Cu^{1+} with dioxygen yielding superoxide (O_2^-) that then reacts readily with Tyr•. However, the full electron transfer may not take place since the Co^{2+} and Ni^{2+} ions are also active for TPQ biogenesis (see Section 5; Okajima et al., 2005). In HPAO, it is proposed that a peroxy intermediate (d) is temporally formed in this step and then converted to DOPA quinone (DPQ; Dove et al., 2000). The lack of accumulation of a spectrometric intermediate in AGAO suggests that this initial step is rate-limiting in the overall reaction of TPQ biogenesis in AGAO. The next crystallographically detected structure, Intermediate e, corresponds to DPQ, revealing that the phenol ring of Tyr 382 is oxidized (oxygenated) first at the C3 position to DPQ through reprotonation and following dehydration, consistent with the proposed mechanism (step d→e).

The DPQ intermediate is chemically trapped by the introduction of an amino group at a nearby position by site-specific mutagenesis of Asp 298, acting as the proton-abstracting base in the catalytic reaction (Chiu et al., 2006), to Lys (Moore et al., 2007). X-ray crystallography of this mutant AGAO revealed that a covalent linkage is formed between the ε-amino side chain of Lys 298 and the C2 position of DPQ formed during TPQ biogenesis. In step e→f, the DPQ quinone ring should rotate around the Cβ–Cγ bond by nearly 180 degrees for incorporation of an oxygen atom into the ring C5 position (corresponding to the C2 position of TPQ) by copper-coordinated water or hydroxide. A water exchange reaction at the bound Cu^{2+} ion is consistent with the previous suggestion that the C2 oxygen is derived from water rather than dioxygen (Nakamura et al., 1996).

By comparison with the structure of Intermediate e, the ring rotation is evident from the structure of Intermediate g in which DPQ has been further converted to TPQ_{red} (step f→g). The C4 hydroxyl group of TPQ_{red} is still coordinated to the bound Cu^{2+} ion in Intermediate g, assuming an on-Cu conformation. Concomitant with the last oxidation step to form TPQ_{ox} (step g→h), the TPQ ring would rotate around both Cα–Cβ and Cβ–Cγ bonds to move into the catalytically active off-Cu conformation. This significant ring movement probably results from the change of the copper coordination structure to afford the axial position for binding and releasing hydrogen peroxide, a product of the TPQ generating reaction. Residues corresponding to Asn 381 and Asp 383 of AGAO, flanking the precursor to TPQ (Tyr 382) in the consensus Asn–Tyr–Asp/Glu sequence, are known to limit the flexibility of the cofactor ring in both TPQ biogenesis and catalysis (Dove et al., 2000). Conformations of the two residues are essentially identical in all AGAO structures, indicating that the orientation and/or position of the TPQ (Tyr) ring is not derived from the movement of these residues (including the main chain) but defined only through the rotations around the Cα–Cβ and Cβ–Cγ bonds of the 382 residue. The side chain of Asn 381 is near the TPQ_{ox} ring only in the structure of the active TPQ-containing enzyme and may be important for stabilizing the active off-Cu conformation of the TPQ ring by an NH-π interaction, together with nearby hydrogen-bonding networks linked to TPQ.

The precursor forms of N381Q and N381D mutants of AGAO show TPQ generation at 50- and 15-fold slower rates, respectively, than that of the wild-type enzyme. For the N381A mutant, TPQ biogenesis takes more than 1 week even under dioxygen-saturated conditions (unpublished data). Thus, the invariant residue Asn 381 also appears to be important to facilitate the movement and/or rotation of the aromatic ring of Tyr/DPQ/TPQ 382. In the superimposed structures of the biogenesis intermediates, the precursor, and the mature forms, the position of the imidazole ring of only His 592 among the three Cu^{2+}-ligating His residues differs significantly by rotation around the Cα–Cβ bond, while maintaining very similar coordination distances to the Cu^{2+} center (2.0 to 2.2 Å). It is assumed that this movement of His 592 results from (or causes) changes in the Cu^{2+} coordination structure during TPQ biogenesis (Matsunami et al., 2004; Wilce et al., 1997).

8.4 METAL ION AND DIOXYGEN PATHWAYS TO ACTIVE SITE

It is well known that TPQ is very unstable when exposed to a solvent by denaturation or protease digestion of the enzyme protein. Like the synthetic TPQ analogs, the exposed TPQ would readily be oxidized by O_2 and self-polymerized. As revealed by the crystal structures of various CAOs (see Chapter 9), the catalytic center is completely buried within the protein interior to protect TPQ from these undesirable oxidative degradations (Figure 8.3). Such localization of TPQ in the protein may be the basis for the self-catalytic process of TPQ biogenesis and renders unlikely the participation of an external modifying system that would be inaccessible to the buried TPQ.

As TPQ generation proceeds in the completely folded protein, channels should be available for the Cu^{2+} ion and dioxygen needed for TPQ biogenesis as cofactor and substrate, respectively, to access the buried active site and for the H_2O_2 departing into the solvent. X-ray crystal structures of the inactive precursor and active mature forms of AGAO (Wilce et al., 1997) show that two channels allow access to the active site; one is a funnel-shaped pocket that has predominantly negative surface charges; the other is a channel leading from an intersubunit space filled with water molecules (central cavity; Figure 8.3a). Negative surface charges of the former channel provide electrostatic guidance for amine substrates that are positively charged at a physiological pH (Wilce et al., 1997).

Chemical modification of Lys 184 and Lys 354 of AGAO, located close to the entrance to the substrate channel, with 4-fluoro-7-nitrobenzo-2-oxa-1,3-diazole (NBD-F) led to rapid inactivation of the enzyme, presumably preventing the substrate amine from entering the channel by steric hindrance (Matsuzaki and Tanizawa, 1998). Interestingly, the Cu^{2+}-dependent generation of TPQ was also retarded in accordance with the amount of NBD incorporated into the precursor enzyme. It is likely that the substrate channel is utilized also by the aquated Cu^{2+} ions whose positive charges are preferable for increasing the local Cu^{2+} concentration in the negatively charged substrate channel. The active site is located at the bottom of this funnel-shaped substrate channel.

As for dioxygen, another substrate in both the catalytic reaction and TPQ biogenesis, the channel connected to the intersubunit space filled with water molecules (central cavity) is assumed to function as its route to the active site (Wilce et al., 1997). To identify the potential binding site for dioxygen approaching the active site, crystals of Cu^{2+}/TPQ-containing AGAO and other CAOs were exposed to pressurized Xe gas; Xe is a rare gas atom with a van der Waals diameter similar to the molecular size of dioxygen that favors hydrophobic environments like dioxygen (Duff et al., 2004).

Since dioxygen is the common substrate in both the TPQ biogenesis and amine-oxidizing catalytic reactions and CAOs show certain K_m values for dioxygen in the latter reaction, a dioxygen-binding site may be present at or near the active site. An x-ray crystallographic analysis of the AGAO–Xe complex showed seven Xe peaks in the anomalous electron density difference map (Duff et al., 2004). One Xe site was closest to the bound Cu^{2+} ion and TPQ and was commonly observed in the Xe complexes of pea seedling CAO and *Pichia pastoris* lysyl oxidase, another CAO.

Since the assigned Xe atom is ~7.5 and ~9.5 Å away from the Cu^{2+} ion and TPQ, respectively, this site may correspond to the pre-binding site for dioxygen proposed by kinetic and mutational studies of HPAO (DuBois and Klinman, 2005). Johnson et al. (2007) recently called the site an "anteroom." The positions of the remaining Xe atoms likely represent the trajectory of dioxygen migrating into the active site (Duff et al., 2004; Johnson et al., 2007). Implicit ligand sampling calculations with the Xe-less AGAO crystal structure gave the energetically favorable routes of dioxygen to the active site; one is from the intersubunit space and another from the molecular surface and through the D3 β sandwich where the Xe atoms are aligned (Johnson et al., 2007; Figure 8.3b). Presumably, multiple pathways of dioxygen are present in CAOs and may differ depending on the individual enzyme structure. The channel connected to the central cavity may also be used as the pathway for H_2O_2, the common product in the catalytic reaction and TPQ biogenesis, for departure from the active site (Johnson et al., 2007).

8.5 METAL ION SPECIFICITY FOR TPQ GENERATION

Although cupric ion is believed to be the sole natural metal ion that assists the TPQ generating reaction, the metal ion specificity for TPQ biogenesis has been investigated in further detail with recombinant AGAO (Okajima et al., 2005). The TPQ-less precursor enzyme was incubated with various other metal ions including Ag^+, Ba^{2+}, Be^{2+}, Ca^{2+}, Cd^{2+}, Co^{2+}, Cr^{2+}, Fe^{2+}, Hg^{2+}, Mg^{2+}, Ni^{2+}, Pd^{2+}, Rb^+, Sn^{2+}, Sr^{2+}, and Zn^{2+} (0.5 *mM*, as chloride salt) under atmospheric conditions. No obvious spectral changes associated with TPQ generation were observed, confirming that these metal ions were all inert for TPQ biogenesis.

The inability to generate TPQ may be due simply to the failure of these ions to bind to the Cu^{2+} binding site. To examine this, the precursor enzyme preincubated with one of these noncupric metal ions was further incubated with 0.5 *mM* Cu^{2+} ion to initiate TPQ biogenesis that should have been observed if the subsequently added Cu^{2+} ion replaced the prebound metal ion at the metal binding site. In most cases, TPQ was generated at essentially the same rate as TPQ biogenesis assisted by Cu^{2+} ion alone, showing that most noncupric metal ions were unbound or only weakly bound. However, TPQ biogenesis in the enzymes preincubated with the three divalent metal ions, Co^{2+}, Ni^{2+}, and Zn^{2+}, was not observed even after the addition of Cu^{2+} ions, strongly suggesting that they were bound to the metal site of the precursor enzyme so tightly that they were not replaced by the subsequent addition of excess Cu^{2+} ions.

Although these noncupric metal ions could not initiate TPQ formation under atmospheric conditions, we noticed that the precursor AGAO solution added with Co^{2+} and Ni^{2+} ions colored pink after several months of storage in a refrigerator.

A more obvious color change was observed by exposure of the Co^{2+}- and Ni^{2+}-containing AGAO solutions to pure O_2 gas, leading to a new finding that these two divalent metal ions can also assist TPQ generation in AGAO in the presence of a high concentration of dissolved dioxygen, although much less efficiently than Cu^{2+} ion.

Resonance Raman spectroscopy, phenylhydrazine titration, metal analyses, and steady state kinetics (Table 8.1) unambiguously show that TPQ is formed in

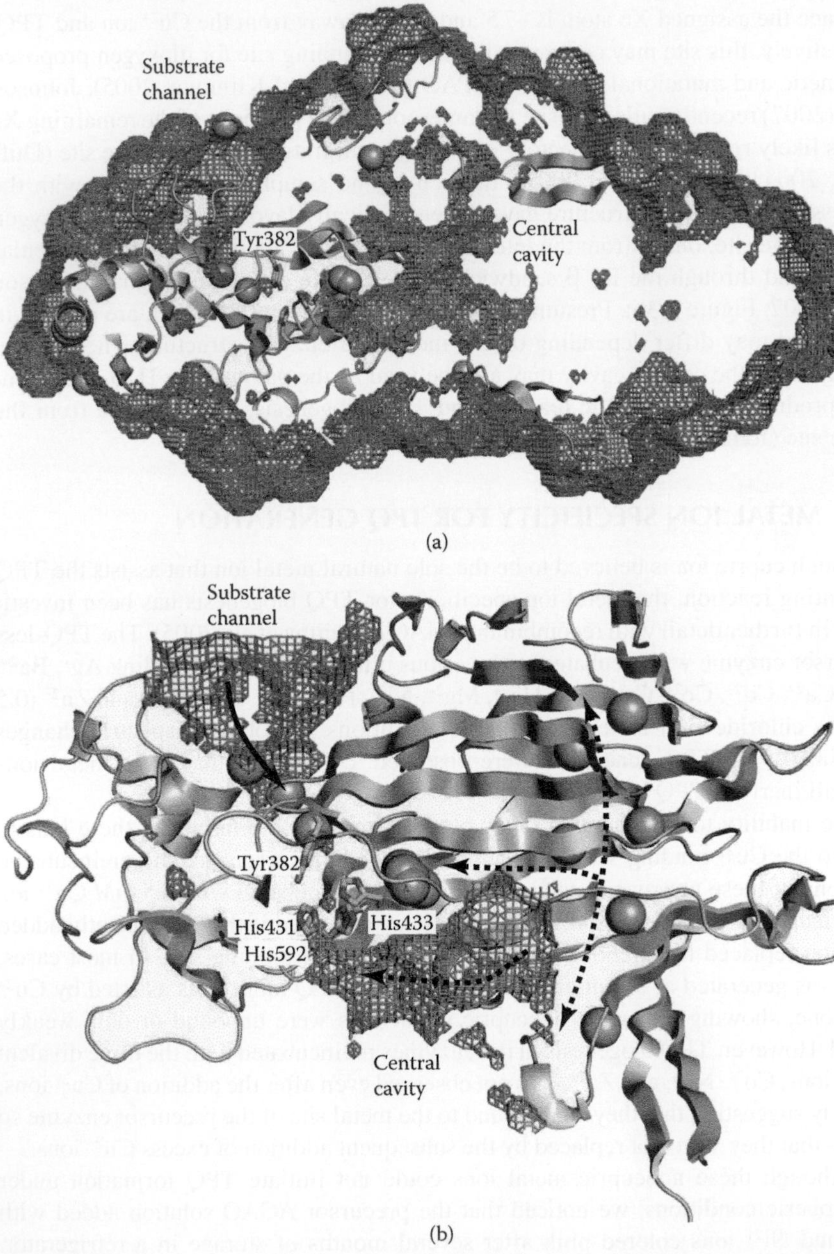

Ni^{2+}- and Co^{2+}-activated enzymes and those enzymes are indistinguishable from the corresponding metal-substituted enzymes (see Chapter 3) prepared from the native copper enzyme (Kishishita et al., 2003). Analyses of Co- and Ni-activated enzymes by x-ray crystallography also revealed an active site structure identical to those of the metal-substituted enzymes (Okajima et al., 2005). However, the rates of TPQ generation (k_{obs}) were very slow for the Co^{2+}- and Ni^{2+}-assisted biogenesis even under O_2-saturated conditions (1.32×10^{-4} min^{-1} and 1.25×10^{-4} min^{-1}, respectively)—about 10^4-fold slower than the rate for the Cu^{2+}-assisted process (Table 8.2). Thus, Cu^{2+} ion is not the sole metal ion assisting TPQ generation; and Co^{2+} and Ni^{2+} are also capable of generating TPQ, although much less efficiently than Cu^{2+}. Similar Ni^{2+}-assisted TPQ biogenesis was reported for HPAO (Samuels and Klinman, 2005), suggesting that intrinsic properties of the metal ions determine the potential to assist TPQ biogenesis. Metal ions for AGAO are classified into three groups:

1. Cu^{2+}, Co^{2+}, and Ni^{2+} tightly bound to the specific metal-binding site and capable of assisting TPQ biogenesis and the catalytic reaction (see Chapter 3) but with different efficiencies
2. Zn^{2+}, also tightly bound to the metal site (even more tightly than Cu^{2+}) but completely inert for TPQ generation
3. Other metal ions that are weakly bound or not bound and are incapable of generating TPQ

These classifications provide insights into the role of metal ions in TPQ biogenesis. First, metal ions must be tightly bound to the metal coordination site formed by three His residues (431, 433, and 592 in AGAO) and the precursor Tyr 382 residue in AGAO with or without a few molecules of H_2O. An x-ray crystal structure of the precursor AGAO bound with a Co^{2+} ion under atmospheric conditions (no TPQ generated, see above;

FIGURE 8.3 (See color insert following page 202.) (Opposite) Channels for Cu^{2+} ion and dioxygen entering the active site. (a) Two channels entering the active site of the precursor form of AGAO are depicted with protein surfaces generated by VOIDOO (Kleywegt and Jones, 1994), primarily with default parameters. The substrate channel dug from the molecular surface and the channel associated with the central cavity are drawn with light blue and pink meshes, respectively. The dimeric structure of AGAO was generated by transforming the deposited monomer coordinates (PDB entry code: 1AVK, chain A) into the other subunit according to the crystal symmetry. Although the channels are calculated with the dimer coordinates, only chain A is presented with the surfaces for showing details. The Xe atoms assigned in the Xe-complexed crystal structures of active AGAO, *Pichia pastoris* lysyl oxidase, pea seedling CAO (Duff et al., 2004), and HPAO (Johnson et al., 2007) are overlaid after superimposing each polypeptide chain and are represented by light brown spheres. The funnel-shaped substrate channel is likely used for both substrate amines and the aquated Cu^{2+} ions entering the active site. Conversely, the channel connected to the central cavity formed between the two subunits is one of the two pathways predicted for dioxygen entering into the active site. (b) Enlarged side view of the channels depicted in panel A shown without protein surface. The bold line with an arrowhead represents the predicted pathway of the Cu^{2+} ion and substrate amines. The dotted lines represent the two potential pathways of dioxygen independently proposed from the array of the Xe-binding sites in the hydrophobic core (Johnson et al., 2007) and from the channel connected to the central cavity (Wilce et al., 1997).

Okajima et al., 2005) indicated that Co^{2+} is coordinated by the same His residues and the unprocessed Tyr residue like the anaerobic complex of Cu^{2+} AGAO (Figure 8.2b).

On the periodic table, Co, Ni, Cu, and Zn are successive elements with the same outer subshell electrons. Presumably, their divalent ions may have proper ionic radii (not too small or large) that can be accommodated within the cavity of the metal center and also an ability to form a specific, stable metal coordination structure (Okajima et al., 2005). Secondly, among physicochemical properties of the divalent metal ions, Lewis acidity may account for the different activities of the three metal ions assisting TPQ generation. When the pK_a values of metal-coordinating water molecule are taken as an assessment of Lewis acidity of metal ions, Cu^{2+} ion (pK_a = 7.5) has a value two pH units lower than that of Co^{2+} ($pK_a = 9.6$) and Ni^{2+} ($pK_a = 9.4$) ions (Table 8.2); Cu^{2+}-coordinating water is 100-fold more dissociated than water molecules coordinating to Co^{2+} and Ni^{2+} at neutral pH.

This tendency of Lewis acidity of the three metal ions corresponds well to the TPQ biogenesis rates. In the biogenesis intermediate (Figure 8.2c), the formation of a Tyr radical (Tyr•) by full $1e^-$ transfer to Cu^{2+} is only feasible with Cu^{2+}; $1e^-$ reduction of Co^{2+} and Ni^{2+} to Co^{1+} and Ni^{1+}, respectively, is energetically unlikely because of their very low reduction potentials. Therefore, the valence change of the bound metal ions is unlikely associated with TPQ biogenesis, although the possibility that it occurs only in the case of Cu^{2+}-assisted biogenesis cannot be excluded.

Nevertheless, the Lewis acidity of metal ions alone does not explain the inability of Zn^{2+} ion to generate TPQ; the pK_a value of Zn^{2+}-coordinating water ($pK_a = 9.6$; Table 8.2) is comparable with that of Co^{2+}-coordinating water. Alternatively, the metal ion specificity may somehow be associated with the versatility of the coordination structure afforded by each metal ion, as typically observed during the Cu^{2+}-assisted process of generating TPQ in AGAO (Kim et al., 2002; see also Section 3). The tetrahedral coordination of the Cu^{2+} ion initially formed with four ligands (Figure 8.2b) is converted to the square pyramidal geometry with five ligands (Figure 8.2h; Wilce et al., 1997), attained through the conformational flexibility of His 592 and also flipping out of the residue 382, allowing the new participation of two water ligands. In the other intermediates (e and g), His 592 has a multiple conformer (Kim et al., 2002; Matsunami et al., 2004; Wilce et al., 1997).

It is interesting to note that the affinity for an anion (e.g., azide) of the Cu^{2+} ion bound to the precursor form of HPAO in the absence of O_2 is two orders of magnitude greater than the same Cu^{2+} ion in the active TPQ-containing enzyme in the presence of O_2 (Schwartz et al., 2000). The anionic affinity of a metal ion is closely related to its Lewis acidity. This observation suggests that Lewis acidity is affected significantly by the metal coordination structure. Although the Lewis acidity of Zn^{2+} is comparable to that of the Co^{2+} ion capable of promoting TPQ biogenesis, the Zn^{2+} ion may lack such versatility of the coordination structure because of its stable electronic structure, thereby being inert for TPQ biogenesis (Table 8.2). In contrast, Co^{2+} ion may have a multiple coordination structure in the metal center of AGAO. At least two forms of Co AGAO have been identified: a tetrahedral coordination in the initial intermediate (Okajima et al., 2005) and an octahedral coordination (Kishishita et al., 2003) in the TPQ-containing active form (Table 8.2).

8.6 CONCLUDING REMARKS

CAO can catalyze the oxygenation and the following oxidation of the precursor Tyr residue in the TPQ biogenesis reaction and also the oxidative deamination of primary amine substrates in the catalytic reaction.

The main role of the Cu^{2+} ion in TPQ biogenesis is the activation of the precursor Tyr residue for the initial reaction with dioxygen; the ion in the catalytic process provide a binding site for $1e^-$- and $2e^-$-reduced dioxygen species produced in the oxidative half-reaction of the catalytic process (Kishishita et al., 2003). The active site of CAO must be optimized for the performance of two distinct reactions involving the same special metal ion, Cu^{2+}.

Although Ni^{2+} and Co^{2+} ions are capable of assisting TPQ biogenesis, the reaction is not a physiological one due to the extremely slow reaction rate and the need for a high concentration of dissolved dioxygen. Cu^{2+} is the best metal ion for efficiently promoting TPQ biogenesis. If the metal ion specificity for catalysis (70- to 100-fold difference) is combined with that for TPQ biogenesis (1200-fold difference), Cu^{2+} ion is about 100,000-fold preferred by AGAO to Co^{2+} and Ni^{2+} ions (Okajima et al., 2005). This could be the basis for the selection of Cu^{2+} among the Co^{2+}, Cu^{2+}, Ni^{2+}, and Zn^{2+} ions with similar natural abundance.

About 15 years after the *in vitro* demonstration of TPQ biogenesis, we have disclosed only part of its sophisticated mechanism. We do not understand clearly why Zn^{2+} is inert for assisting TPQ biogenesis and whether the dioxygen binding site is near the precursor Tyr residue and plays a role in TPQ biogenesis, although kinetic experiments with HPAO suggest its presence (Dove et al., 2000). Further studies are necessary for refining the mechanisms of TPQ biogenesis and fully elucidate the molecular details. In addition to various kinetic, spectroscopic, and x-ray crystallographic studies, quantum mechanistic studies (e.g., QM/MM calculations) are also required to explain unequivocally the role of the metal ion and the structures of intermediates that have not been identified by the previous flash-freeze technique.

ACKNOWLEDGMENTS

We wish to thank a large number of collaborators in our AGAO studies over the past 15 years and also many friends in the CAO community for their valuable comments and discussion (in alphabetical order): K. Chigusa, Y.-C. Chiu, Y.-H. Choi, D.M. Dooley, H.C. Freeman (with our deepest sympathy for his passing), I. Frébort, T. Fukui, J.M. Guss, H. Hayashi, K. Hirose, S. Hirota, H. Hori, H. Itaya, T. Iwamoto, N. Kamiya, A. Kawamori, Y. Kawano, M. Kim, S. Kishishita, J.P. Klinman, P.F. Knowles, T. Koyanagi, S. Kuroda, H. Matsunami, R. Matsuzaki, W.S. McIntire, M.J. McPherson, T. Murakawa, M. Mure, T. Nakamoto, N. Nakamura, S. Nakanishi, N. Ohtani, Y. Ozaki, S.E. Phillips, C.E. Ruggiero, J. Sanders-Loehr, E. Shimizu, M.A. Spies, S. Suzuki, M. Taki, M. Uchida, M.C.J. Wilce, C.M. Wilmot, H. Yamaguchi, K. Yamaguchi, Y. Yamamoto, T. Yorifuji, and M. Yoshimura.

The original studies reviewed were supported by grants-in-aid for scientific research from Japan Society for the Promotion of Science (Research for the Future), an Osaka University Center of Excellence program (Creation of Highly Harmonized

Functional Materials) to K.T., a research grant from the Japan Foundation for Applied Enzymology to K.T., grants-in-aid for scientific research from the Ministry of Education, Culture, Sports, Science and Technology of Japan (Priority Area 13125204; 21st Century Center of Excellence Program) to K.T., and from the Japan Society for the Promotion of Science (Nos. 07224210, 07680687, 8214210, 8249224, 12480180, and 18370043 to K.T., and 14560066 and 18350085 to T.O.).

9 Copper Amine Oxidase Crystal Structures

J. Mitchell Guss, Giuseppe Zanotti, and Tiina A. Salminen

CONTENTS

Crystal structures of copper-containing amine oxidases (CAOs) have been solved from eubacteria, plants, yeasts and mammals including humans, but interestingly not from archaea or mycobacteria (Table 9.1). The structures are remarkably conserved

TABLE 9.1
Crystal Structures of Copper-Containing Amine Oxidases Deposited in the Protein Data Bank as of March 9, 2008

PDB ID	Source	Structure Description	Resolution (Å)	Reference
	Eubacteria			
1oac	*Escherichia coli*	Native	2.0	Parsons et al., 1995
1d6u	*Escherichia coli*	Anaerobically substrate reduced	2.4	Wilmot et al., 1999
1d6y	*Escherichia coli*	Anaerobically substrate reduced plus NO	2.4	Wilmot et al., 1999
1d6z	*Escherichia coli*	Aerobically trapped equilibrium species	2.1	Wilmot et al., 1999
1dyu	*Escherichia coli*	Native	2.0	Murray et al., 1999
1lvn	*Escherichia coli*	Complex with tranylcypromine	2.4	Wilmot et al., 2004
1spu	*Escherichia coli*	Complex with 2-hydrazinopyridine	2.0	Wilmot et al., 1997
1jrq	*Escherichia coli*	Y369F mutant	2.1	Murray et al., 2001
1qak	*Escherichia coli*	D383A mutant	2.0	Murray et al., 1999
1qaf	*Escherichia coli*	D383E mutant	2.2	Murray et al., 1999
1qal	*Escherichia coli*	D383N mutant	2.2	Murray et al., 1999
1av4	*Arthrobacter globiformis*	Active holo	2.3	Wilce et al., 1997
1avi	*Arthrobacter globiformis*	Inactive holo	2.8	Wilce et al., 1997
1avk	*Arthrobacter globiformis*	Apo	2.2	Wilce et al., 1997
1iqx	*Arthrobacter globiformis*	Co-substituted	2.0	Kishishita et al., 2003
1iqy	*Arthrobacter globiformis*	Ni-substituted	1.8	Kishishita et al., 2003
1iu7	*Arthrobacter globiformis*	Native	1.8	Kishishita et al., 2003
1ivu	*Arthrobacter globiformis*	TPQ biogenesis Tyr + Cu	1.9	Kim et al., 2002
1ivv	*Arthrobacter globiformis*	TPQ biogenesis QPQ + Cu	2.1	Kim et al., 2002
1ivw	*Arthrobacter globiformis*	TPQ biogenesis, inactive	1.9	Kim et al., 2002
1ivx	*Arthrobacter globiformis*	TPQ biogenesis, active	2.2	Kim et al., 2002
1wmn	*Arthrobacter globiformis*	Co-substituted	1.8	Okajima et al., 2005
1wno	*Arthrobacter globiformis*	Ni-substituted	1.8	Okajima et al., 2005
1wmp	*Arthrobacter globiformis*	TPQ-biogenesis. Co-assisted + Tyr	2.0	Okajima et al., 2005
2e2t	*Arthrobacter globiformis*	Complex with phenylhydrazine	2.05	Murakawa, T., et al., To be published
2cwt	*Arthrobacter globiformis*	D298A mutant	1.82	Chiu et al., 2006
2cwu	*Arthrobacter globiformis*	D298A mutant + 2PEA (1 hr)	1.85	Chiu et al., 2006
2cwv	*Arthrobacter globiformis*	D298A mutant + 2PEA (1 wk)	1.85	Chiu et al., 2006

(*Continued*)

TABLE 9.1 (*Continued*)

PDB ID	Source	Structure Description	Resolution (Å)	Reference
2d1w	*Arthrobacter globiformis*	D298A + tyramine (substrate Schiff base)	1.74	Murakawa et al., 2006
1rjo	*Arthrobacter globiformis*	Complex with xenon	1.67	Duff et al., 2004
1sih	*Arthrobacter globiformis*	Complex with MOBA	1.73	O'Connell et al., 2004
1sii	*Arthrobacter globiformis*	Complex with NOBA	1.70	O'Connell et al., 2004
1ui7	*Arthrobacter globiformis*	H433A mutant	2.0	Matsunami et al., 2004
1ui8	*Arthrobacter globiformis*	H592A mutant	1.8	Matsunami et al., 2004
1w6g	*Arthrobacter globiformis*	Native	1.55	Langley et al., 2006
1w6c	*Arthrobacter globiformis*	Native (small cell)	2.2	Langley et al., 2006
2bt3	*Arthrobacter globiformis*	Complex with C4-racemic wire	1.73	Contakes et al., 2005
2cfd	*Arthrobacter globiformis*	Complex with C4-lambda wire	1.60	Langley et al., 2008a
2cfg	*Arthrobacter globiformis*	Complex with C4-delta wire	1.55	Langley et al., 2008a
2cfk	*Arthrobacter globiformis*	Complex with C5 wire	1.8	Langley et al., 2008a
2cfl	*Arthrobacter globiformis*	Complex with C6 wire	1.8	Langley et al., 2008a
2cfw	*Arthrobacter globiformis*	Complex with C7 wire	1.74	Langley et al., 2008a
2cg0	*Arthrobacter globiformis*	Complex with C9 wire	1.8	Langley et al., 2008a
2cg1	*Arthrobacter globiformis*	Complex with C11 wire	1.67	Langley et al., 2008a
1w4n	*Arthrobacter globiformis*	Complex with tranylcypromine	1.65	Langley, Trambaiolo et al., 2008b
1w5z	*Arthrobacter globiformis*	Complex with benzylhydrazine	1.86	Langley, Trambaiolo et al., 2008b
2e2u	*Arthrobacter globiformis*	Complex with 4-hydroxybenzyl hydrazine	1.68	Murakawa et al., To be published
2e2v	*Arthrobacter globiformis*	Complex with benzylhydrazine	1.80	Murakawa et al., To be published
	Yeasts			
1a2v	*Hansenula polymorpha*	Native	2.4	Li et al., 1998
1ekm	*Hansenula polymorpha*	Zn-substituted	2.5	Chen et al., 2000
2oov	*Hansenula polymorpha*	Xe complex	1.7	Johnson et al., 2007
2oqe	*Hansenula polymorpha*	Xe complex	1.6	Johnson et al., 2007
1w7c	*Pichia pastoris*	Native	1.23	Duff, Cohen et al., 2006a
1n9e	*Pichia pastoris*	Native	1.65	Duff et al., 2003
1rky	*Pichia pastoris*	Xe complex	1.68	Duff et al., 2004
	Plants			
1ksi	*Pisum sativum* (garden pea)	Native	2.2	Kumar et al., 1996
1w2z	*Pisum sativum* (garden pea)	Xe complex	2.24	Duff, Shepard et al., 2006b

(*Continued*)

TABLE 9.1 (*Continued*)

PDB ID	Source	Structure Description	Resolution (Å)	Reference
	Mammals			
1us1	*Homo sapiens*	Native	2.9	Airenne et al., 2005
1pu4	*Homo sapiens*	Native	3.2	Airenne et al., 2005
2c10	*Homo sapiens*	Native	2.5	Jakobsson et al., 2005
2c11	*Homo sapiens*	Complex with 2-hydrazinopyridine	2.9	Jakobsson et al., 2005
1tu5	*Bos taurus*	Native	2.37	Lunelli et al., 2005
2pnc	*Bos taurus*	Complex with clonidine	2.4	Holt et al., 2008

TABLE 9.2
Pair-Wise Structural Comparison of Copper-Containing Amine Oxidases[a]

	ECAO (1oac)	AGAO (1w6g)	HPAO (2oqe)	PPLO (1w7c)	PSAO (1ksi)	BPAO (1tu5)	VAP-1 (2c10)
ECAO (1oac)	–	30%	28%	20%	29%	24%	25%
AGAO (1w6g)	1.4 (588)	–	34%	18%	25%	21%	21%
HPAO (2oqe)	1.7 (583)	1.5 (600)	–	19%	26%	23%	22%
PPLO (1w7c)	2.3 (554)	2.2 (533)	2.2 (548)	–	19%	28%	26%
PSAO (1ksi)	1.4 (597)	1.6 (589)	1.8 (592)	2.3 (559)	–	20%	20%
BPAO (1tu5)	2.0 (534)	2.3 (534)	2.2 (523)	1.9 (566)	2.1 (545)	–	84%
VAP-1 (2c10)	2.1 (559)	2.2 (553)	2.2 (558)	2.0 (628)	2.2 (576)	0.7 (625)	–

Notes: [a] Protein Structure Comparison Service SSM at European Bioinformatics Institute (http://www.ebi.ac.uk/msd-srv/ssm), authored by E. Krissinel and K. Henrick, 2004; Lower left triangle: root mean square deviation (Å) for superposed C^{α}; number of paired residues in parenthesis. Upper right triangle: percentage of paired residues that are identical; ECAO = *E. coli* amine oxidase; AGAO = *A. globiformis* amine oxidase; HPAO = *H. polymorpha* amine oxidase; PPLO = *P. pastoris* amine oxidase; PSAO = pea seedling (*P. sativum*) amine oxidase; BPAO = bovine plasma amine oxidase; VAP-1 = vascular adhesion protein-1 (*H. sapiens*).

(Table 9.2) both in the core catalytic domains and also in the peripheral domains that presently have no known function.

The structures display a relatively low level of sequence similarity (Table 9.2; Figure 9.1). The first to be solved was that of the *Escherichia coli* enzyme by Simon Phillips and his colleagues in Leeds who likened it to a mushroom (Parsons et al.,

1995). See Figure 9.2. Every structure is a dimer that buries a large surface in the interface in which each subunit embraces the other with two extended arms. In keeping with the observation of a very strong dimer in the crystals, no evidence indicates a stable monomeric species in solution. Every subunit has ~700 amino acid residues.

Each subunit of the *E. coli* copper amine oxidase (ECAO) structure has four domains, with the N-terminal domain (Dl) forming the stalk of the mushroom. The second and third domains (D2 and D3) form the periphery of the mushroom cap when viewed down the axis of the stalk. While the topology of these two domains is conserved in all the structures, their orientation to the core of the cap formed by the dimer of the D4 domain varies slightly. An interesting feature common to all the structures is a central cavity lying in the core of the dimer between the D4 domains. This "lake" laps against the back of one of the histidine copper ligands and is in turn connected to the external solvent surrounding the molecule by a series of narrow channels along the two-fold axis of the dimer. This rear approach to the active site via the lake has been suggested as a route for small substrates and products of both TPQ biogenesis and enzymatic reactions (O_2 and H_2O_2) to enter and leave the active site.

One enzyme active site, defined by the presence of the topaquinone cofactor and the copper ion, is buried deeply in each subunit. The active site in one subunit is connected to the other subunit by one of the embracing arms, residues from which form part of the rim of the active site funnel. The active sites are linked to the surface of the enzyme by a channel, the width and length of which vary. This channel may serve as an entry and exit route for the amine substrates and aldehyde products of the enzymatic reaction.

The TPQ is generally found in one of two states (Figure 9.3). In the first, the O4 atom coordinates directly to the Cu ion in such a way that the TPQ is inaccessible to a substrate travelling the channel. In this conformation, the substrate channel is often blocked by a so-called gate residue. This conformation of the TPQ is called "on-copper" or inactive. In the alternate conformation, the TPQ is swung away from the Cu into the base of the active site channel. This is known as the "off-copper" or active conformation. In many crystal structures, the TPQ is disordered despite relatively high resolution and clear electron density for the remainder of the structure including the active site. This observation supports the notion that the TPQ is flexible and moves during biogenesis and possibly the enzymatic reaction.

9.1 FROM BACTERIA

9.1.1 Function and Occurrence

As stated by McIntire and Hartmann (1992), "The function of amine oxidases in prokaryotes is obvious. These enzymes allow the organism to use the appropriate alkyl- or arylamine as a carbon source for growth." This statement has not been confirmed by the appropriate knock-outs nor disproven. Therefore, in the absence of evidence to the contrary, it provides a good hypothesis for the role of these enzymes in bacteria.

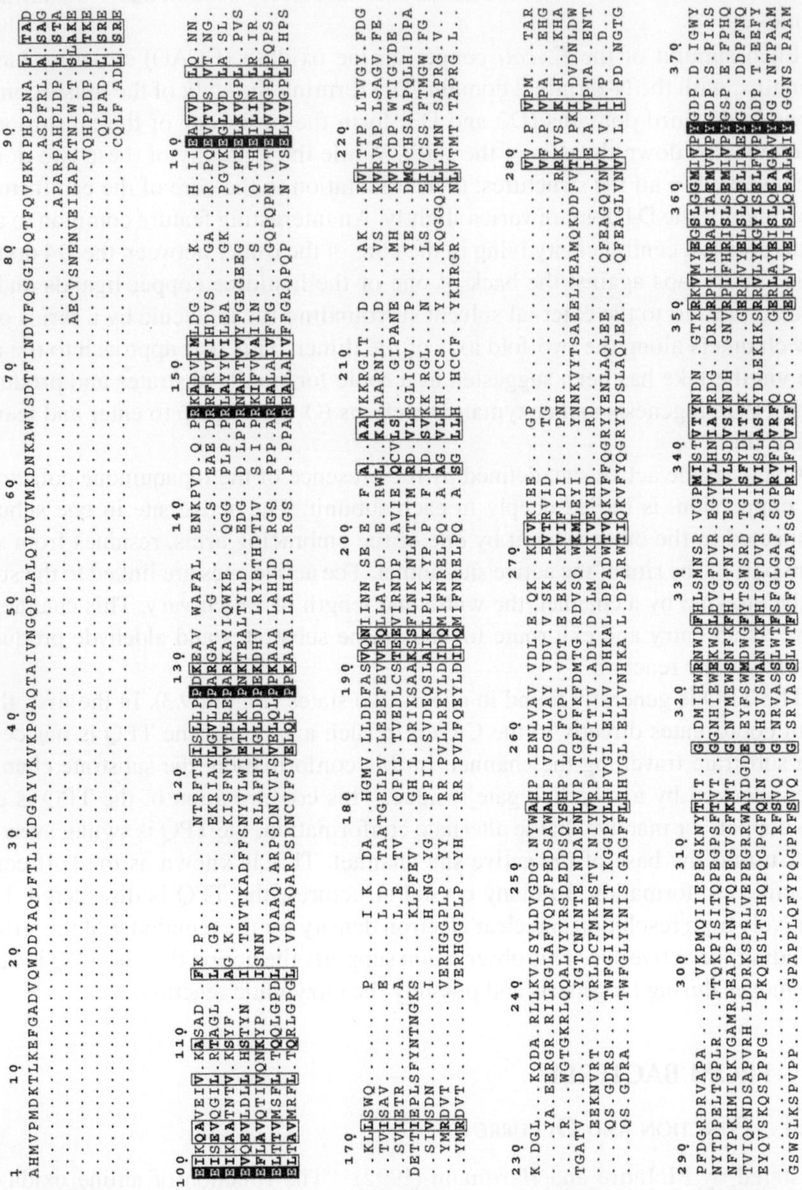

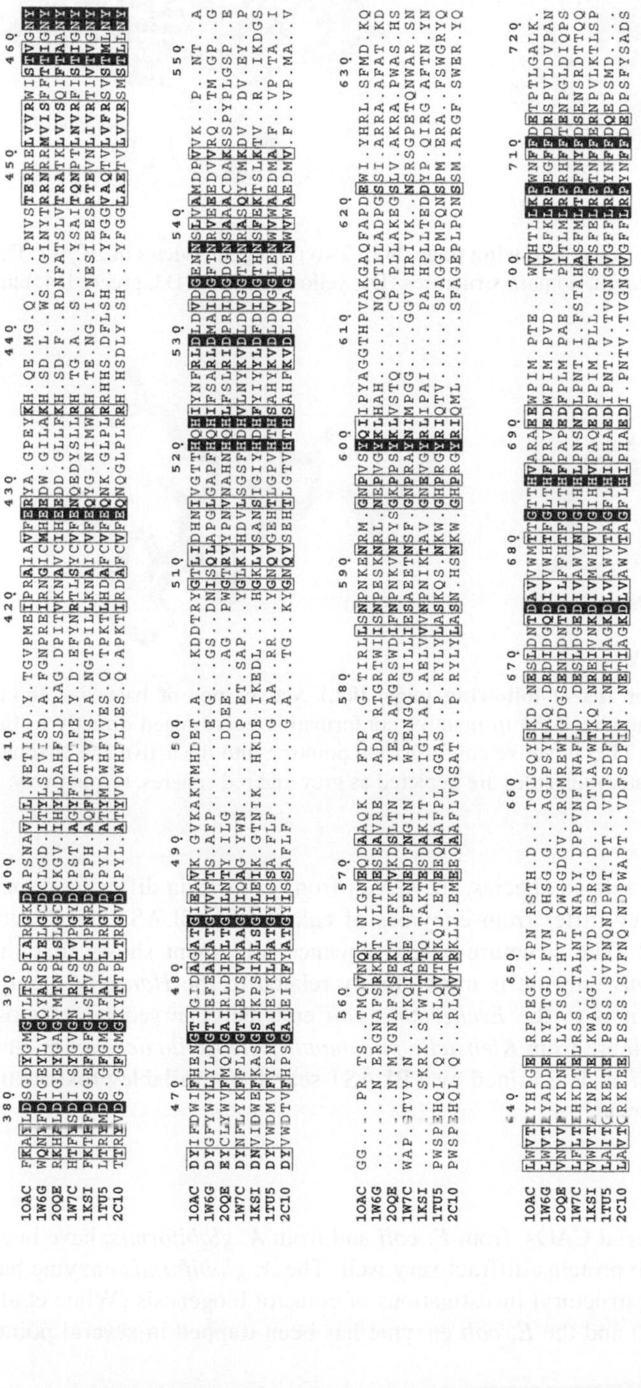

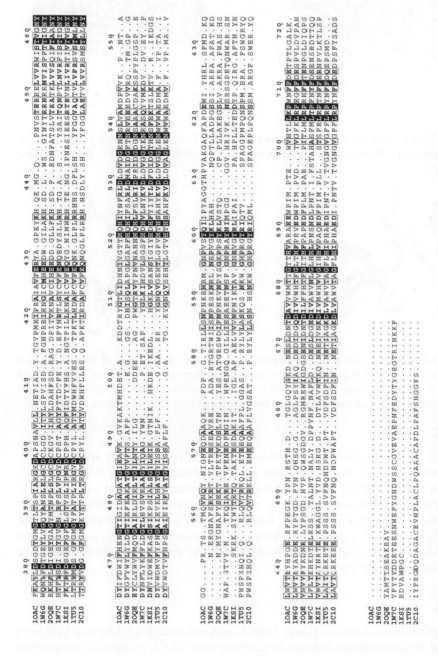

FIGURE 9.1 Structure-based sequence alignments of CAOs. The A chain from the highest resolution structure for each species has been used. The structures are 1oac (*E. coli*), 1w6g (*A. globiformis*), 2oqe (*H. polymorpha*), 1w7c (*P. pastoris*), 1ksi (*P. sativum*), 1tu5 (*bovine*), 2c10 (*human*, VAP-1).

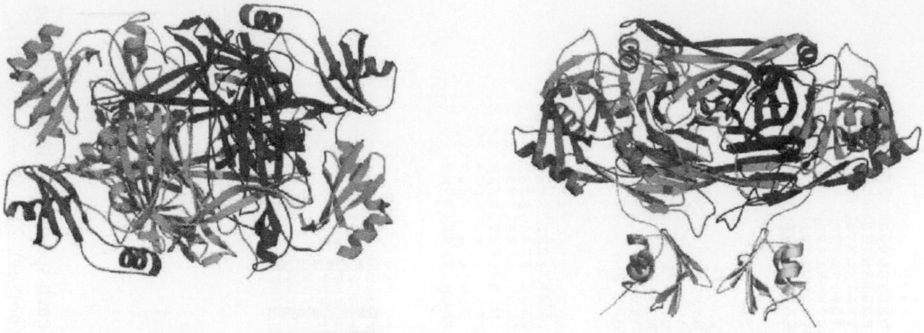

FIGURE 9.2 **(See color insert following page 202.)** Two orthogonal views of ECAO. The chains are colored to highlight domain structures (D1, yellow; D2, red; D3, green; D4, blue and cyan).

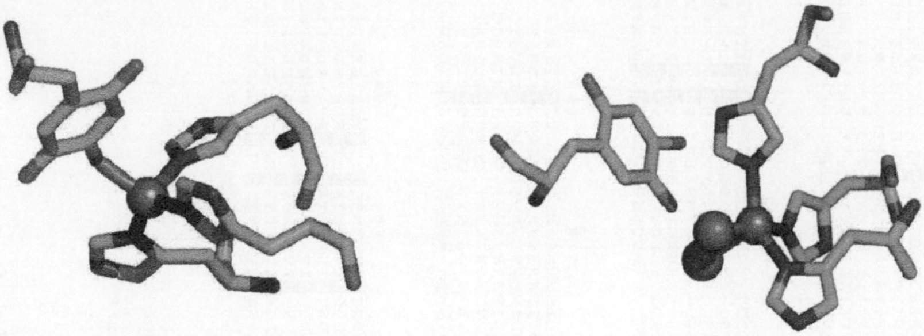

FIGURE 9.3 **(See color insert following page 202.)** Active sites of bacterial amine oxidases. Left: ECAO showing TPQ in inactive conformation coordinated directly to the TPQ. Right: AGAO with TPQ in active conformation pointing into the active site channel. The copper atoms and water molecules are depicted as grey and red spheres, respectively.

9.1.2 Sequence

Except in very closely related species, the CAOs from eubacteria differ from one another as much as they differ from enzymes of eukaryotes (BLAST results not shown). See Table 9.2. The structure-based sequence alignment shows that the *Arthrobacter globiformis* enzyme is more closely related to the *Hansenula polymorpha* enzyme than it is to the *Escherichia coli* enzyme. The sequences most closely related to *E. coli* are from *Klebsiella pneumonia*, *Klebsiella aerogenes,* and *Acinobacter baumannii* as determined by a BLAST search of available prokaryotic sequences (data not shown).

9.1.3 Structure

Structures of two bacterial CAOs, from *E. coli* and from *A. globiformis*, have been solved. Crystals of both proteins diffract very well. The *A. globiformis* enzyme has been used in extensive structural investigations of cofactor biogenesis (Wilce et al., 1997; Kim et al., 2002) and the *E. coli* enzyme has been trapped in several points of the catalytic cycle (Wilmot et al., 1999). In total, 11 *E. coli* and 37 *A. globiformis* coordinate sets have been deposited in the Protein Data Bank (PDB; Berman et al., 2003). As noted earlier, the first structure of a CAO to be solved was from *E. coli* and it thus defines the archetypal fold (Figure 9.2) to which subsequent structures from other organisms are related. Of all the CAO structures, only ECAO has the N-terminal domain D1. In a BLAST search of all sequences based on ECAO, only the CAOs with the closest sequence similarity to ECAO, *Klebsiella pneumonia*, *Klebsiella aerogenes*, and *Acinobacter baumannii*, are predicted to contain this extra domain.

The active sites in ECAO and in AGAO are shown in Figure 9.3. The structures were chosen to illustrate the two major states of the active site. In the ECAO structure, the TPQ is coordinated directly to the Cu, creating a distorted tetrahedral environment. In the other state, illustrated by the AGAO structure, the TPQ has swung away from the Cu into the active site channel placing its O5 atom in a position to interact with the incoming amine substrate. In many structures of CAOs, the TPQ is disordered, further emphasizing the conformational flexibility of this ligand. At several points during the TPQ biogenesis reaction, direct coordination of the tyrosine precursor and intermediates to the Cu has been confirmed by structure analyses (Chapter 8).

The three His ligands of the Cu are conserved in all CAOs. When the TPQ is not coordinated directly to the Cu, an additional one or two water ligands are generally observed bound to the Cu. Depending on the resolution of the structure analysis, one or both of these solvent molecules may not be observed as a result of disorder or partial occupancy. An interesting feature of a number of AGAO structures is that one of the copper ligands, His 592, has two distinct conformations. This behavior is apparently not connected with the position of the TPQ.

Despite the relatively low sequence identity between the two structurally characterized CAOs of bacterial origin, ECAO and AGAO, the residues lining the active site channel are identical and in equivalent locations, with one or two differences (Table 9.3). Initial inspection of the first structures of CAOs from ECAO (Parsons et al., 1995) and PSAO (Kumar et al., 1996) failed to reveal a facile route for the entry of amine substrates into the active site. This led to the hypothesis that large domain movements were required to allow the entry of substrates and the exit of products from the active sites. The first structure of AGAO clearly revealed the presence of an active site channel. Re-examination of the ECAO and PSAO structures showed equivalent channels that had been obscured by a residue lying across the channel (the "gate"). Structures of complexes of ECAO and AGAO with inhibitors (Table 9.1),

TABLE 9.3
Equivalent Residues in Active Site Channels of Two Bacterial CuAOs

AGAO	Phe	Ala	Pro	Leu	Trp	Tyr	Tyr	Tyr	Gly	Leu
	105	135	136	137	168	296	302	307	380	358
ECAO	Phe	Thr	Pro	Leu	Trp	Tyr	Tyr	Leu	Gly	Met
	192	223	224	225	257	381	387	392	464	443

whether or not covalently attached to the TPQ, always have the gate open and the TPQ in active conformation (Figure 9.3). This led to the mouse-trap model whereby the entry of an inhibitor or substrate into the active site channel requires the gate to be open and the TPQ to snap into the active site ready for interaction.

9.2 FROM YEASTS

9.2.1 Function and Occurrence

The structures of CAOs from two yeasts, HPAO from *Hansenula polymorpha* (also known as *Pichia angusta*) and PPLO from *P. pastoris*, have been solved, each with three entries in the PDB.

9.2.2 Sequence

Although HPAO and PPLO come from related organisms, their structures and sequences are as different from each other as they are from CAOs from prokaryotic and other eukaryotic organisms (Table 9.2). In fact, the structure of PPLO appears to be an outlier by having the fewest close structural homologues (Table 9.2). Despite these differences both HPAO and PPLO have the same overall structures and domain organizations as the other CAOs.

9.2.3 Structure

Unlike most other CAO structures that have a monomer or dimer in the asymmetric unit, the original structure of HPAO in space group $P2_12_12_1$ has three tight dimers in the asymmetric unit of the crystal arranged to form a ring of six monomers (Li et al., 1998). Both HPAO and PPLO are glycoproteins. HPAO has one potential N-linked glycosylation site and a single sugar residue is located at this site in the original structure at 2.5 Å resolution. PPLO, on the other hand, has five potential glycosylation sites and sugar residues were modelled at all five sites in the structure. Despite an estimated carbohydrate content exceeding 20% by weight, crystals of PPLO obtained subsequently diffracted to 1.23 Å resolution, the highest currently obtained for any CAO.

The outlier status of PPLO relates to its enzyme activity. PPLO alone among the structurally characterized CAOs has the ability to catalyse the oxidation of a peptidyl–lysine residue to the corresponding allylsine aldehyde. In this respect only, PPLO mimics the enzyme activity of the unrelated mammalian lysyl oxidase enzyme. A true lysyl oxidase has a lysyl–tyrosyl–quinone (LTQ) cofactor replacing the TPQ of the CAO family. Lysyl oxidases are also monomers; CAOs are always dimers in their active forms. This unusual activity of PPLO gives it its common name, *Pichia pastoris* lysyl oxidase. This extraordinary activity is explained by the presence in PPLO of a much wider and open active site channel (Figure 9.4) that can accommodate a peptide substrate, albeit one that is bent into a turn conformation (Duff et al., 2003).

A dramatic illustration of the potential activity of PPLO was shown in the structure from the 1.23 Å refinement that revealed an intramolecular cross link between two lysine residues (Duff et al., 2006a). It was proposed that the cross link formed

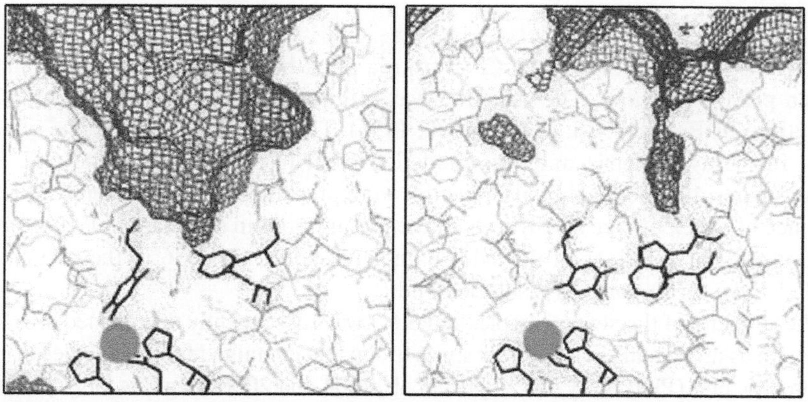

FIGURE 9.4 Comparison of active site channels of PPLO (left) and HPAO (right). The blue mesh represents the surface of the enzyme; Cu atoms (cyan spheres), Cu ligands, and TPQ are shown. Gate residues are shown as sticks. (Adapted from Duff et al., 2003, Figure 7.)

spontaneously following oxidation of one of the two lysine residues and the subsequent condensation of the resulting aldehyde with the other lysine residue.

9.3 FROM PLANTS

9.3.1 Function and Occurrence

The roles and distributions of CAOs in plants are still not fully understood despite many years of study. Histochemical and immuno-staining of CAOs in plants grown under different conditions and exposed to fungal pathogens suggests that they play a role in cell growth, cell differentiation, and pathogen resistance (Laurenzi et al., 2001). The CAO was proposed to function by the regulation of the levels of di- and polyamines and/or by the production of reactive oxygen species. Inhibition of CAO following wounding of plants resulted in lower H_2O_2 accumulation at wound sites and at more distal locations (Rea et al., 2002). The production of H_2O_2 generated by CO is the signal generated by absicic acid-induced stromal closure in *Vicia faba* plants (An et al., 2008). Thus, CAOs function in plants to both degrade and thus regulate the levels of cellular polyamines and participate in important physiological processes via the production of H_2O_2 (Cona et al., 2006).

9.3.2 Sequence

In a multiple sequence alignment of four CAOs from higher plants (pea seedling, lentil seedling, chickpea, and soybean) using ClustalW, 70% of the sequence was invariant. In addition to the copper–ligand residues, the TPQ and gate residue (Phe in plant CAOs)—all the residues lining the active site channel—are conserved. This indicates that these proteins are likely to act on the same or very similar substrates. In the absence of the structure of a complex of a substrate or inhibitor of a plant CAO, it is not possible to state with certainty which residues pursue specific interactions with substrates and thus determine substrate specificity.

9.3.3 Structure

Only one structure (PSAO) of a CAO from a plant has been determined from the garden pea (*P. sativum*). The structure was originally solved and refined at 2.2 Å resolution (PDB entry 1ksi; Kumar et al., 1996). The polypeptide backbone of each chain in the asymmetric unit was modelled from residue 6 to residue 647. The remaining residues at the N- and C-termini were presumed to be disordered. Despite evidence of a lack of free sulfhydryls in the mature protein (Kumar et al., 1996), the positions and orientations of the side chains of residues Cys 636 in the two chains precluded the formation of a disulfide bond.

The structure of the xenon complex of PSAO was subsequently determined at 2.24 Å resolution using crystals in a different space group that contained four molecules in each asymmetric unit (PDB entry 1w2z; Duff et al., 2004 and 2006b). In this latter structure the C-terminus containing the disulfide bond is clearly resolved (chains A, B, C, and D; residues 6 through 647) and is asymmetrically related to the molecular two-fold axis of the dimer. The missing disulfide bond in the previous structure may be explained by crystal packing forcing the superposition of two different orientations of the C-terminus. The disulfide bond forms a knot linking the polypeptide chains of the two subunits.

The four Asn–Xaa–Ser/Thr tripeptide consensus "sequons" per monomer of PSAO may possibly have N-linked glycosylation. Carbohydrate residues could be modelled at all six independent chains in the two structures attached to Asn 131 and Asn 558 (Figure 9.5).

No electron density was seen for any linked carbohydrate at Asn 334 or Asn 364, despite the fact that these residues also lie on the surface of the protein. Three sites (131, 364 and 558) are conserved in the lentil enzyme; 364 was proposed not to be glycosylated based on hydropathy profiles (Rossi et al., 1992). These observations are consistent with deglycosylation of pea and lentil seedling enzymes that suggest a total glycosylation amounting to 4% of total molecular weight (Tipping and McPherson, 1995). None of these actual or putative sites for N-linked glycosylation lies close to the active site or substrate binding channel, suggesting that their role is to enhance the stability of the protein or participate in protein–protein interaction.

In both crystal forms of PSAO, the Cu atoms and the three ligand histidine residues are well resolved and the TPQ is in the off-Cu or active conformation. However,

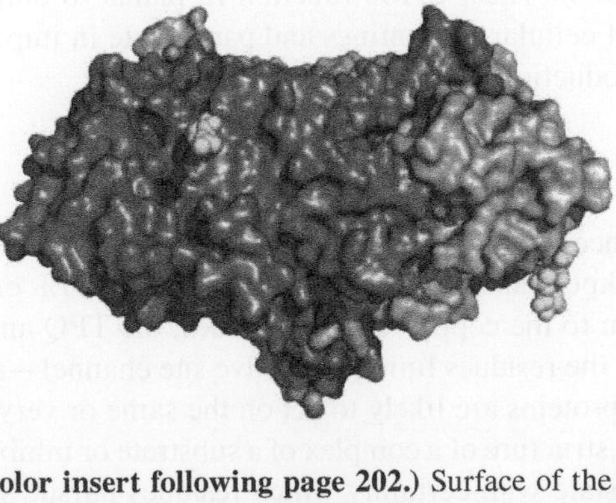

FIGURE 9.5 **(See color insert following page 202.)** Surface of the dimer of PSAO with subunits shown in blue and pink. The attached carbohydrate is represented by yellow space-filling atoms.

the temperature parameters for the TPQ are generally higher than those for the surrounding residues and the electron density indicates that the TPQ has at least two conformations related by a 180-degree rotation about the C1–C4 axis of the six-membered ring (Kumar et al., 1996). The volume of the substrate channel of PSAO is similar to those of the other structurally characterized amine oxidases with the exception of the *P. pastoris* enzyme that is much larger (Duff et al., 2006a).

The residues that line the lower part of the substrate channel (Gln 108, Ser 138, Phe 140, Phe 298, Phe 304, Ser 309, and Gly 385) are conserved in plant CAOs (for which sequences are available) but not in the CAOs of other organisms. Tyr 168 is replaced by a Trp in the sequence of the soybean enzyme. In a study of 13 substrates of 3 CAOs from 2 plants, *Lathyrus cicera* (LCAO) and PSAO, and from BSAO (Pietrangeli et al., 2007), the plant enzymes revealed very similar activities to the same substrates in accord with the proposed structural conservation of the active sites based on sequence similarities. The plant enzymes clearly preferred diamine over monoamine substrates with the highest k_{cat} for putrescine.

9.4 FROM BOVINE SERUM

The crystal structure of BSAO has been determined, alone, in the presence of Xe (Calderone et al., 2003; Lunelli et al., 2005), and in complex with a clonidine inhibitor (Holt et al., 2008) at a resolution of ~2.4 Å.

9.4.1 Overall Folding of BSAO Monomer

The BSAO monomer is composed of three domains labeled D2, D3, and D4. The primary sequence of BSAO is 746 amino acids long, comparable to ECAO and definitely longer than other members of the family. However, it lacks the D1 domain present in *E. coli*. Nevertheless, in BSAO the first 40 amino acids are disordered and cannot be seen in the crystal structure. D2 and D3 domains are relatively small (composed of four or five strands and two α-helices each) and present similar but not identical quaternary organizations. Domain D2 comprises residues from 57 to 161 (the first 56 amino acids are possibly flexible and cannot be seen on an electron density map). It is connected to domain D3 (residues 171 to 282) by a ten-residue stretch.

The structural similarity of domains D2 and D3 in various species led to the hypothesis of gene duplication (Parsons et al., 1995), despite the low sequence identity (13%) about D2 and D3 in the bovine enzyme. Domain D3 is connected to the largest domain, D4, through a long stretch (residues 283 to 321), partially disordered. The end of domain D3 and the beginning of domain D4 are ~50 Å apart, and this flexible stretch runs on the protein surface. Domain D4 comprises about 400 residues characterized by a complex topology. It is also the most conserved domain among the AOs.

9.4.2 Quaternary Organization

BSAO, like the amine oxidases of other species, is a homodimer composed of two monomers related by a twofold axis. This makes the entire molecule roughly a parallelepiped of approximate size 105 × 60 × 45 Å (Figure 9.6a). Dimerization takes place through the two D4 domains, whose β-sandwich structures pack together,

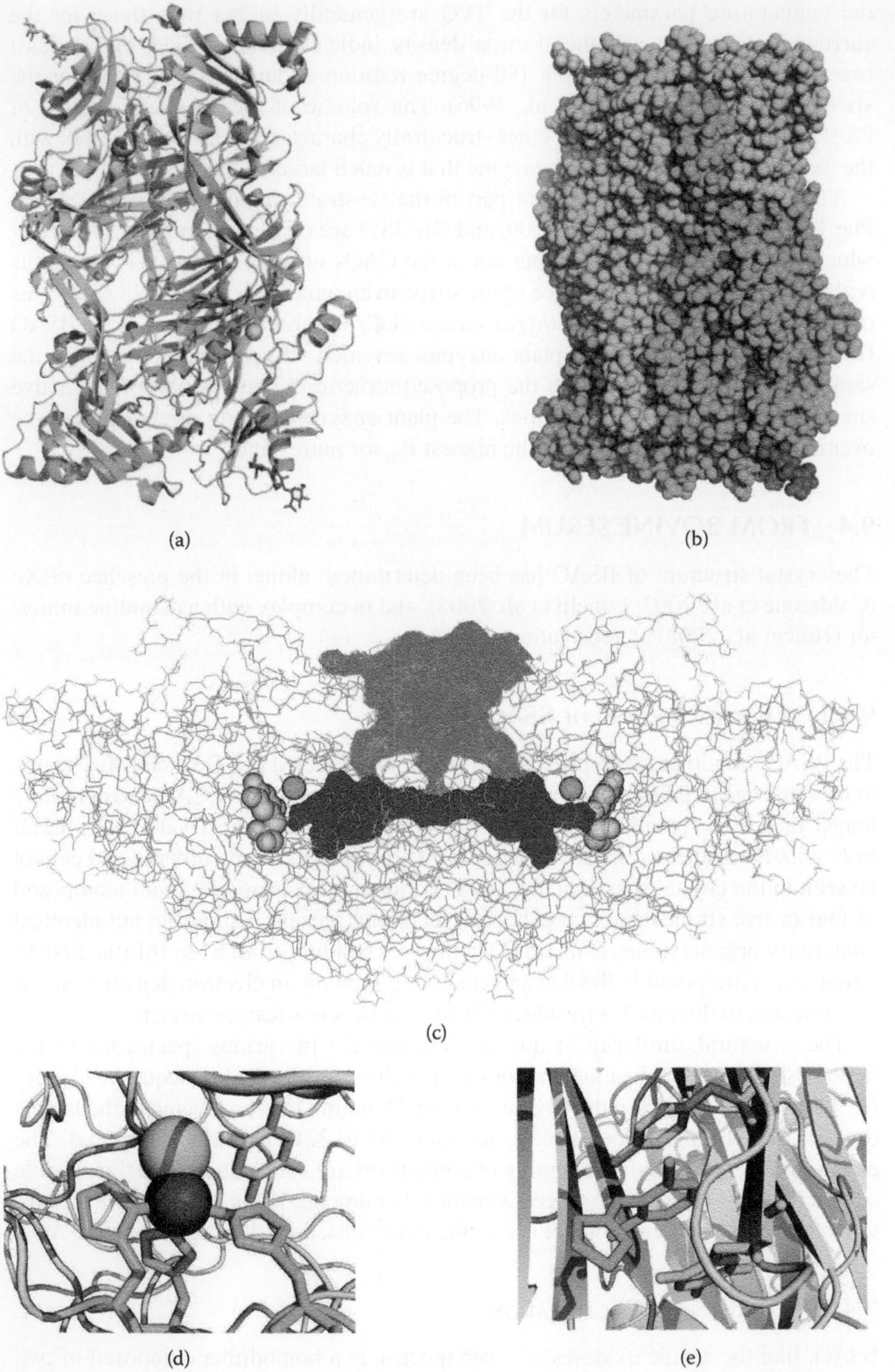
(a)
(b)
(c)
(d)
(e)

forming a sort of disk. The surface buried through dimerization is large (~5500 Å^2), but definitely smaller than those of other CAOs, mainly due to a large internal cavity with a volume of 1400 Å^3 in BSAO. The internal cavity is connected to the external solvent by three small channels that open into an external wide mouth along the twofold dimer axis (Figure 9.6c).

This mouth is 10 Å deep and 7 × 18 Å wide at the protein surface, between residues G 579 and R 584 of the two subunits. The central small channel presents hydrophobic walls; the two lateral ones are hydrophilic and hydrophobic. The internal cavity reaches the copper site, delimited by residue H 683, which coordinates the copper ion. It may therefore provide a pathway for the substrate O_2 and for the H_2O_2 reaction product, as proposed for other CAOs (Li et al., 1998). The possibility that this system of cavities may serve as an alternative pathway for the amine can be excluded because the internal cavity does not communicate with TPQ; and the channels have small dimensions. Furthermore, the electrostatic potential calculated on the surface of the mouth is positive everywhere and thus repulsive for cations.

The two subunits are held together also by two long arms, each protruding from a domain D4 of one monomer to wrap around the other (Figure 9.6b). Both arms are disordered at their ends. These two arms achieve direct contact of one subunit with the active site of the other. Moreover, in our structure the two subunits were virtually identical, the root mean square deviation of equivalent Cα atoms being 0.17 Å.

9.4.3 Secondary Cation Binding Sites

Besides the Cu ions of the active site, two additional cations are present in each subunit. Although their chemical natures are undefined, they have been refined as Ca(II) in the bovine structure via anomalous diffraction data. They are both near the molecular surface and far from the active site copper (30 to 36 Å). The two ions in one subunit are also ~13 Å distant from each other. The first cationic site presents an octahedral geometry coordinated by six oxygen atoms (D 528, D 530, D 672, carbonyl oxygens of L 529 and L 673, and a water molecule). The presence of this site

FIGURE 9.6 (See color insert following page 202.) (Opposite) (a) Cartoon representation of BSAO dimer. The two monomers (green and light blue) are related by a twofold axis approximately perpendicular to the plane of the paper. Copper ions are shown as red spheres, secondary cations as yellow spheres, bound sugars as yellow and magenta sticks. (b) Space-filling model of the BSAO dimer seen in the same view as (a). Two arms embracing each other that form one monomer are clearly visible. (c) Internal cavity and access pathways. The volume of the central cavity between monomers is represented in blue, the three small channels and the large mouth connecting the central cavity to the external solvent in orange. Protein atoms are shaded in grey. TPQ (yellow) and the copper ions (green) are represented as CPK models. (From Lunelli et al., 2005. With permission.). (d) BSAO active site. Cu(II) (red) is coordinated by three histidine residues (519, 521 and 683, green) and by a water molecule (light blue). TPQ in the off-Cu position, along with the chain trace of the enzyme, is shown in yellow. (e) Clonidine (green) inhibitor bound to the active site of BSAO (yellow). The phenyl ring of clonidine is stacked with respect to TPQ 470 and Tyr 472 (light blue).

in all CAOs including *E. coli* and the mostly conserved natures of the coordinating residues suggest a physiological role for this site. The ligands of the second cationic site are seven oxygen atoms: N 664, E 666, E 571 (bidentate), the carbonyl oxygen of F 662, and two water molecules. Its coordination is not conserved in most other species and its physiological relevance is questionable.

9.4.4 Glycosylation Sites

The bovine protein was crystallized only upon extensive deglycosylation (Calderone et al., 2003). Despite that, carbohydrates are still present in the glycosylation sites. In particular, the chain of three sugar residues is clearly visible, in correspondence to the N 136 site in each BSAO monomer. Three N-acetyl-d-glucosamine (NAcGlc) residues were fitted and refined. In addition, one NAcGlc monomer is bound to N 665 and another to N 231, but the latter is visible only in one of the two independent subunits in the asymmetric unit.

9.4.5 Xe Derivative

Xe, which binds preferentially to hydrophobic sites, may be used to mimic the binding of oxygen molecules. Xe derivative data were collected to help to solve the phase problem, and a relatively high Xe pressure was used (2 MPa for 5 min). For this reason, about 70 peaks larger than 4 σ could be observed in the anomalous difference map. After the first four maxima corresponding to the positions of the secondary cations, the others corresponded to Cu and S of the Cys or Met residues. Thirty-five peaks could be attributed to Xe atoms with relatively low occupancy (generally ~0.3). The highest peak attributed to Xe was 8.6 Å from the Cu of one subunit (labeled B in the deposited PDB file) located in a direction opposite the TPQ. This broadened peak corresponds to a small hydrophobic cavity near the side chain of L 488 and surrounded by other hydrophobic residues (F 681, V 700, A 490, A 523, A 367, Y 472, and W 474), with the side chain of F 681 separating this site from Cu.

This may mimic the position of the O_2 molecule approaching the Cu site. Indeed the central cavity observed between the subunits extends to this small hydrophobic cavity and supports the hypothesis that it may correspond to the O_2 pathway. In subunit A, a smaller peak (3σ) is observed. This site is adjacent to the oxygen binding site suggested for HPAO (Li et al., 1998), corresponding to residues Y 472, A 490, and T 693 in BSAO, between the Cu ion and TPQ—the position where the molecular oxygen has been postulated to be bound in the ECAO crystal (Wilmot et al., 1999). Other peaks are found in other hydrophobic cavities of the molecule.

Interestingly, the presence of Xe atoms makes the TPQ less mobile, making clear the off-Cu orientation, but in a productive orientation, pointing its O5 toward D 385 and the O2 toward the copper site.

9.4.6 Active Sites

The two active sites of BSAO are located between the two β-sheets of the β-sandwich and contain the cofactor TPQ and Cu(II) ion (Figure 9.6d). The latter is coordinated by three His residues (519, 521, and 683 in the numbering system of the bovine enzyme), conserved in all CAO sequences. The position of each His is stabilized by an H-bond involving the nonmetal-coordinated imidazole N atom (H 519 N^{δ}–P 517 O, 2.90 and 2.95 Å in the A and B subunits, respectively; H 521 N^{δ}–Y 472 OH, 3.07 and 3.08 Å in the A and B subunits, respectively; H 683 N^{ε}–D 689 O^{δ}, 2.93 and 2.81 Å in the A and B subunits, respectively).

Possibly the fourth coordination position in the absence of a cosubstrate is occupied by a water molecule, but the limited resolution of the structure prevents a definitive conclusion. In our crystal form, the TPQ was in the off-Cu positions in both subunits, i.e., its side chain pointing away from the Cu(II) ion. Moreover, the electron density for the TPQ ring was not well defined, suggesting that it may rotate around its central axis.

The two arms of CAOs protrude from one subunit into the other; their functional relevance is the subject of debate. In fact, they strengthen the interactions of the two monomers and may also affect accessibility to the active sites. The end of the arm of one subunit is positioned at the mouth of the active site channel of the other subunit, suggesting the possibility of interaction between the two active sites. In the case of BSAO, reactivities of half of the sites have been reported (De Biase et al., 1996; Morpurgo et al., 1988), suggesting negative cooperativity.

A key residue in this sense is L 468, located 6 Å from TPQ, and the only residue of the protein that has the backbone dihedral torsion angles φ and ψ falling in a forbidden region of the Ramachandran plot (about 60 and –105 degrees, respectively, in both subunits). The position of its side chain may determine substrate accessibility to the active site. In the other available CAO structures, residue 468 corresponds to a glycine, except in the case of HPAO, where 468 is an alanine residue. In all cases, φ and ψ torsion angles of these residues are similar to those of L 468 in BSAO. The L 468 of BSAO can form a hydrogen bond with H 442, belonging to the mobile arm coming from the other subunit, since the distance between N^{ε} of this histidine and the O of the carbonyl group of L 468 is 3.3 Å in one subunit and 3.5 Å in the other. Furthermore, the ~180-degree rotation of the imidazole ring of H 442 would bring its N^{δ} to be hydrogen bonded to D 471, the aspartate following TPQ. It is therefore plausible to cast these residues in a role that exerts some control on the reaction mechanism; its exact nature is not clear and warrants further studies.

From a structural view, the positions of the amino acids in the two active sites of BSAO are almost identical, thus supporting their equivalence. Some differences are visible, such as the electron density associated with solvent molecules in the active site region. Water molecules involved in interactions among TPQ and protein residues and also the water molecules surrounding the Cu(II) ions appear different in the two subunits. Moreover, the relatively low resolution of this crystal structure does not allow definite conclusions. Higher resolution studies are necessary to clarify this point.

9.4.7 Entry Channel for Active Site

TPQ is located at the bottom of a funnel-shaped cavity, whose entrance at the protein surface is about 25 × 9 Å. This cavity is located between domains D3 and D4 and is more similar to that found in PPLO than to the narrow channels of other CAOs. The opening of the cavity of one subunit is partially occupied by the end of one hairpin from the other subunit, suggesting communication between the two active sites. At the bottom of the cavity, the side chain of L 468, which is about 6 Å from TPQ toward the protein surface, along with Y 238 and N 211 belonging to the D3 domain, reduces the lower part of the cavity to a small channel. On the opposite wall of this small channel is the side chain of D 385, the active site base, whose carboxyl group is 5 Å from the O5 of TPQ.

Finally, the channel is delimited by F 388, M 467, and Y 383; the latter corresponds to the residue-defined active site gates in other CAOs, and lies in a position that allows the substrate to reach the cofactor. Several positively or negatively charged residues are distributed around the cavity on the surface, but only one negative residue (D 445) is present in the cavity wall. Moreover, the presence at the bottom of the cavity of the catalytic aspartate (D 385) and TPQ creates an electrostatic gradient, possibly driving the positively charged amino group of the substrate to the active site in a correct orientation for the catalytic reaction.

The affinity of BSAO is higher for physiological polyamines such as spermine and spermidine (that are more highly charged), long hydrophobic amines (Di Paolo et al., 2003), and benzylamine with hydrophobic substituents (Hartmann and Klinman, 1991). Although the overall architectures of the known CAO models appear similar, significant differences exist both in the active site structures and in charge distributions. In particular, the active site cavity of BSAO is a large hydrophobic surface consisting of residues belonging to the D4 domain (F 388, L 468, M 467, F 393, A 394, P 396, L 424, W 412, F 414, Y 392), with the negatively charged D 445 at one end and the TPQ and catalytic D 385 at the other end. The area between TPQ and D 385 can accommodate the charged amino group on the head of spermine, spermidine, and benzylamine, while the positive charges on the tails of the polyamines may interact with and be stabilized by D 445. This residue is at the end of the flexible β-hairpin of the other subunit and its high mobility may provide a flexible docking system for the substrate. This structural feature can explain the broad range of possible substrates characterized by different chain lengths and charge distributions that are characteristic of BSAO (Bardsley et al., 1970; Di Paolo et al., 2003).

The presence of an area with negative electrostatic potential in the active site cavity makes this region a potential binding site for cations such as Ca(II) and Mg(II) demonstrated to be competitive inhibitors of highly charged polyamines (Di Paolo et al., 2002).

9.4.8 Inhibitor Clonidine

The crystal structure with the clonidine inhibitor is the only BSAO complex with an inhibitor determined to date (Holt et al., 2008). The binding of the inhibitor does not alter the quaternary architecture of the enzyme. A superposition of Cα atoms of the free and complexed enzymes gives an overall root mean standard deviation value of 0.35 Å, with major changes involving the residues defining the catalytic pocket and substrate entrance channel. Loops of Tyr 230–His 240 and Met 467–Asp 471 exhibit the highest Cα displacements. Asn 469 rotates approximately 180 degrees away from the internal pocket to face the mouth of the entrance channel, thereby permitting the docking of clonidine. In addition, several residues (Gly 580–Ala 581, Ala 610–Pro 613 and Val 696–Gly 701) forming an internal channel at the interface between monomers show small but significant displacements with respect to native BSAO.

One molecule of clonidine binds similarly in each of the two active sites of the BSAO dimer. Clonidine bound at the bottom of the catalytic cavity causes a displacement of TPQ, inducing a rotation of the TPQ side chain to the on-Cu nonproductive orientation, in which the hydroxyl group of TPQ coordinates the copper ion. The phenyl ring of clonidine lies almost perpendicular to the imidazolidine portion of the molecule and is clearly trapped by stacking interactions in a sandwich-like orientation within the aromatic side chains of TPQ 470 and Tyr 472 residues (Figure 9.6e).

Nitrogen atoms N2 and N3 in the imidazolidine moiety are at a suitable orientation and distance to interact through hydrogen bonding with Asp 385, the general catalytic base (Hartmann et al., 1993; Mure et al., 2002). The N3 nitrogen in the heterocyclic ring can also strengthen the inhibitor–enzyme interaction through a NH-π interaction with Tyr 383.

9.5 VASCULAR ADHESION PROTEIN (VAP)-1

9.5.1 Overall Structure

Human vascular adhesion protein-1 (VAP-1) is a 170- to 180-kDa homodimeric sialoglycoprotein that exists in membrane-bound (Salmi and Jalkanen 1992; Smith et al., 1998) and soluble forms (Abella et al., 2004; Stolen et al., 2004b). The membrane-bound form is anchored by an N-terminal helix and has a very short N-terminal cytoplasmic tail. The structure of the extracellular part of the VAP dimer was first predicted by homology modeling (Salminen et al., 1998; Yegutkin et al., 2004; Maula et al., 2005). The x-ray structures of VAP-1 revealed by two independent studies (Airenne et al., 2005; Jakobsson et al., 2005) later confirmed that it consists of CAO domains D2, D3, and D4 that form a heart-shaped dimer (Figure 9.7a).

The 2.9 Å x-ray structure (Airenne et al., 2005; PDB ID 1US1) was solved using the full-length protein including the N-terminal transmembrane helix; the 2.5 Å structure (Jakobsson et al., 2005; PDB ID 2C10) was solved using a truncated form (residues 29 through 763). The 2.9 Å x-ray structure of VAP-1 consists of residues A 55 through A 761 in monomer A and B 57 through B 761 in monomer B but lacks the N-terminal residues, which did not show any electron density, and the residues A 202 through A 204, B 203, and B 742 through B 746 whose electron density was obscure (Airenne et al., 2005). In the 2.5 Å structure of the truncated form, the N-terminus was disordered as well, and only residues A 41, A 58 through A 761, B 39 through B 43 and B 58 through B 761 built in (Jakobsson et al., 2005). In both structures that are very similar, the last two C-terminal residues were not interpretable. Similarly to BSAO (Lunelli et al., 2005; PDB ID 1TU5), the conserved core of the D2 domain (residues 55 through 162; dark red and dark cyan in Figure 9.7b) consists of four β-strands and two α-helices but has two additional short β-strands and a short α-helix.

Domain D2 is linked to D3 (residues 172 through 283; magenta and light cyan in Figure 9.7b) by nine residues. The D3 domain shares the same conserved core as D2

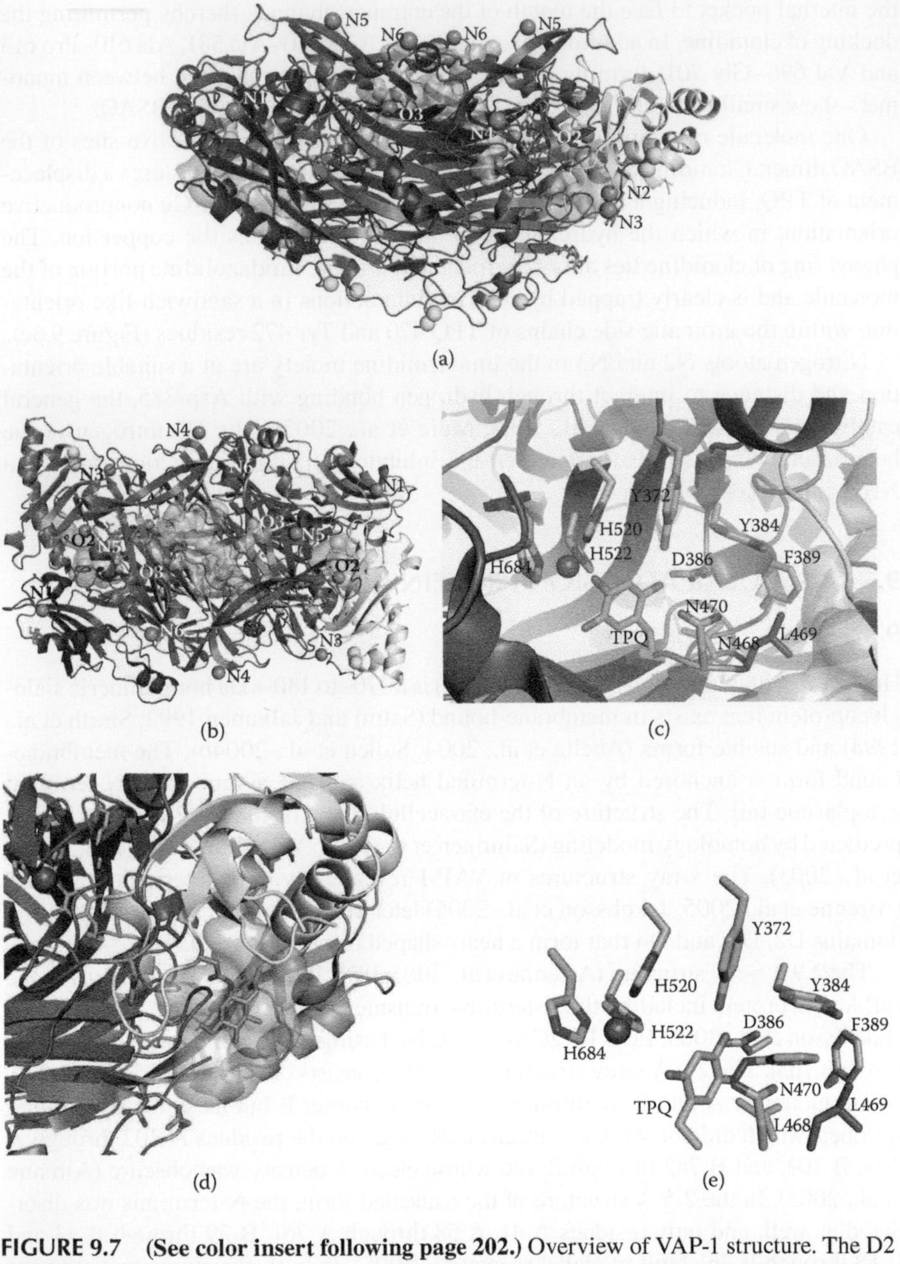

FIGURE 9.7 (See color insert following page 202.) Overview of VAP-1 structure. The D2 domains are shown in dark red and cyan, the D3 domains in magenta and light cyan, and the D4 domains in red and blue. The N-glycosylation sites and attached carbohydrates are shown in orange as spheres and as sticks, respectively. The cysteines are shown as yellow spheres. The RGD loop is shown as sticks with yellow carbon atoms. The active site cavity is shown as light violet surface and the central cavity as light pink surface. The active site residues are shown as lines with green carbon atoms and the copper ion as a dark yellow sphere. Leu 469 is shown in red. (a) Side view. (b) View from top toward membrane. (c) Active site. (d) Active site cavity. (e) Active site of 2HP adduct structure. The figures were created with the PyMOL Molecular Graphics System (DeLano Scientific).

but has one additional β-strand and two α-helices. As a result of the additional secondary elements, the D2 and D3 domains differ considerably in VAP-1. The linker between D3 and the largest D4 domain consists of residues 284 through 322. The catalytic D4 domain (residues 323 through 761; red and blue in Figure 9.7b) contains the residues involved in TPQ generation and the catalytic reaction, including TPQ 471, the catalytic base Asp 386, Tyr 372, and Asn 470 positioning the TPQ along with the three histidines (520, 522, and 684) coordinating the copper ion (Figure 9.7c). As in the other known CAO structures, the two long hairpin arms I (residues 423 through 468) and II (residues 536 through 569) extend from the D4 domain of one monomer to embrace the other one. The percentage sequence identities of VAP-1 with the other structurally known CAOs based on structure-based sequence alignment are listed in Table 9.4.

9.5.2 Disulfide Bridges

VAP-1 contains ten cysteine residues in total (yellow in Figure 9.7a). Six are involved in disulfide bridges in all VAP-1 structures, one is unpaired (Cys 95), and another (Cys 22) located in the putative N-terminal transmembrane helix that is not visible in the full-length VAP-1 structure and does not exist in the truncated form. Cys 198 and Cys 199 form an intradomain disulfide bond in D3 that may stabilize the turn at the C-terminal end of helix α6 (Airenne et al., 2005) or be involved in redox switch mechanisms (Jakobssen et al., 2005). The D4 domain is stabilized by the conserved intradomain disulfide bridge between Cys 404 and Cys 430. The C-terminal part of VAP-1 has three cysteines. Cys 734 and Cys 741 form an intramonomeric bond that positions the hairpin loop that has an RGD cell adhesion motif (Arg 726, Gly 727, and Asp 728).

The remaining disulfide bridges differ in the structures of the full-length VAP-1 and its truncated form. Jakobsson et al. (2005) interpreted an intramonomeric disulfide bond between Cys 41 and Cys 748 whereas Airenne et al. (2005) modeled an intermonomeric bond between Cys 748s and Cys 41 could not be built in the structure. The intramonomeric Cys 41–Cys 748 bond links the N- and C-termini similarly as observed in the x-ray structure of PPLO (Duff et al., 2003) whereas the intermonomeric Cys 748A–Cys 748B bond links the monomer similarly as in PSAO (Duff et al., 2006b).

TABLE 9.4
Comparison of Sequence Identity of VAP-1 with Other Structurally Known CAOs Based on Structure-Based Sequence Alignment

Enzyme	% Identity
Bos taurus serum amine oxidase (Lunelli et al., 2005)	78.9
Pichia pastoris lysyl oxidase (Duff et al., 2003)	23.9
Escherichia coli amine oxidase (Parsons et al., 1995)	20.6
Hansenula polymorpha amine oxidase (Li et al., 1998)	18.5
Arthrobacter globiformis amine oxidase (Wilce et al., 1997)	18.5
Pisum sativum amine oxidase (Kumar et al., 1996)	16.1

The differences in the topologies of the disulfide bridges are interesting and the reasons for the differences remain unknown. Jakobsson et al. (2005) speculated that the Cys 41–Cys 748 bond in their VAP-1 structure may arise from protein purification. They noted that an additional disulfide bond between Cys 22 and Cys 41 is unlikely since Cys 22 is located in the putative transmembrane part whereas an intermonomeric disulfide bridge between Cys 41s would be possible when Cys 748s are paired with each other. They suggested that this putative disulfide bond between Cys 41s from both monomers may play a role in exposing the protease cleavage site that likely exists between the transmembrane helix and the D2 domain and is requisite for the creation of the soluble form of VAP-1.

The flexibility of the C-terminal part is an intriguing feature of mammalian CAOs. In the complex structure of 2-hydrazinopyridine (2HP; an irreversibly bound substrate Schiff-based reaction intermediate) and VAP-1 (Jakobsson et al., 2005; PDB ID 2C11), the C-terminal part is disordered like the C-terminal residues 718 through 762 of BSAO, which has TPQ in the active conformation. Interestingly, these are the only mammalian CAO structures in which TPQ is in the active conformation. This flexibility of the C-terminal part suggests that the disulfide bridges may have a special role. Jakobsson et al. (2005) hypothesized that cysteines in their reduced form may be oxidized by hydrogen peroxide, one of the reaction products, resulting in conformational changes on the VAP-1 surface. For example, the reduction of disulfide bridge Cys 198–Cys 199 near the entrance to the active site would modulate the shape of the protein surface and possibly affect the mechanism of enzymatic action. The functional roles of the disulfide bridges in VAP-1 remain unclear and must be further characterized by site-specific mutations of cysteines.

9.5.3 Active Site

The active site cavity of VAP-1 (Figure 9.7d) resembles more the funnels of BSAO and PPLO (Duff et al., 2003) than the narrow substrate channels of the other known CAOs. The cavity is about 20 Å deep and its opening has approximate dimensions of 10 × 15 Å. The cavity walls are lined by residues from the D2 domain (Ala 87), from the least conserved D3 domain (Phe 173, Tyr 176, Leu 177, Asp 180, Thr 210, Met 211, and Phe 227) and from the D4 domain (Thr 395, Thr 396, Pro 397, Phe 415, Leu 416, Leu 417, Glu 418, Ser 419, Ala 421, Pro 422, Lys 423, Thr 424, Ile 425, and Arg 426). Furthermore, the tip of the long β-hairpin arm I from the other monomer (His 444, Asp 446, Leu 447, Tyr 448, and His 450) is located on the edge of the active site cavity (red in Figure 9.7d) and the main chain oxygen atoms of the C-terminal residues (Gly 758, Gly 759, Phe 760, and Ser 761) shape the cavity. Interestingly, the tip of arm I has a different conformation in VAP-1 compared to BSAO.

The differences may result from the fact that in the VAP-1 structure an additional β-strand (residues 723 through 726; red in Figure 9.7d) from the other monomer interacts with arm I while in BSAO the C-terminal part (starting from residue 717) is unstructured. In VAP-1 the additional β-strand is followed by the RGD loop (yellow in Figure 9.7d) and a short α-helix. This part of the active site entrance that protrudes from the other monomer is further stabilized by the C-terminus on the surface. The sugar units attached to Asn 232 (orange in Figure 9.7d) interact with arm I and the RGD loop and thus modulate the charge and shape of the active site entrance.

Residues from the D4 domain (Thr 212, Lys 393, Tyr 394, Thr 396, Pro 397, Phe 398, Phe 415, Leu 417, Leu 425, Thr 467, Leu 468, and Leu 469) form the bottom of the cavity. The active site cavity of VAP-1 is mainly formed by aromatic and hydrophobic residues with the exception of the side chains of Asp 180, Glu 418, Lys 393, Lys 423, Thr 212, Thr 396, Thr 467, Ser 419, and Ser 761. The active site channel is very narrow (4.5 × 4.5 Å) near the TPQ residue and the side chains of Leu 468, Leu 469, and Tyr 384 (Met 467, Leu 468, and Tyr 383 in BSAO) narrow the entrance to the active sites in structures in which the TPQ residue is in on-Cu conformation (Figure 9.7c). The conserved leucine (Leu 469 in VAP-1 and Leu 468 in BSAO corresponding to a glycine in most CAO structures) blocks the entrance to the TPQ in the VAP-1 structures; in the BSAO structure, it adopts a different conformation and the active site is open. In the complex structure of 2HP and VAP-1, the TPQ residue is in the off-Cu conformation (Figure 9.7e). The 2HP adduct is in hydrazone form and forms hydrogen bonds with Asp 386. Its pyridine ring stacks with Tyr 384 and has hydrophobic interactions with Leu 468, Leu 469, and Phe 384.

9.5.4 Glycosylation Sites and RGD Motif

VAP-1 is a heavily glycosylated protein; its putative glycosylation sites have been studied by site-specific mutagenesis in Chinese hamster ovary cells. Based on these studies, Ser 43 and Thr 679 (O3 in Figure 9.7) are O-glycosylated and all the predicted N-glycosylation sites (Asn 137, Asn 232, Asn 294, Asn 592, Asn 618, and Asn 666 [N1 through N6 in Figure 9.7]) indeed have carbohydrates attached. An additional O-glycosylation site, Thr 212, is suggested by the extra electron density attached to the residue in the 2.9 Å x-ray structure (O2 in Figure 9.7b). The putative O-glycan attached to Thr 212 lies next to Leu 469 and may thus influence ligand binding.

In the VAP-1 x-ray structures, one or more attached carbohydrate moieties are seen in all the other N-glycosylation sites except N5. Four of the N-glycosylation sites (N1, N4, N5, and N6) are located on top of the VAP-1 molecule and three (N4 through N6) are proposed to have a role in cell adhesion (Maula et al., 2005). The N4 and N6 glycosylation sites line the entrance of central cavity between the monomers. Thus, the carbohydrates attached to these sites may function by regulating the entry and/or exit of molecular oxygen, hydrogen peroxide, and ammonia. The carbohydrates attached to N2 and N3 might exert effects on both enzymatic activity and cell adhesion since they are found at the active site entrance (orange in Figure 9.7d) in the vicinity of the RGD cell adhesion motif.

The RGD site in VAP-1 is in a finger-tip like conformation that can be recognized by the ligands of the motif including several members of the family of cell surface integrins (Ruoslahti and Pierschbacher, 1987) involved in cell–cell and cell–matrix interactions. The deletion of the RGD motif from VAP-1 decreases lymphocyte adhesion (Salmi et al., 2000) suggesting a possible functional role. Jakobsson et al. (2005) proposed that the RGD motif may play a role in the release of soluble VAP-1 from the membrane-bound form. Integrins may target the metalloprotease cleaving

VAP-1 to the appropriate site by interacting with the RGD motif. Since the RGD motif of monomer A is located next to the active site entrance of monomer B (and vice versa), the binding of a ligand to one site may change the structure of the VAP and therefore affect its enzymatic activity and other functions. VAP-1 structural data clearly support the idea that the RGD motif has a biological function but further studies are needed to validate the hypotheses of *in vivo* functions.

ACKNOWLEDGMENT

We acknowledge the assistance of Heidi Kidron, Ph.D. who prepared the figures in the section on VAP-1.

10 Plasma Amine Oxidases in Various Clinical Conditions and in Apoptosis

Frans Boomsma, Anton H. van den Meiracker, and Antonio Toninello

CONTENTS

10.1 INTRODUCTION

About 40 years ago, a number of papers described abnormal activities of what was then called plasma (serum) amine oxidase in patients with various diseases (Nilsson et al., 1968; Tryding et al., 1969; McEwen and Cohen, 1963; McEwen and Castell, 1967). This plasma amine oxidase is now generally referred to as semicarbazide-sensitive amine oxidase (SSAO). SSAO is also known as vascular adhesion protein-1 (VAP-1) since the discovery that VAP-1 and SSAO are the same; the two names simply reflect the different functional aspects of the same protein (Smith et al., 1998).

SSAO is the product of two or three genes: AOC2, AOC3, and in animals other than rodents and humans, AOC4. It is an ectopic membrane-bound protein. The SSAO found in plasma is believed to result from proteolytic cleavage of the membrane-bound form (Abella et al., 2004; Stolen et al., 2004b). Plasma SSAO activity varies widely among species, with low activity in rats and mice, moderate activity in humans, and far greater activity in dogs, rabbits, pigs and horses, and especially in cows, sheep, and goats (Boomsma et al., 2000a). Discussion in this chapter is limited to SSAO in human plasma.

In healthy adults, plasma SSAO activity stays within a relatively narrow range, and the within-subject activity varies little from one day to another. No differences in activity exists between arterial or venous plasma or between men and women. In children from birth until 16 years of age, SSAO activity is 30 to 40% higher; after age 50, activity slowly increases again (Boomsma et al., 2003). For proper activity measurements, plasma SSAO should be measured within 1 to 2 days after blood sampling or after storage at –70°C prior to measurement. Certain medications act as SSAO inhibitors, e.g., hydralazine, isoniazid, benzerazide, and carbidopa.

10.1.1 Plasma SSAO Activities in Clinical Conditions and Diseases

Increased levels of plasma SSAO have been reported in patients with Type 1 and Type 2 diabetes mellitus (Boomsma et al., 1995 and 1999; Mészáros et al., 1999b; Garpenstrand et al., 1999a; Salmi et al., 2002; Göktürk et al., 2003b). The increases were greater in patients with complications such as retinopathy and nephropathy than in patients without such complications. In children at first diagnosis of diabetes, SSAO was already increased compared to controls. In patients with congestive heart failure, SSAO was also reported to be increased, again in accordance with the severity of the disease as indicated by the New York Heart Association classification (Boomsma et al., 1997). Interestingly, the highest SSAO activities were found in patients who had both congestive heart failure and diabetes mellitus.

Plasma SSAO served as an independent marker for survival in patients with congestive heart failure (Boomsma et al., 2000b). In patients with various forms of liver disease, strongly elevated plasma SSAO values were found (McEwen and Castell, 1967; Kurkijärvi et al., 1998; Boomsma et al., 2003). Elevations in plasma SSAO were reported during pregnancy (Sikkema et al., 2002) and in malignant (but not essential or renovascular) hypertension (Boomsma et al., 2003), atherosclerosis (Mészáros et al., 1999a; Karadi et al., 2002; Boomsma et al., 2003), and severe obesity (Weiss et al., 2003). Stroke reports vary: a decrease in patients with cerebral bleeding (Ishizaki, 1990), no change in patients with cerebral thrombosis or embolism (Garpenstrand et al., 1999b), and an increase in those with multiple cerebral infarction (Ishizaki, 1990). Slightly increased activities were found in unspecified stroke (Boomsma et al., 2003). Increases in the first 6 hr after acute stroke were also reported (Airas et al., 2007). In patients with moderate to severe Alzheimer's disease and in patients with active multiple sclerosis increased plasma SSAO values were reported (Hernandez et al., 2005; Airas et al., 2006).

In Sjögren's syndrome, rheumatoid arthritis, colitis ulcerosa, chronic hepatitis C without liver cirrhosis, and in children with attention deficit/hyperactivity disorder

(ADHD), no deviations from normal plasma SSAO levels were found (Kurkijärvi et al., 1998; Boomsma et al., 2003; Roessner et al., 2006a).

Decreased plasma SSAO activities were reported in cases of major and anxious depression, children born with hernia diaphragmatica, severely burnt patients, and patients with cancer (Roessner et al., 2006b; Uzbekov et al., 2006; Boomsma et al., 2003; Lewinsohn, 1977). Increased values were found in patients with skeletal metastases of prostate cancer (Ekblom et al., 1999). In chronic renal failure, both increased and decreased plasma SSAO levels were found (Kurkijärvi et al., 1998; Boomsma et al., 2003; Boomsma et al., 2005a; Nemcsik et al., 2007). The discrepancy is interesting, and may be related to methodologies.

SSAO is usually measured functionally by the generation of benzaldehyde from benzylamine or immunologically as VAP-1. In renal failure, Nemcsik et al. (2007) found decreased levels of plasma SSAO when SSAO was measured by the generation of benzaldehyde from benzylamine, in agreement with earlier reports (Boomsma et al., 2003 and 2005a). Increased levels were found in the same plasma samples when SSAO was measured by the generation of hydrogen peroxide, another product of the same reaction, in agreement with another report of an immunological method (Kurkijärvi et al., 1998). The authors suggest that elevated concentrations of methylamine in renal failure may compete with benzylamine in the enzymatic conversion, thereby yielding falsely low values when benzaldehyde is measured and proper values when hydrogen peroxide (always generated no matter the substrate) is measured. Whether the methylamine plasma concentrations in these patients were high enough to cause such an effect must be investigated, however. Table 10.1 shows SSAO changes in various clinical conditions.

10.1.2 Why Do Plasma SSAO Levels Change in Some Clinical Conditions and Diseases?

It is difficult to answer this question in view of the scarcity of knowledge about the regulation of plasma SSAO. Insulin has been proposed as a regulator of plasma SSAO (Salmi et al., 2002) but others found only a minor effect (Dullaart et al., 2006). An important clue may be the decrease of very high plasma SSAO levels in children to lower adult levels around age 16, suggesting a link with the growing process, but this has not been followed up (Boomsma et al., 2003).

Changes in amounts of activators and inhibitors have been suggested, but are not substantiated; C peptide and insulin were shown not to influence plasma SSAO activity (Boomsma et al. 2001). A common factor may be involved in the regulation of plasma SSAO and plasma angiotensin I-converting enzyme (Boomsma et al., 2005b); both are membrane-bound enzymes that also circulate in soluble forms in plasma. Follow-up studies in children born with hernia diaphragmatica seem to suggest that properly developed lungs are essential for attaining normal plasma SSAO activities (Boomsma et al., 2003).

SSAO is expressed ubiquitously, but highest activities are found in adipocytes and vascular smooth muscle cells. It is present almost everywhere except for the brain (but present in blood vessels and meninges). Abnormalities in many organs and tissues may thus lead to abnormal plasma SSAO levels. An increase may be

TABLE 10.1
Overview of Changes in Plasma SSAO in Various Clinical Conditions

Clinical Condition or Disease	Change in Plasma SSAO
Age	↑ < 16 and > 50 years
Pregnancy	↑
Diabetes mellitus (Types 1 and 2)	↑
Congestive heart failure	↑
Liver disease	↑
Hypertension, malignant	↑
Hypertension, essential	=
Hypertension, renovascular	=
Atherosclerosis	↑
Obesity	↑
Stroke	↑, ↓, =
Alzheimer's disease	↑
Multiple sclerosis	↑
Sjögren's syndrome	=
Rheumatoid arthritis	=
Colitis ulcerosa	=
Chronic hepatitis C	=
ADHD	=
Depression	↓
Hernia diaphragmatica	↓
Burns	↓
Cancer	↓, ↑
Chronic renal failure	↓, ↑

Notes: ↑, increased; ↓, decreased; =, unchanged compared to plasma SSAO of healthy adults.

caused by higher expression in some organs, greater cleavage from cell membranes by secretases, more efficient post-translational modifications, an increase in conversion efficiency or stability of the enzyme, decreased breakdown, etc. Decreases in plasma SSAO levels may of course be caused by a reversal of the same processes.

Under different conditions, different mechanisms may be responsible. Plasma SSAO is higher in patients with heart failure with concomitant diabetes mellitus than in those without the diabetes (Boomsma et al., 1997). Similarly, children have higher plasma SSAO levels; those with diabetes mellitus have still higher levels (Boomsma et al., 2003). It thus appears that the rise of plasma SSAO in diabetes mellitus is caused by a different mechanism than the rise in heart failure or during the process of growing up.

In agreement with the well-established role of SSAO in the inflammatory process, increased expression has been found at sites of inflammation. Whether this leads to increased plasma levels depends on the extent of the increase and the site of inflammation. Therefore, the finding that patients with inflammatory diseases like

Sjögren's syndrome and colitis ulcerosa did not have elevated plasma SSAO levels does not mean that no local upregulation occurs. We are starting to recognize that in diabetes mellitus, heart failure and atherosclerosis, inflammation may be an important trigger for the disease.

In view of the insulin-mimicking effects of SSAO, one might speculate that in diabetes mellitus SSAO could be upregulated in an attempt to compensate for the lack of insulin. In a mouse model of diabetes, however, increases in SSAO were accompanied by a decrease in SSAO mRNA while K_m did not change (Göktürk et al., 2004). These workers suggested that the increased glucose fluxes in diabetes mellitus may enhance glycosylation of proteins, a process known to protect them from degradation and thereby increasing the plasma concentration. In pregnancy, the increase in plasma SSAO may simply be the result of the development of the placenta, a known rich source of SSAO.

In a few instances where the probable cause of elevated plasma SSAO (liver cirrhosis, kidney failure, and pregnancy) was eliminated by transplantation of liver or kidney or by giving birth, plasma SSAO values returned to normal (Boomsma et al., 2003; Kurkijärvi et al., 2001).

10.1.3 Do Decreased or Increased Plasma SSAO Levels Matter?

Again, no unambiguous answer is possible based on our present state of knowledge, although abnormal plasma SSAO appears to be a biomarker for some disease states. Assuming that plasma SSAO can be viewed as indicative for the total (soluble and bound) SSAO in the body, reactions with natural substrates (methylamine and aminoacetone) would lead to more or less of the three reaction products: the corresponding aldehydes (formaldehyde and methylglyoxal), hydrogen peroxide, and ammonia.

The aldehydes are very reactive, and cause cross linking of proteins. To some extent this is advantageous. Animal studies clearly indicate that lack of SSAO leads to problems with proper formation of blood vessels. Too much cross linking creates other problems, e.g., via the formation of advanced glycation end products or plaques. Similarly, some hydrogen peroxide formation is advantageous, since at some sites it may serve as a signaling molecule, e.g. for recruitment of GLUT4 to the cell membrane for uptake of glucose. Too much hydrogen peroxide generation may lead to increased oxidative stress. The third reaction product, ammonia, is generally considered toxic, but has not been investigated thoroughly.

10.1.4 Concluding Remarks

Much progress has been made in bringing to light the many faces and functions of SSAO in (patho)physiology, but much remains to be discovered. Clearly, plasma SSAO levels in healthy people are kept within a relatively narrow range in order to balance its positive and negative effects. Abnormal plasma SSAO levels are indicative of some disease states. Whether this is a positive or a negative adaptation remains to be established for each disease separately. In diabetes mellitus, for example, elevated SSAO may be advantageous in channeling more glucose from the blood into cells, but disadvantageous through its effect on the formation of advanced glycation end products. In inflammatory diseases elevated SSAO may help leukocytes reach the site of inflammation, but too much transmigration of leukocytes may have deleterious effects. The continuing development of selective SSAO inhibitors and good SSAO substrates means we can soon expect clinical study results that may answer some of these questions and show whether manipulation of SSAO activity may serve as s a new tool to treat various diseases.

10.2 APOPTOSIS

A study of liver diseases, (Kerr et al., 1972) revealed a type of cell death that appeared different from the necrotic hepatocyte death induced by toxins. The dead cells showed round bodies, surrounded by membranes and containing intact organelles along with parts of nuclei. These bodies were the results of progressive cell condensation and budding and were often engulfed by phagocytic cells. This type of cell death was named apoptosis and the morphological changes that further characterize this process were designated chromatin condensation and nuclear fragmentation (Kroemer et al., 2005).

The phenomenon of apoptosis has been proposed as responsible for an important defense system that eliminates superfluous, damaged, and mutated cells. Apoptosis can also exhibit a therapeutic behavior by eliminating neoplastic cells. However, disabled apoptosis can favor pathogenic events that contribute to oncogenesis and cancer progression, while acute massive apoptosis participates in the pathophysiology of infectious diseases, septic shock, and intoxication (Reed, 2002).

Apoptosis is a programmed cell death, a predetermined mechanism that may be triggered by a large number of inducer agents that activate several molecular pathways. The best characterized are called the extrinsic and intrinsic pathways.

The extrinsic pathway is characterized by activation of death receptors, located on cell surfaces and induced by particular ligands; it is also known as the death receptor pathway. Conversely, the intrinsic pathway involves an intracellular cascade of events in which mitochondria play a prominent role by means of mitochondrial permeability transition (MPT); it is also known as mitochondrial pathway.

The ligand-induced activation of death receptors in the extrinsic pathway induces the assembly of the death-inducing signaling complex on the cytoplasmic side of the plasma membrane. This activates a specific class of proteases, the caspases. The first, caspase-8, is in turn able to activate other caspases or so-called effector caspases (Reed, 2002; Susin et al., 1998; Kokoszka et al., 2004) to produce an apoptotic phenotype. Caspase-8 can also activate the pro-apoptotic Bid factor that induces MPT and leads to transition pore opening (Jacotot et al., 1999). This final effect represents the main link between extrinsic and intrinsic apoptosis.

In the intrinsic pathway, several intracellular signals including DNA damage are directed to mitochondria with the result of the transition pore opening. This process causes the release of the pro-apoptotic factors (cytochrome c, apoptosis-inducing factor (AIF), second mitochondria-derived activator of caspases/direct IAP binding protein with low pI (Smac/DIABLO), and Omi stress-regulated endoprotease/high temperature requirement protein A2 (Omi HtrA2) from the intermembrane space.

Cytochrome c, for example, induces the assembly of ATP/deoxyATP with the apoptosis protease-activating factor 1 and the formation of apoptosome which activates caspase-9. Caspase-9, in turn, activates the effector caspases that finally lead to apoptosis. As noted above, MPT plays a crucial role in the mitochondrial pathway of apoptosis.

10.2.1 Mitochondrial Permeability Transition

The phenomena of MPT and the opening of the transition pore serve as the links between mitochondria and intrinsic apoptosis and occur in all types of mitochondria. This change in mitochondrial permeability strongly affects the peculiar insulating property of the mitochondrial inner membrane, that is, the energy-transducing membrane discovered in the late 1970s. Hunter et al. (1976) first observed that isolated mitochondria incubated with supraphysiological Ca^{2+} concentrations in combination with a wide variety of inducing agents or conditions underwent a dramatic increase in inner membrane permeability, with a nonspecific behavior designated MPT (see Susin et al., 1998).

This phenomenon also occurred in cultured cells and, subsequently, was confirmed *in vivo* by a large number of studies. MPT occurs on the opening of a proteinaceous pore, the so-called transition pore, that has an estimated diameter of 3 nm and permits bidirectional traffic of metabolites with molecular masses up to 1.5 kDa. The pore is most probably located at the gap junction between the two mitochondrial membranes. Different diameter values for the transition pore have been experimentally demonstrated by different research. This suggests that the pore or channel is capable of different opening levels or consists of additional structures of different molecular compositions.

The precise architecture and molecular structure of the pore have not been completely determined. Evidence indicates that it includes cytosolic proteins such as hexokinase and creatine kinase, components of outer membrane such as a voltage-dependent anion channel, and an inner membrane such as adenine nucleotide translocase (AdNT) serving as the main component, and a matrix protein such as cyclophylin D. However, some genetic experiments have raised doubts about the role of AdNT as the main component of the pore structure and the question is strongly debated (Kokoszka et al., 2004.).

The opening of the transition pore, which exhibits several levels of conductance and negligible solute selectivity, leads to colloid osmotic matrix swelling, collapse of the electrical and chemical gradients, loss of glutathione and endogenous cations, release of adenine and pyridine nucleotides, and oxidation of several components including sulfhydryl groups, glutathione and pyridine nucleotides. All these events compromise cellular energy metabolism and Ca^{2+} homeostasis to the point of provoking necrotic cell death. However, pore opening causes large amplitude swelling of mitochondria with the rupture of the outer membrane and the release in the cytosol of cytochrome c, AIF, Smac/DIABLO—so-called pro-apoptotic factors that trigger the caspase cascade and induce apoptosis. The type of death, necrosis or apoptosis, may be related to the energy balance of the cell and the duration of pore opening. Apoptosis requires the presence of ATP (Susin et al., 1998).

10.2.2 Reactive Oxygen Species in Mitochondrial Permeability Transition and Apoptosis

The involvement of reactive oxygen species (ROS), that is hydrogen peroxide, superoxide anion and hydroxyl radical, in the induction of MPT and, consequently, in apoptosis was proposed first by Vercesi and collaborators (1997). They obtained experimental evidence of the presence of critical sulfhydryl groups located on the AdNT whose oxidation is one of the main events responsible for the opening of the transition pore.

The most important facts supporting this hypothesis are the increased production of intracellular ROS prior the induction of the MPT and its prevention by several antioxidant agents (Vercesi et al., 1997). The involvement of ROS in MPT is considered highly relevant and has been confirmed in several cell lines under different pathophysiological conditions. Indeed, natural and pharmacological compounds such as isoflavons, cyclic triterpens, salicylates, and polyphenols can induce MPT through the increase in ROS generation by interaction with the electron transport chain (Battaglia et al., 2005).

All the naturally occurring polyamines exhibit significant protection against MPT in isolated mitochondria when the phenomenon is induced by different agents (Toninello et al., 2004). Spermine can be considered one of the most powerful physiological inhibitors of MPT occurrence and several mechanisms have been proposed to elucidate this effect. An electrostatic interaction between the positive charges of this amine with the negatively charged head of the cardiolipin annular domain of AdNT was previously proposed. This involves a direct correlation between number of positive charges and efficacy of the protection: spermine > spermidine > putrescine. Another proposal was that polyamines act by activating or inhibiting protein phosphorylation–dephosphorylation, taking into account that MPT may be regulated by this posttranslational event. The increase in the affinity of ADP for an inhibitor site of MPT induction favored by polyamines was also taken into account along with a direct binding to a specific site in mitochondria responsible for MPT prevention.

A recent paper demonstrated that spermine directly acts against ROS like a free radical scavenger (Sava et al., 2006). Spermine prevented oxidation of several mitochondrial molecules such as glutathione and pyridine nucleotides and particular components such as sulfhydryl groups, all involved in MPT induction. Indeed, spermine exhibited a protective effect against lipid peroxidation and protein oxidation in mitochondria when these oxidations were due to a specific ROS, namely the hydroxyl radical.

While all the natural polyamines exert protective effects against MPT in isolated mitochondria, the monoamines exhibit particular properties that differ from the activities of the others. Several monoamines such as tyramine, octopamine, and benzylamine at appropriate concentrations (≈ 200 to 500 μM) are able to trigger or amplify MPT in the presence of Ca^{2+} and phosphate (Toninello et al., 2004). A similar effect under the same experimental conditions was observable with the diamine agmatine (Battaglia et al., 2007). At low concentrations (10 to 100 μM), tyramine and agmatine exhibited a protective effect on MPT induced by supraphysiological Ca^{2+}

concentrations (Battaglia et al., 2007). This different behavior is likely ascribable to the fact that at low concentrations these amines act as ROS producers, due to the activity of mitochondrial amine oxidases that form aldehydes and H_2O_2. At high concentrations, the amines not yet oxidized by the oxidases act as scavengers of the ROS they form (Battaglia et al., 2007). These observations clearly demonstrate a link of amine oxidase activity and MPT induction and subsequent triggering of the apoptotic pathway.

10.2.3 Role of Amine Oxidases in Apoptosis

As noted above, amines may exhibit paradoxical roles by inducing and preventing apoptosis. A review by Schipper et al. (2000) describes these controversial effects of polyamines in apoptosis. However, direct *in vivo* observations of the protective activities of polyamines on apoptosis have not yet been reported.

The main pathway in amine-mediated apoptosis is the oxidation of these molecules by bovine serum amine oxidase (BSAO), by intracellular polyamine oxidase (PAO), or by the mitochondrial monoamine oxidase (MAO). The hydrogen peroxide and aldehyde reaction products of these oxidations are strong inducers of apoptosis (Maruyama et al., 2000). These observations agree with results demonstrating that the catabolic reaction products of amine oxidation are powerful inducers of MPT. The role of H_2O_2 and its derivatives in amine-induced apoptosis is strongly supported by the ability of catalase to inhibit apoptosis induced by polyamines, their analogues, and other biogenic amines (Toninello et al., 2004). A further confirmation that oxidative stress is involved in MPT induction is the observation that the clorgyline and pargyline MAO inhibitors are able to inhibit the triggering of this phenomenon (Toninello et al., 2004).

All these investigations, when considered in the context of whole cell behavior, strongly support the proposal that amine oxidases, via their catalytic effects on biogenic amines, play an important role in controlling apoptosis by the induction of MPT and inducing fluctuations in concentrations. This proposal is further supported by the inhibition exhibited by these MAO inhibitors on melanoma cell line M14 (Malorni et al., 1998) apoptosis induced by serum starvation and by the reactive oxygen species produced by MAO activity in neuronal cell cultures (Buckman et al., 1993). Other structurally related MAO inhibitors such as rasagiline and L-deprenyl (both propargylamines) also hamper apoptosis in dopaminergic neuroblastoma SH-SY5Y cells, triggered by the endogenous N-methyl salsolinol toxin. It is noteworthy that clorgyline and pargyline are ineffective in this regard. This can be explained by taking into account that MPT induction and prevention are stereochemical in nature (Maruyama et al., 2000) and deprenyl and rasagiline induce their effect by suppressing the collapse of mitochondrial $\Delta\Psi$. The inefficacy of clorgyline and pargilyne is explained by their lack of stereospecificity for pore structures.

10.2.4 Amine Oxidation Products

Other observations on amine oxidases indicate that the oxidation products of polyamines rather than the polyamine alone are responsible for altering mitochondrial structures, with loss of metabolites and important protein factors leading to apoptosis or necrosis. Maccarrone et al. (2001) reported that spermine induces cytochrome c release from mitochondria, most probably accompanied also by AIF and Smac/DIABLO, in parallel with $\Delta\Psi$ collapse. These damaging effects were prevented by pargyline and catalase and amplified by BSAO. These findings confirm that hydrogen peroxide and also its very toxic derivative hydroxyl radical are largely responsible for the highly damaging effects and further confirm their ability to induce the loss of pro-apoptotic factors, caspase activation, and subsequent programmed cell death in mammalians (Hockenbery et al., 1993; Maccarrone et al., 1997).

In these experiments, the release of cytochrome c was selective and not due to an unspecific leakage. Cyclosporin A was not able to protect against $\Delta\Psi$ drop. This fact led the authors to propose that the transition pore does not open in this process (Maccarrone et al., 2001) and that under certain conditions, the induction of MPT is not sensitive to cyclosporin A. The same authors performed other investigations of the activity of BSAO and polyamine oxidation. They provoked activation of amine oxidase, leading to mitochondrial dysfunction and bioenergetic collapse, accompanied by loss of pro-apoptotic factors, and programmed cell death. These effects were counteracted by pargyline. The main conclusion raised by these investigations was that the activation of amine oxidases is the cause of the apoptotic programs and that the reaction products of intracellular polyamine oxidation are responsible of apoptosis induced by UVB (Maccarrone et al., 2001). The activity of amine oxidases in conditions in which there is a strong increase of intracellular accumulation of amine is also responsible for increases of hydrogen peroxide and its derivatives during pathophysiological conditions (Toninello et al., 2004; Stefanelli et al., 1998 and 1999). Most likely under these conditions the production of toxic compounds is not completely abolished by the endogenous safety systems by which apoptosis is induced. Triggering of apoptosis by excessive accumulation of polyamines was also evidenced by other investigators (Stefanelli et al., 1998; Poulin et al., 1995).

10.2.5 Loss of Pro-Apoptotic Factors

A crucial event to induce caspase activation—a fundamental step for the onset of apoptosis—is the loss of pro-apoptotic factors like cytochrome c, AIF, and Smac/DIABLO. The mechanism of the release of these factors is not yet fully understood. Some investigations have focused on cytochrome c release, showing that spermine causes leakage of this protein into the cytosol in intact cells and cell extracts, without provoking damage of mitochondria (Stefanelli et al., 1998, 1999, and 2000).

Unlike other investigations showing that cytochrome c can be lost through the effects of caspases (Marzo et al., 1998a), by the pro-apoptotic protein BAX, or by other means, pore opening was not involved (Stefanelli et al., 2000; Marzo et al., 1998b). In particular, the released cytochrome c appears to belong to a different pool of the protein not operating in electron transport. As a result, it does not affect the mitochondrial respiration and ATP synthesis necessary to promote a cell death program. To induce this effect, spermine binds to mitochondria at the level of the so-called S_1 site—the first step of polyamine transport in these organelles. The S_1 site is also responsible for the prevention of MPT (Sava et al., 2006). This accounts for the lack of pore opening and membrane damage. These observations also demonstrate that the S_1 site is located in the inner mitochondrial membrane (Toninello et al., 2000) near the site of cytochrome c binding. The physiological relevance of these results remains obscure. However, the outflow of cytochrome c from mitochondria induced by spermine may be considered a prerequisite for the establishment of different forms of apoptosis.

If we consider that cytochrome c is a fundamental component of apoptosome, a multiprotein complex whose activity is necessary for caspase activation, the effect described above may acquire additional importance, depending on the amount of protein lost. Leakage of cytochrome c into HL60 leukemia cells by spermine is responsible for apoptosis (Stefanelli et al., 1998). A similar effect of spermine was observed in embryonic cardiomyocytes in which caspase activation was equally triggered by the released cytochrome c from mitochondria (Stefanelli et al., 2000). This observation suggests the involvement of polyamine in certain pathological forms of apoptosis in cardiomyocytes. Cardiac hypertrophy consequent to hypertension is characterized by consistent augments of polyamine levels, and is strictly linked to excessive apoptosis (Ibrahim et al., 1995; Pegg and Hibasami, 1980). One possible explanation is that an intense and abnormal accumulation of polyamines in cardiomyocytes, instead of stimulating cell growth, is directed to triggering apoptosis by the loss of pro-apoptotic factors that in turn stimulate caspase activity or other stimuli favoring apoptosis.

Based on previous investigations (Hockenbery et al., 1993), amine oxidases, by oxidizing their biogenic amine substrates via the production of hydrogen peroxide, can induce apoptosis in a large number of cell types (Parchment et al., 1990; Lindsay and Wallace, 1999). The toxicity of biogenic amines is normally observed when evaluated in the presence of serum containing amine oxidase activity. However, purified BSAO and other amine oxidases can also cause apoptosis when biogenic amines are present. The reaction catalyzed by BSAO also involves water and molecular oxygen and produces ammonia, hydrogen peroxide, and aldehydes. If the substrate is spermine, the reaction products are monoaldehydes, an unstable dialdehyde, and an acrolein breakdown product that is very toxic (Alarcon, 1970).

10.2.6 Amine Oxidases and Human Cancer Cells

When pig kidney diamine oxidase (a copper amine oxidase [CAO]) is injected into the peritoneal cavities of Swiss mice after a viable transplantation of Ehrlich ascites cells, a remarkable inhibition of tumor growth is observed (Mondovì et al., 1982). Likewise, chick embryo fibroblasts transformed by Rous's sarcoma virus undergo cell death after microinjection with diamine oxidase or BSAO. When normal cells were injected, no significant changes in their morphology were noted (Bachrach et al., 1987). The main factor responsible for the observed cytotoxicity induced by these amine oxidases is most probably hydrogen peroxide, since the damaging effects are prevented by catalase. Aldehydes, however, are also responsible for cytotoxicity since incubation with aldehyde dehydrogenase prevents the damaging effects (Averil-Bates et al., 1994).

The observation that the reaction products of spermine oxidation by BSAO induce cell death in tumor cells led to a proposed strategy to overcome multidrug resistance of human cancer cells. Investigations were performed on both drug-sensitive (LoWoWT) and drug resistant (LoVoDX) colon adenocarcinoma cells (Calcabrini et al., 2002) that contain elevated levels of polyamines. The amine oxidases were used as antitumor agents. Results indicated that hydrogen peroxide, the enzymatic product of spermine oxidation by amine oxidases, was responsible for most of the observed cytotoxic effects on both cell lines. The aldehydes, particularly acrolein, contributed to the toxicity. Drug-resistant cells were much more sensitive to the toxic products than the wild types. The higher sensitivity of LoWoDX cells to H_2O_2 and aldehydes was not prevented by the antioxidant effect exhibited by glutathione and other physiological protective agents.

These results were supported by electron microscopy studies showing that drug-resistant cells exposed to both BSAO and spermine exhibited more evident morphologic and ultrastructural modifications than drug-sensitive cells. The subcellular organelles more sensitive to this treatment were mitochondria of drug-resistant cells that exhibited massive intracristal swelling and inner membrane depolarization. Similar damaging effects were observed when BSAO and spermine were substituted with exogenous hydrogen peroxide. These results may be explained by considering the fact that resistant cells exhibit more intense activity of the mitochondrial respiratory chain than wild-type cells. In fact, resistant cells require very high bioenergetic capacities to ensure the functionality of the phosphoglycoproteins that characterize them (Jia et al., 1997). These observations were noted also by other authors (Jia et al., 1996) who pointed out the requirement for different ATP production activities of resistant and sensitive cells and the possibility that overexpression of phosphoglycoproteins may regulate mitochondrial energy transduction.

Indeed, resistant cell lines with augmented flows of electrons along the respiratory chain demonstrated greater susceptibility to apoptosis induced by different effectors than the sensitive parental cells (Jia et al., 1996). In conclusion, all these investigations indicate that LoVoDX cells are more sensitive to amine oxidase activity because of the enhanced activities of their mitochondrial respiratory complexes that further produce ROS (Sohal, 1997). Similar correlations were demonstrated by several other studies utilizing pro-apoptotic agents. In particular, a new anticancer compound named BMD188 activated apoptosis by increasing the electron transport rate of mitochondrial respiratory complex with consequent enhancement of mitochondrial membrane potential and overproduction of ROS (Joshi et al., 1999). Collapse of electrical membrane potential and uncoupling of mitochondrial functions and apoptotic phenotypes are generally preceded by consistent augmentation of $\Delta\Psi$ values and high amounts of ROS, as noted by other investigations, implying the cytotoxicity of tumor necrosis factor (Schulze-Osthoff et al., 1993).

The activity of amine oxidases in oxidizing biogenic amines with the production of reactive oxygen species is generally accepted as the main event leading to the opening of the transition pore. The alterations in mitochondrial structure and function following this phenomenon trigger an increase in the oxidation level of sulfhydryl groups, glutathione and pyridine nucleotides, accompanied by a drop of the energy capacity of the organelles and the release of pro-apoptotic factors. However, a severe and persistent drop in energy production shifts the apoptotic programmed cell death toward the triggering of the necrotic pathway (Eguchi et al., 1997).

Amine oxidase activity can induce different cell death pathways, depending on the utilization of the polyamines as their substrates in particular cell lines. The characteristic aspects of apoptosis (chromatin condensation, nuclear fragmentation, DNA cleavage, poly(ADP ribose) polymerase cleavage) were not evidenced in L1210 leukemia cells subjected to the toxic effects of oxidation products of spermine by BSAO activity. The necrotic morphology was evidenced but the cells killed by deregulated polyamine uptake showed biochemical and morphological aspects typical of apoptotic cell death (Bonneau and Poulin, 2000).

An impairment in the regulation of polyamine biosynthesis due to changes in their accumulation or catabolism can also trigger the apoptotic pathway (Pegg and Hibasami, 1980; Grassilli et al., 1995). The inefficacy of caspase inhibitors to hinder cell death induced by polyamine oxidase activity and the delayed effect on this process provoked by catalase indicate that the reaction products of these oxidases contribute to the necrosis of L1210 leukemia cells (Bonneau and Poulin, 2000). The necrotic death of these cells is characteristic of a primary necrosis, but also observable is a redistribution of plasma membrane phosphatidylserine to the outer leaflet, suggesting the involvement of a signal for phagocytosis. The authors proposed that immunosuppression by polyamines reflecting their induction of a fast necrotic pathway involving amine oxidase activity and a rapid bioenergetic drop led to phosphatidylserine exposure and elimination by phagocytes. This proposal considered previous observations that the oxidation of spermine by polyamine oxidase exhibited strong immunosuppression through an unknown mechanism of cell death.

The induction of cell death by polyamines has also been observed in the absence of ROS generation by using the MDL72527 polyamine oxidase inhibitor (Poulin et al., 1993 and 1995; Mitchell et al., 1992) and by studying the effects of spermine–spermidine acetyltransferase (SSAT) overproduction in transgenic mice (Alhonen et al., 2000). These investigations suggest that in some cell types the induction of apoptosis by polyamines does not require their oxidation. One explanation may be that the massive accumulation or depletion of polyamines may exert several damaging effects on many cellular functions. These effects include DNA–protein interactions, protein–protein interactions, and impacts on mitochondrial integrity (Stefanelli et al., 2000; Thomas et al., 1999), all events linked to apoptosis induction. A similar process not requiring oxidized polyamine catabolites is observable in the apoptotic program triggered by 2-deoxy-D-ribose in HL60 cells in which a relationship between induction of the G1 phase transition by spermine and the onset of apoptosis induced by deoxyribose was demonstrated.

Studies of polyamine analogues yielded important results concerning the induction of intrinsic apoptosis. Experiments with these analogues evidenced cell death induction in cell lung carcinoma line by showing the typical signs of apoptosis, i.e., cytochrome c release, collapse of $\Delta\Psi$, activation of caspase-3, PARP cleavage, and DNA fragmentation (Ha et al., 1998). Bcl-2 overexpression exhibited inhibition of all these effects without completely preventing cell death, suggesting a caspase-independent pathway activated by spermine analogues. Other analogues exhibited anticancer activities and pro-apoptotic effects in some cell types (Chen et al., 2001).

Treatment of SK-MEL-28 human melanoma cells showed very high SSAT activation and depletion of polyamine accompanied by $\Delta\Psi$ collapse and loss of cytochrome c. Mitochondrial alterations were followed by caspase-3 activation and PARP cleavage. Caspase inhibitors caused the cells to be resistant to spermine analogues and reduced the extent of apoptosis.

A study of different spermine analogues indicated they produced different induction effects on SSAT activity, but were able to deplete polyamines to the same extent, demonstrating a correlation between the induction of this enzyme and the mitochondrial signaling pathway of apoptosis.

Depletion of polyamines in rat intestinal epithelial (IEF-6) cells by treatment with 2-difluoromethyl ornithine (DFMO), an inhibitor of ornithine decarboxylase (ODC), was demonstrated to delay apoptosis induced by campothecin, an antineoplastic compound. The depletion of polyamines prevented the loss of cytochrome c from mitochondria, induced by the agent, with the result of diminishing the activities of caspase-3 and caspase-9. The expression of the antiapoptotic factors Bcl-2 and Bcl-X_L was also induced, while caspase-8 activity and Bid cleavage were diminished (Yuan et al., 2002). The conclusion is that polyamine depletion protects against apoptosis by hindering the recruitment of BAX by mitochondria, resulting in inhibition of cytochrome c release.

Another study emphasized that the loss of polyamines triggers the pro-apoptotic pathway by mitochondrial mediation (Nitta et al., 2002). A combination of inhibitors of polyamine biosynthetic enzymes (ODC, S-adenosylmethionine decarboxylase, and spermidine synthase) was used to deplete WEHI 231 murine B cells of polyamine content. Chromatin condensation and DNA fragmentation were the results of the treatment, demonstrating the typical aspects of apoptosis that were completely reversed by administration of exogenous polyamines. Similar effects following polyamine depletion were observed in other tumor cell lines such as human B cells and human T Jurkat cells (Nitta et al., 2002). In all these cells, the lack of intracellular polyamine content led to a collapse of mitochondrial electrical gradient and activation of caspase-3. This demonstrates the loss of impermeability of mitochondrial inner membrane and consequent collapse of bioenergetic capacities of isolated organelles and also *in situ*. The inability to prevent $\Delta\Psi$ drop by a caspase-3 inhibitor also demonstrated that disruption of this gradient is an early event in the apoptotic process.

Several other examples indicate that the depletion of polyamine content is correlated with apoptosis as observed in rat thymocytes (Desiderio et al., 1995) and in different tumor cell lines (Penning et al., 1998). The agmatine diamine is a very effective agent in inducing polyamine depletion in hepatocytes. Agmatine is produced by arginine decarboxylation catalyzed by arginine decarboxylase (ADC). It strongly induces SSAT which acetylates polyamines by favoring their oxidation by the respective oxidases. Agmatine is widely distributed in mammalian tissues and organs (Raasch et al., 1995) and can act as a neurotransmitter.

The strong reduction in polyamine content induced by agmatine in hepatocytes may be responsible for apoptosis in rat hepatocytes (Gardini et al., 2001). However, it has also been proposed that agmatine can be oxidized by a diamine oxidase, whose reaction products, in particular hydrogen peroxide and its derivatives, induce oxidative stress leading to the opening of the mitochondrial transition pore. This event, as noted above, provokes the release of pro-apoptotic factors and activation of the caspase cascade.

Most likely agmatine acts in two synergic ways. It induces hyperproduction of SSAT and increases hydrogen peroxide generation. Indeed, the activity of SSAT in producing acetylated polyamines, the diamine oxidase catalysis of oxidative deamination of agmatine, and the inhibition of ODC and nitric oxide synthase are critical aspects of programmed cell death in different cancer and normal cells (Yuan et al., 2002).

The most significant investigations of apoptosis induced by monoamines were performed with dopamine. Oxidative stress generated by MAO activity in the normal metabolism of dopamine is involved in Parkinson's disease based on the observation of selective degeneration of dopaminergic neurons (Spina and Cohen, 1989). MAO mediates the oxidation of the catecholamine to form quinines and ROS that contribute to mitochondria impairment and apoptosis induction (Cohen and Heikkila, 1974).

Oxidation of dopamine and 6-hydroxydopamine or their reaction products can induce apoptosis in a large number of cell types (Coronas et al., 1997). The mechanism of dopamine toxicity observed in Parkinson's disease most likely involves defects in Complex I activity that can predispose cells to this effect. Dopamine can also induce the loss of cytochrome c which is dependent on p38 MAP kinase activation and precedes the cleavage of caspases (Junn and Mouradian, 2001). These findings indicate that dopamine is involved in apoptosis induction in SH-SY 5Y neuroblastoma cells due to ROS generation upon catecholamine oxidation.

This oxidation is accompanied by p38 MAP kinase activation, cytochrome c release, and caspase activation. P38MAP kinase belongs to the mitogen-activated protein (MAP) kinase superfamily and is involved in several cellular processes including inflammation and apoptosis (Han et al., 1997). The treatment of SH-SY 5Y neuroblastoma cells with dopamine and other catecholamines activates the above-mentioned kinase by phosphorylation of tyrosine and threonine residues in the sequence Thr–Glu–Tyr by dual MAP kinase kinase 3 and SAPK/ERK kinase 1 (Han et al., 1997; Tibbles et al., 1996).

p38 MAP kinase is phosphorylated by a mechanism requiring ROS production by MAO activity (Junn and Mouradian, 2001); this process is inhibited by the antioxidant N-acetyl-L-cysteine. Indeed, the activation of p38 MAP kinase is responsible for the release of cytochrome c and activation of executor caspases in dopamine-induced apoptosis (Junn and Mouradian, 2001).

These observations led to the proposal that this catecholamine has an important role in dopaminergic neurons most likely responsible for mitochondrial impairment in Parkinson's disease. Dopamine oxidation by MAO with its reaction products provokes mitochondrial dysfunction with the loss of pro-apoptotic factors. These reaction products are increased in the substantia nigra of postmortem brain tissues of Parkinson's patients (Spencer et al., 1998). In conclusion, all these observations suggest that targeting p38 MAP kinase and preventing cytochrome c loss may provide a mechanism for maintaining the viability of dopaminergic neurons. A very interesting investigation demonstrated that the mitochondrial voltage dependent anion channels (VDACs) are contributors to cell death induced by dopamine (Premkumar and Simantov, 2002). The effect of dopamine should be exerted by downregulation of VDAC expression. VDACs participate in the protein assembly involved in transition pore architecture. The injection of anti-VDAC antibodies prevents apoptotic cell death (Shimizu et al., 2001).

10.2.7 Conclusions and Future Perspectives

This chapter describes the physiological role of amine oxidases and their substrates in controlling or triggering apoptotic cell death and, in many cases, neoplastic growth. These effects depend on cell type fluctuations in amine concentrations and amine oxidase activity and turnover.

Because cell death pathways are often affected in cancer cells, studies of the roles of amine oxidases in controlling apoptosis will undoubtedly enhance our understanding of the processes of malignant transformation. The use of MAO inhibitors is particularly relevant for therapeutic intervention, particularly in neurological pathologies such as Alzheimer's and Parkinson's diseases. In recent years, nonspecific and irreversible MAO inhibitors were utilized in neuropsychiatry as antidepressants and were later discarded because of toxicity and adverse side effects. The two known forms of MAO, A and B, exhibit different substrate specificities and tissue and cell distributions in the central nervous system and peripheral organs. This indicates that the two forms have distinct physiological functions that may be targeted by specific inhibitors. A number of reversible and irreversible inhibitors with relative specificities and different inhibition mechanisms for MAO A and B have been described. The important issue concerns the mild side effects of these inhibitors. The type of effort used to determine MAO activity should be applied to the other amine oxidases to individualize their negative or positive effects and ultimately use their activities to inhibit apoptosis induction. Future investigations should address the effects of amine oxidase activity on the regulation of signal transduction pathways, taking into account the discovery of Src kinases, tyrosine phosphatase, and their substrates in mitochondria (Lewandrowski et al., 2008).

ACKNOWLEDGMENT

We wish to thank Valentina Battaglia, Ph.D. for helpful collaboration.

11 Copper Amine Oxidases in Intestinal Diseases

Wieslawa Agnieszka Fogel, Ewa Toporowska-Kowalska, and Anna Stasiak

CONTENTS

11.1 INTESTINAL DIAMINE OXIDASE AND ITS MAIN SUBSTRATES, HISTAMINE AND PUTRESCINE

In healthy individuals of most mammalian species, high levels of activity of DAO (histaminases; Enzyme Commission [EC] 1.4.3.6) are present in the intestinal tract. In humans, marked DAO activity is found only in kidney and mesenteric lymph nodes (Kusche et al., 1980). The longitudinal distribution of DAO in the gastrointestinal tract measured by gastric, jejunal, ileal, and colonic biopsies is similar in humans and in animals (Bieganski et al., 1983; Rokkas et al., 1990a). The enzyme is localized in the mucosa and within cell, usually in the cytoplasm, except in guinea pig intestine where it appears to be associated with the particulate fraction (Bieganski et al., 1983).

Analysis of subcellular localization of pig intestinal DAO by confocal laser scanning fluorescence microscopy suggests that the enzyme is present in vesicular structures that form a network near the plasma membranes of cells (Schwelberger et al., 1998). The villus cells, predominantly those from the middle and apex (but not crypt cells), contain intestinal DAO. This implies that any event leading to a damage or loss of villus structure will exert great impact on enzyme levels in the intestinal mucosa. DAO is not associated with the brush border and is exported from parenchymal cells to vascular binding sites of intestinal microvasculature where it is released by heparin. The low basal plasma level of DAO is considered to be of the intestinal origin (Shakir et al., 1977; Robinson-White et al., 1985). Intestinal DAO does not accept monoamines but degrades short-chain aliphatic diamines, putrescine, cadaverine, histamine, N(τ)-methylhistamine, and spermidine (Bieganski et al., 1983).

This is not surprising because the mammalian digestive tract mucosa is in permanent contact with a vast array of amine substrates; histamine, putrescine, polyamines, and other substrates are normal dietary constituents. Foods produced with technology that includes fermentation processes (cheeses, certain sausages, yogurts, etc.) are especially rich in biogenic amines. Under healthy conditions, the enzyme activity is sufficient although food-induced histaminosis under DAO blockade has been clearly demonstrated.

Some biogenic amines are formed by bacteria in the lower part of the large bowel (Fogel et al., 2005a and 2007). Enterocytes acquire biogenic amines from the lumen through passive diffusion and metabolize and/or store them (Milovic, 2001).

Important endogenous depots of histamine and other amines in the gastrointestinal tract are enterochromaffin-like cells in the stomach, APUD (amine precursor uptake and decarboxylation) endocrine cells, mast cells, and intramural nerves (Fogel, 2005a). During chronic inflammation, the mucosal epithelium is exposed, in addition to the lumen, to histamine excesses released from internal sources such as mast and other infiltrating cells. The intraepithelial lymphocytes and attracted circulating cells (basophils, macrophages, and platelets) make significant contributions to the histamine pool, as cytokines: TNFα, IL-1, GM-CSF, and others, by triggering histidine decarboxylase (HDC, EC 41.1.22) gene expression in these cells (Murata et al., 2005). While histamine can exert a large variety of effects extracellularly, for example, evoking smooth muscle contraction, stimulating mucus, water, and electrolyte secretions, and causing edema. It can influence neuronal transmissions, neural–enterochromaffin cell interactions, and immune responses (Pearce, 1991). Histamine effects are mediated by four cell surface histamine receptors designated H_1 through H_4 on various cells of the digestive system including epithelial and smooth muscle cells, endothelial and immune cells, and intestinal neurons (Fogel et al., 2005a). Generally, histamine H_1 receptors participate in inflammatory effects while H_2 and H_4 receptors mediate histamine effects release and thereby neurotransmission and/or neuromodulation. In rodents, the effects mediated by histamine H_3 receptors involve presynaptic inhibition of neurotransmitter release from the myenteric plexus (Poli et al., 2001). However, no expression of H_3 receptors in the immune system (induction of suppressor cells, inhibition of lymphocyte proliferation, chemotaxis of neutrophils, mast cells, and macrophages) resulted in locally decreased immune response against cancer or exacerbation of inflammation (Jutel et al., 2006). Acting through H_3 receptors, histamine influences the neurotransmitters of the human gut (Sander et al., 2006; Boér et al., 2008). Close contact of gut mucosal mast cells and enteric nerves produces interactions between resident immune cells and the enteric nervous system which in inflammatory diseases may lead to altered gut physiology and increased sensory perception (Mantyh et al., 1991). The mast cells release mediators that stimulate intestinal ion transport through direct epithelial action and via nerves (Fogel et al., 2005a).

Due to a high affinity to histamine, intestinal DAO is considered to play a crucial role in controlling its level and in termination of amine action at appropriate receptor. If DAO activity is reduced or absent (Table 11.1, Figures 11.1 and 11.2), this

TABLE 11.1
Clinical Use of DAO Activity Measurement

Subjects and References	Tissue/Mucosal DAO Activity (nmol/min/g protein)	Clinical Relevance
Healthy controls; GI tract; Rokkas et al., 1990c	Stomach [n = 12]: 1.73 ± 2.08 (1.47 to 7.0) Jejunum [n = 16]: 19.0 ± 9.48 (8.4 to 58.6) Ileum [n = 6]: 33.4 ± 6.56 (24.2 to 40.2) Colon [n = 18]: 7.43 ± 3.56 (1.5 to 13.2)	• Proximal-to-distal gradient in DAO activity from stomach to ileum; low enzyme level in colon
Crohn's disease (CD) ↓ DAO activity		• Positive correlation with histologic scores in CD • Reflects involvement of small intestine with CD
CD + intestinal resection; Schmidt et al., 1990	Normal proximal resection [n = 20]: 25.93 (12.67 to 65.50) Normal distal resection [n = 20]: 22 (9.75 to 49.17) Inflamed area [n = 20]: 12.42 (0.83 to 43.67)	• Very low DAO activity → factor predicting risk of recurrence or complications after intestinal resection • Marker for assessing extent of inflamed area in CD that may influence surgical strategies
Ulcerative colitis (UC); Mennigen et al., 1990 ↓ DAO activity	UC [n = 12]: *ca.* 27 UC in remission [n = 3]: 1393 (1030 to 2108) Controls [n = 30]: 228 (37 to 1365)	• Indicator of rectal mucosal antiproliferative function in UC • DAO: suitable indicator of activity grade of UC • ↑ DAO in UC remission → antiproliferative rebound effect
Celiac disease (CLD) ↓ DAO activity		• Decreased mucosal DAO activity reflects mucosal atrophy (correlation with low serum enzyme activity; see Table 11.2)
CLD; children; Forget et al., 1986	Controls: 81.0 ± 6.5 CLD: 20.2 ± 3.7	
CLD and gluten-free diet; adults; Rokkas et al., 1990a	*pre*-GFT [n = 5]: 11.7 ± 2.91 CLD + 6 mo GFT [n = 5]: 25.7 ± 11.54	

(*Continued*)

TABLE 11.1 (*Continued*)

Subjects and References	Tissue/Mucosal DAO Activity (nmol/min/g protein)	Clinical Relevance
Large bowel polyps ↓ DAO activity		• Reduction of intestinal DAO activity may depend on malignant or premalignant states or general disturbances of mucosal integrity
Mennigen et al., 1988	LBP tissue [n = 20]: 134 (0 to 530) Mucosa, 5 to 10 cm distal to LBP [n = 14]: 375 (50 to 680) Normal mucosa [n = 30]: 228 (37 to 1365)	
Hesterberg et al., 1981	Normal rectal mucosa [n = 41]: 410 (280 to 794) LBP tissue [n = 7]: 77 (10 to 480)	
Large bowel cancer; Mennigen et al., 1988 ↓ DAO activity	Normal mucosa [n = 30]: 228 (37 to 1365) Cancerous mucosa [n = 17]: 177 (60 to 470) Cancer tissue [n = 42]: 26 (0 to 456)	• Mucosal hyperproliferation associated with reduced DAO activity • DAO activity: marker of abnormal mucosal proliferation

Notes: Values = mean ± SD and ranges (in round brackets) for number of patients (square brackets). Data recalculated and DAO activity uniformly expressed in nmol/min/g protein, based on information provided by authors (definitions of enzyme units, sample sizes, etc.). Measures were standardized: 1 g fresh/wet tissue = 100 mg protein. Tissue DAO activity in all presented cases was measured by radiochemical methods based on procedure of Okuyama and Kobayashi (1961), with ^{14}C-putrescine as substrate.

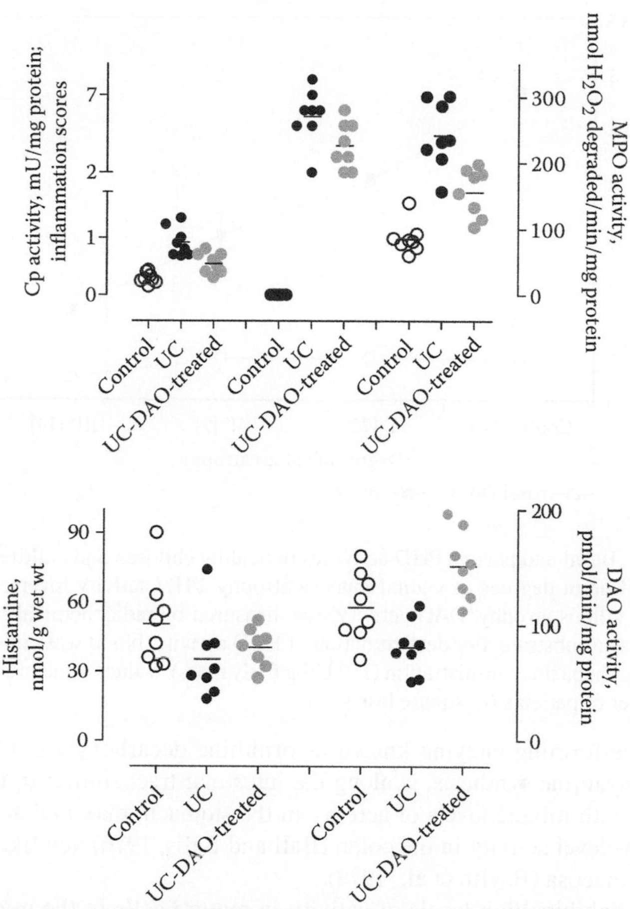

FIGURE 11.1 Therapeutic effects of exogenous DAO on reparative processes in rats with UC. Male Wistar rats (260 to 300 g) were randomly allocated to three groups of eight animals each: control, UC, and UC–DAO-treated. UC was induced by intrarectal administration of 4% (v/v) acetic acid for 15 sec. Hog kidney DAO immobilized on concanavalin A–Sepharose was given to rats as a suspension of 25 mU/day intraperitoneally, until the fifth day when the animals were sacrificed. The severity of colonic lesions was evaluated using a nine-point rating scale and inflammation by measurements of CP and colonic MPO activities. Histamine was separated by Cellex P chromatography and assayed spectrofluorimetrically; DAO activity was measured with ^{14}C -putrescine (Fogel et al., 2005b and 2006).

protective role is attenuated or missing and undesired effects of histamine may occur as observed in anaphylactic and allergic reactions, mesenteric infarction, ischemic bowel syndrome, colorectal cancer, chronic nonspecific inflammatory bowel disorders, Crohn's disease, and ulcerative colitis.

In addition to its role in histamine metabolism, intestinal DAO is implicated in the metabolism of putrescine, a precursor of polyamines that serve as universal regulators of all normal, regenerative, and neoplastic growth processes. The distribution

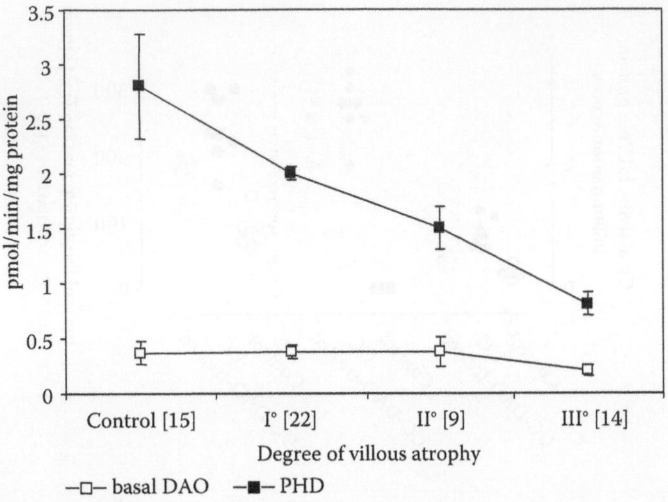

FIGURE 11.2 Basal and plasma PHD activities in healthy children and children with celiac disease with different degrees of jejunal mucosa atrophy. PHD activity inversely correlated with degree of villous atrophy. DAO activity was measured by radiochemical method using ^{14}C-putrescine as a substrate. For determination of PHD activity, blood was collected 60 min after intravenous heparin administration (150 U/kg body mass). Values = mean ± SD obtained from the number of patients (in square brackets).

of a putrescine-forming enzyme known as ornithine decarboxylase (ODC), a key enzyme in polyamine synthesis, is along the intestinal tract similar to the distribution of DAO, with minute levels of activity in the stomach, maximal activity in the ileum, and low level activity in the colon (Ball and Balis, 1976) and likewise across the intestinal mucosa (Baylin et al., 1978).

ODC also exhibits high levels of activity in mature cells in the mid villus and tip. It has been clearly demonstrated with the use of α-difluoromethyl ornithine, a highly selective, irreversible ODC inhibitor, that an increase in ODC activity preceding an increase in polyamine content is indispensable to achieve intestinal mucosal maturation and regeneration in rats (Luk et al., 1980a). Evidence also indicates that DAO is more important than ODC in putrescine level regulation and DAO prevents feedback inhibition of ODC by putrescine. After DAO depletion by heparin, putrescine concentration increased while ODC activity decreased in the jejunum and ileum. At the same time, DAO depletion exerted no significant effects on spermine and spermidine concentrations (Rokkas et al., 1990b). Based on high levels of ODC and DAO activities, the gastrointestinal tract is a potential source of putrescine-derived GABA since the first product of oxidative deamination of putrescine by DAO is γ-aminobutyraldehyde, later oxidized to GABA. Still open to speculation is the question whether oxidative deamination by DAO, while limiting levels of growth-promoting putrescine, polyamines, and histamine, gives rise first to inhibitors of cell proliferation (aminoaldehydes) and subsequently inducers of differentiation or maturation (GABA, aminovaleric acid, and imidazoleacetic acid; Fogel, 1986).

TABLE 11.2

Basal and Postheparin Plasma Diamine Oxidase Activity as Clinical Diagnostic Tool

Subjects and References	Plasma/Serum DAO Activity (nmol/h/ml) Basal	PHD	DAO Assay	Clinical Relevance
Crohn's disease (CD); Thompson et al., 1991 ↓ DAO activity	CD [n = 7]: 3.4 ± 2.0 Control [n = 30]: 4.3 ± 2.5	CD [n = 37]: 17.4 ± 3.0 Control [n = 30]: 32.8 ± 30.8	• Radiochemical with ^{14}C-putrescine • Heparin: 3000 U *iv* T_{30}[1]	• DAO activity reflects intestinal involvement in CD • Correlation between changes in DAO activity and CDAI[2] • Effective therapy: ↑ PHD activity and ↓ CDAI
Small bowel Crohn's disease (SB-CD); D'Agostino et al., 1988a, b	SB-CD [n = 51]: 0.08 ± 0.04 Control [n = 20]: 0.11 ± 0.08	SB-CD [n = 51]: < 3 Control [n = 20]: *ca.* 5	• Radiochemical with ^{14}C-putrescine • Heparin: 15000 U *iv* T_{60}	• Low PHD values in direct proportion to small bowel mucosa damage • Anti-inflammatory therapy led to significant increase in PHD
Celiac disease (CLD) ↓ DAO activity				• Useful for assessing mucosal damage in patients with malapsorption • Helpful in quantitating remaining mature enterocyte mass
Adults: CLD + gluten-free diet (GFD); D'Agostino et al. 1987a, b	CLD [n = 12]: 0.06 ± 0.03 Control [n = 20]: 0.09 ± 0.05	CLD [n = 25]: *ca.* 2.5 CLD + 3 mo GFD [n = 14]: *ca.* 3.5 CLD + 6 mo GFD [n = 14]: *ca.* 5 to 6 Control [n = 16]: *ca.* 5	• Radiochemical with ^{14}C-putrescine • Heparin: 15,000 U *iv* T_{60}	

(*Continued*)

TABLE 11.2 (Continued)

Subjects and References	Plasma/Serum DAO Activity (nmol/h/ml)		DAO Assay	Clinical Relevance
	Basal	PHD		
Children; Forget et al., 1986	CLD: 0.013 ± 0.002 Control: 0.039 ± 0.012		• Radiochemical with ^{14}C-putrescine	
Children with protracted diarrhea (PD) and degrees I through III (partial to total) villous atrophy	CLD+PD: I° [n = 22]: 1.59 ± 1.26 II° [n = 9]: 1.59 ± 1.64 III° [n = 14]: 0.88 ± 0.76 Control [n = 15]: 1.55 ± 1.68	CLD + PD: I° [n = 22]: 6.80 ± 1.26 II° [n = 9]: 4.70 ± 2.39 III° [n = 14]: 2.52 ± 1.64 Control [n = 15]: 10.20 ± 7.76	• Radiochemical with ^{14}C-putrescine • Heparin: 150 U/kg body mass *iv* T_{60}	• PHD activity is good indicator of jejunal mucosa atrophy during PD (correlated with degree of villous atrophy)
Small bowel lymphoma (SBL); D'Agostino et al., 1987a and 1988b ↓ DAO activity	SBL [n = 3]: 0.15 ± 0.07 Control [n = 20]: 0.09 ± 0.05	SBL [n = 5]: <1 Treated SBL [n = 4]: *ca.* 1.5 SBL in remission: >3.7 Control [n = 16]: *ca.* 5	• Radiochemical with ^{14}C-putrescine • Heparin: 15000 U *iv* T_{60}	• Monitoring recovery of small bowel enterocytic mass • Useful in monitoring recovery of mucosal lesions induced by chemotherapy
Ileus + nutritional management: total parental (TPN), enteral (EN), soluble dietary fiber (SDF), oral feeding (OF); children; Tanaka et al., 2003	Ileus + TPN: 4.73 ± 1.2 Ileus + EN: 6.84 ± 1.18 Ileus + SDF + EN: 7.62 ± 0.67 Ileus + OF: 8.82 ± 1.26 Control: 6.36 ± 0.98		• Colorimetric with cadaverine and DA-67[3]	• Plasma DAO activity may be sensitive marker of intestine function in children • EN and OF prevent intestinal atrophy
Acute gastroenteritis (GE); children; Forget et al., 1985	Control [n = 20]: 0.036 ± 0.006 GE acute phase [n = 11]: 0.031 ± 0.005 GE healing phase [n = 21]: 0.018 ± 0.002		• Radiochemical with ^{14}C-putrescine	• Marker of total mass of functional enterocytes

Gastric cancer; total gastrectomy with esophagojejunostomy and total parental (TPN) or enteral (EN) nutrition; Kamei et al., 2005	TPN [n = 21]: 5.8 ± 2.3 EN [n = 27]: 8.1 ± 2.3		• Colorimetric with cadaverine and DA-67[3]	• EN prevents intestinal atrophy • DAO activity is circulating marker of effect of nutritional therapy on intestinal mucosa
Chronic idiopathic urticaria (CIU); oligoallergenic diet; Guida et al., 2000		CIU [n = 10]: 2.79 ± 1.56 post-diet: 2.90 ± 1.22 Control: > 3.7	• Radiochemical with ^{14}C-putrescine • Heparin: 15000 U *iv* T_{60}	• Subclinical impairment of small bowel enterocyte function in CIU (normal intestinal permeability index) • Abdominal symptoms: plasma DAO activity correlated with mucosal activity
Biliary cancer; bile replacement (BR) following percutaneous transhepatic biliary drainage; Kamiya et al., 2004 BR → ↑ DAO	pre-BR [n = 25]: 3.9 ± 1.4 post-BR [n = 25]: 5.1 ± 1.6 Control: 4.9		• Colorimetric with cadaverine and DA-67[3]	• Correlation between magnitude of alteration in serum DAO activity and length of BR • BR helps restore intestinal barrier function in patients with biliary obstruction
Intestinal ischemia; → coronary artery bypass grafting (CABG) + cardiopulmonary bypass, (CPB); Tsunooka et al., 2004 ↑ DAO activity		pre-CABG CPB [n = 12]: 6.2 ± 2.2 CABG CPB [n = 12]: 34.7 ± 22.2 CABG [n = 10]; no rise Control: 5.0 ± 1.5	• Colorimetric with cadaverine and DA-67[3] • Heparinized patients	• Index of intestinal ischemia • Heparin may contribute to rise in serum DAO during CPB

(*Continued*)

TABLE 11.2 (*Continued*)

Subjects and References	Plasma/Serum DAO Activity (nmol/h/ml)		DAO Assay	Clinical Relevance
	Basal	PHD		
Chemotherapy (Ch) ↓↑ DAO				• Indicator of mucosal injury during chemotherapy
Leukemia; bone marrow transplant (BMT); Tsujikawa et al., 1999	pre-Ch: 12.1 ± 3.7 Ch induction: *ca.* 7.4 Ch maintenance: *ca.* 11.3 High dose Ch + BMT: *ca.* 0.85		• Colorimetric with cadaverine and DA-67[3]	
Neuroblastoma; children; Tanaka et al., 2003	Control [n = 138]: 6.36 ± 0.98 Ch [n = 19]: significant ↓ DAO activity up to first 4 days Ch		• Colorimetric with cadaverine and DA-67[3]	

Notes: Values = mean ± SD or ranges for number of patients (square brackets). Data recalculated and DAO activity uniformly expressed in nanomol per hour per milliliter, based on information provided by authors (definitions of enzyme units, sample sizes, etc.). Measures standardized: 1 ml plasma/serum = 70 mg protein. Radiochemical methods of DAO activity measurement based on procedure of Okuyama and Kobayashi (1961).[1] Time of blood collection following heparin administration; [2] CDAI = Crohn's disease activity index; [3] DA 67: 10-(carboxymethyl-aminocarbonyl)-3,7-bis (dimethylamino) phenothiazine sodium salt is a chromogen from Wako Pure Chemical Industries.

11.2 INTESTINAL DAO AS ANTIPROLIFERATIVE PRINCIPLE

In patients with adenocarcinoma of the large bowel or stomach, DAO activity within the tumors was reduced (Kusche et al., 1980; Boér et al., 2004) but was enhanced in the adjacent mucosa as compared to healthy tissue (Kusche et al., 1980). Biopsy material indicated that DAO activity was markedly lower in rectal mucosa of patients with rectal polyps and still lower values were obtained for patients with rectal carcinomas (Hesterberg et al., 1981). Compatible with these data are results of a recent study demonstrating upregulation of histidine decarboxylase activity and histamine concentration, highly correlating with tumor stage in colorectal cancer (Masini et al., 2005a).

Significant reduction of intestinal DAO was recorded in studies of azoxymethane carcinogenesis in rat digestive systems. The clinical and experimental data prompted the hypothesis that DAO inhibition, by enabling hyperproliferation, may increase the risk of intestinal carcinoma promotion (Kusche et al., 1986). Experiments have shown that aminoguanidine, a DAO inhibitor, combined with azoxymethane treatment resulted in considerable promotion of large bowel tumor, supporting involvement of DAO in a feedback regulatory mechanism terminating proliferation (Kusche et al., 1988). This finding has important clinical aspects for many DAO inhibitor drugs (Sattler and Lorenz, 1990) since their prolonged use may increase cancer risk. However, another study reported that suppression of DAO with aminoguanidine during the adaptive period to subtotal jejuno-ileal resection in rats resulted in enhanced mucosal proliferation but had no effect on mucosal functional differentiation. Mucosal proliferation, as measured by mucosal mass, protein content, and DNA content, was greater in resected animals receiving aminoguanidine but sucrase activity was similar to levels in controls (Erdman et al., 1989). Similar results were obtained by other authors who suggested that inhibition of DAO activity may be important therapeutically for patients with short bowel syndrome. The stimulatory effect of histamine on proliferation and resultant type of cell growth appears related to the local patterns of histamine receptors. In human colorectal tumors, downregulation of H_1R and H_4R expression (Boér et al., 2008) that may produce favorable conditions for neoplastic cell growth from H_2R-mediated negative regulation of Th1 and Th2 lymphocyte responses was reported (Jutel et al. 2006).

11.3 DAO AND IMMUNE-MEDIATED DISORDERS OF DIGESTIVE TRACT

Various adverse reactions to food mediated immunologically via IgE or by T cells or both mechanisms share the same symptoms, i.e., vomiting, abdominal pain, and diarrhea (Mansueto et al., 2006). Celiac disease, an immune-mediated intestinal disorder, is a type of gluten-sensitive enteropathy and one of the most common genetic diseases in Europe (prevalence = 1 in 300 persons). In susceptible individuals, gliadin, a dietary gluten component, initiates an immune response that involves T cell activation and cytokine production, a cascade of downstream events resulting finally in mucosa damage characterized by villous blunting, crypt hyperplasia, and lamina propria expansion (Dickson et al., 2006).

Chronic inflammation and atrophy of intestinal villi result in malabsorption and a variety of clinical manifestations. Along with the classic symptoms cited above, anemia, fatigue, neurologic dysfunction, bone changes, and skin rashes associated with deficits of necessary iron, calcium, or vitamin D nutrients may occur. Patients with food allergies exhibit reduced gut mucosal DAO and significantly higher histamine concentrations in tissues (Raithel et al., 1995; Fogel et al., 2005a). Histamine H_1 and H_2 receptors are upregulated in food allergies and in inflammatory bowel diseases (IBDs; Sander et al., 2006).

11.4 DAO IN INFLAMMATORY BOWEL DISEASES

IBDs include nonspecific, noninfectious inflammatory and ulcerative disorders of small and large intestines such as Crohn's disease (CD) and ulcerative colitis (UC). The pathogenesis of IBD is considered multifactorial, having genetic, immunologic and environmental components, but is still far from clear. CD can affect any area of the gastrointestinal tract, but the terminal ileum is most often involved. The disease is a segmental, destructive inflammation with a patchy and transmural tissue damage. Deep necrosis of mucosal cells can be followed by fibrosis and stenosis. UC is a chronic inflammatory disease, affecting mainly the mucosa of the rectum or distal part of the colon. Superficial mucosal ulceration may appear in varying degrees.

Both CD and UC are characterized by periods of remission and exacerbation of a different duration. In the active phases, in inflamed mucosa and submucosa, large increases of infiltrating neutrophils and macrophages are seen along with mast cell hyperplasia. In CD, responding T cells exhibit Th1 phenotypes; UC is believed to be Th2-mediated disease (Dvorak et al., 1979; Xavier and Podolsky, 2007). Patients with IBDs exhibit higher frequencies of DAO Dra I allele 2 homozygotes, whether they suffer from UC or CD. In patients with Crohn's ileitis, DAO activity was diminished and correlated with the severity of histologic changes in the mucosa. Patients with very low DAO activities soon demonstrated disease recurrence. In Crohn's colitis or UC, however, enzyme activity was near normal. These observations prompted the suggestion that tissue DAO activity may prove useful in predicting the risk of recurrence or anastomotic complications after resection in CD patients (Thompson et al., 1988a).

Another study demonstrated that DAO activity in the mucosa of inflamed areas of bowel in patients with CD were significantly lower in comparison to healthy tissues; this also applied to DAO throughout the intestinal wall (Schmidt et al., 1990). CD is associated with abnormal intestinal motility. Altered receptor-mediated contraction was disclosed in *ex vivo* examinations.

Small intestinal circular muscles resected from CD patients showed 2.5-fold increase in maximal response to histamine that was blocked by mepyramine, an H_1 antagonist, but was not altered by a DAO inhibitor, thus further indicating DAO deficits in these patients (Vermillion et al., 1993). Despite reduced DAO activity, patients with CD had tissue histamine concentrations within normal ranges (Raithel et al., 1995). This may be due to an enhancement of histamine catabolism rate on the methylation pathway.

In acute phase CD, the urinary excretion of the N-methylhistamine metabolite was increased and showed positive correlation with the degree of inflammation (Winterkamp et al., 2002). Menningen et al. (1990) studied DAO in biopsy materials and in a group of patients with UC, DAO activity levels were about 1/10 of normal; enzyme activity in patients in remission was 5 to 10 times normal, indicating antiproliferative rebound effects in these subjects. The patients displayed also strongly reduced intestinal MAO activity (Menningen et al., 1990). Unlike patients with CD, tissue concentrations of histamine in patients with UC were increased despite higher spontaneous histamine release expressed by mast cells (Raithel et al., 1995).

The importance of histamine excess for disease symptomatology was addressed in rat models of UC. Drugs used included histamine H_1 (ketotifen), H_2 (ranitidine), and dual H_3–H_4 (thioperamide) receptor antagonists (Fogel et al., 2005b) as well as an exogenous (hog kidney) DAO preparation (Fogel and Lewinski, 2006). Treatment started immediately after UC induction and continued for 4 days. Both strategies aimed to eliminate histamine from any kind of interaction—receptor-mediated or occurring directly—positively modified the disease course and improved the animal state as illustrated in Figure 11.1 showing therapeutic effects of DAO preparation. Inflammation markers, plasma ceruloplasmin (CP) and intestinal myeloperoxidase (MPO) activities, and inflammation scores were lowered.

The DAO-treated UC rats showed significantly more DAO activity in the large intestine ($p < 0.05$ versus control; $p < 0.001$ versus UC), indicating that the tissue was able to bind the enzyme (Figure 11.1). The benefits of histamine receptor antagonists may arise from interference with multiple effects of histamine on the immune system (Jutel et al., 2006).

H_1 and H_2 receptors are present on T cells and dendritic cells. H_4 histamine receptors are expressed on mast cells, basophils, and eosinophils (Hofstra et al., 2003). Histamine may affect antigen receptor-mediated immune responses of T and B cells via signals from histamine H_1 receptors. By activation of H_2 receptors present on immature dendritic cells, histamine may change the repertoire of cytokines and chemokines secreted by mature dendritic cells. Increased IL-10 (interleukin-10) production and reduced IL-12 (interleukin-12) secretion by histamine-matured dendritic cells polarized naive CD4+ T cells toward a Th2 phenotype. Th2 cells favor IgE production, and the final result was increased histamine secretion by mast cells. Histamine creates a positive feedback loop that may contribute to the severity of inflammatory and allergic diseases (Jutel et al., 2006).

Histamine is a potent chemoattractant for mast cells, eosinophils, and neutrophils. JNJ 7777120, a selective H_4 antagonist, significantly blocked neutrophil infiltration in zymosan-induced peritonitis in mice (Thurmond et al., 2004), histamine-induced hemotaxis, calcium influx in mouse bone marrow-derived mast cells, and histamine-induced migration of tracheal mast cells (Hofstra et al., 2003), indicating H_4 receptor involvement. Experimental colitis in rats evoked by TNSB (trinitrobenzene sulfonic acid) instillation was shown to respond positively to therapy with selective H_4 receptor antagonists JNJ 10191584 and JNJ 7777120 (Varga et al., 2005).

Ketotifen was reported to inhibit eicosanoid accumulation by cultured colonic mucosa of UC patients (Eliakim et al., 1992) while another H_1 blocker, ebastine, inhibited T cell migration and production of Th2 type cytokines and pro-inflammatory cytokines by cultured lymphocytes from healthy donors (Nori et al., 2003). Jones et al. (1998) observed certain health improvement from ketotifen (4 mg/day) therapy to children with moderate UC. Recent work suggests that in the digestive systems of patients with IBD, upregulation of H_1 and H_2 receptors occurs (Sander et al., 2006).

Positive effects of DAO applications have been demonstrated in several animal models of human diseases: anaphylactic shock (Mondovì et al., 1975), asthma (Masini et al., 2004), growth of Ehrlich ascites cells in mice (Mondovì et al., 1982). and cardiac (Masini et al., 2003), kidney, and intestinal ischemia and/or reperfusion injury (Masini et al., 2007).

A positive protective effect of DAO pretreatment was manifested by less severe hypotension, decreased mortality rate. and normalizations of several examined parameters related to tissue leukocyte infiltration and free radical-mediated tissue injury.

To date, DAO in animal experiments has been administered by intravenous or intraperitoneal injection. In clinical use, at least to treat digestive tract diseases, the best route would be oral administration. The introduction of DAO gene into any probiotic strain characterized by anti-inflammatory properties and the ability to modify the courses of immune-dependent anti-inflammatory processes could serve as a highly efficacious treatment of a variety of inflammatory diseases in humans, e.g., *Lactobacillus casei* strain Shirota (Matsumoto et al., 2005) or genetically modified probiotic strains secreting potent immunomodulatory proteins such as *Lactococcus lactis* (Foligne et al., 2007).

11.5 SEMICARBAZIDE-SENSITIVE AMINE OXIDASE IN INFLAMMATORY BOWEL DISEASE

The two existing forms of SSAO are (1) a tissue-bound form expressed in humans mainly on smooth muscle cells of the vasculature and (2) a plasma-soluble enzyme (Stolen et al., 2004b; O'Sullivan et al., 2004). The soluble form has proven of no value in the assessment of disease activity or severity of inflammation in patients with IBDs (Koutroubakis et al., 2002; Boomsma et al., 2003). The serum concentrations of a soluble VAP-1 (vascular adhesion protein-1) were measured by commercially available ELISA in patients with UC [n = 90], CD [n = 71], and with both diseases [n = 40] in both active and inactive stages and compared to concentrations in healthy controls [n = 93]. No significant differences between IBD patients and controls and no association with disease activity or its location were found (Koutroubakis et al., 2002). When benzylamine was used as a substrate, no differences in plasma SSAO activities of healthy adults and patients with colitis ulecerosa were detected (Boomsma et al., 2003).

Membrane-bound SSAO is an inflammation-inducible endothelial cell adhesion molecule. The importance of its function as an adhesion protein is related to the control of leukocyte traffic *in vivo*, enabling early recruitment of polymorphonuclear leukocytes and later lymphocytes to the area of inflammation (Salmi et al., 2001). SSAO-mediated adhesion and transendothelial migration of leukocytes is warranted by its enzymatic activity and can be abrogated by the enzyme inhibitors (Koskinen et al., 2004). In line with these findings, AOC3 gene knockout mice, characterized by the absence of SSAO, showed greatly reduced influx of leukocytes in non-infectious inflammatory diseases while antimicrobial responses in these animals were not compromised (Stolen et al., 2005). Recent studies using different animal models of inflammation including oxazolone-induced colitis in mice indicate that downregulation of tissue inflammatory markers and reductions in tissue damage may be achieved by therapeutic or prophylactic treatment of animals with an orally active, potent, and selective inhibitor of SSAO known as LJP 1207. The inhibitor is a hydrazine derivative: N-(2-phenyl-allyl)-hydrazine hydrochloride. Treatment with LJP 1207 had a significant effect in preventing weight loss and doubled the survival rates (80 versus 40% survival in the vehicle-treated group). Further results included significant suppression of inflammation, injury, and ulceration scores (Salter-Cid et al., 2005).

A survey of the literature clearly shows both representatives of copper-dependent amine oxidases, DAO and SSAO, serve as focal points for inflammatory processes and are negatively coupled to each other. While DAO blockade may be disadvantageous and intensify inflammatory processes, SSAO inhibition may produce benefits by limiting them.

11.6 DAO ACTIVITY IN PLASMA AND SERUM AS A DIAGNOSTIC TOOL

The diagnostic utility of measuring circulating DAO activity in intestinal pathologies is based on the fact that the intestine is the chief source of the blood enzyme activity (Rokkas et al., 1990a). The distribution of DAO in well-differentiated mature epithelial cells in the apical villi makes it a suitable measure of mucosa maturity and integrity (Luk et al., 1980b) since any injury and loss of villi structure will decrease mucosal and subsequent plasma DAO activity (Forget et al., 1986; Rokkas et al., 1990a, c).

The magnitude of a decrease in plasma enzyme activity is associated with the degree of reduction in mucosal DAO levels (Luk et al., 1980b). However, DAO is normally present in very small amounts in the circulation (Luk et al., 1980b; D'Agostino et al., 1987a, 1988a, b). Radiometric methods may not be sensitive enough to register the enzyme activity decrement. Of great help was a finding that enterocytic DAO may be released into the blood stream by some liberators, among which heparin is especially active. After intravenous, intraperitoneal, or subcutaneous heparin bolus, DAO is released from capillaries of the lamina propria into the peripheral circulation (Shakir et al., 1977; Robinson-White et al., 1985) as first shown by Hansson et al. (1966). Measurement of plasma postheparin DAO (PHD) enhances assay sensitivity, making it a reliable screening test for different enteropathies (celiac disease and other malabsorption syndromes, CD, UC, short bowel syndrome, mucosal injury induced by chemotherapy, and other diseases affecting mucosal architecture). In contrast to the intestine, heparin does not exert a significant effect on DAO activity in other organs (Rokkas et al., 1990a, c).

Plasma and serum DAO activity serves as potential marker of a variety of bowel disorders, including inflammation, atrophy and ischemia and may also be used to evaluate alterations in the integrity and functioning of digestive tract mucosa associated with diseases affecting other systems and organs (Tsujikawa et al., 1999; Guida et al., 2000; Tsunooka et al., 2004; Kamiya et al., 2004). The diagnostic value of measuring plasma DAO activity was confirmed in experimental and clinical studies. Generally, PHD activity is significantly higher than basal activity according to all published reports (Table 11.2). The measurement of PHD activity is recommended as a noninvasive alternative to intestine biopsy (Luk et al., 1980b).

Conversely, the evaluation of tissue (mucosal) DAO activity in biopsy specimens or material obtained via resection also has diagnostic significance (Table 11.1). Despite clinical data clearly confirming the usefulness and accuracy of DAO activity testing, its applicability remains restricted to research. The main reason appears to be the lack of commercially available kits—a decisive factor under clinical conditions.

Tables 11.1 and 11.2 show selected values of DAO activity associated with different disorders involving gastrointestinal function alterations. Obviously, it is difficult to compare the data from different laboratories. The values may be divergent for many reasons; discrepancies may arise from sampling techniques (biopsy, resection, micro and/or macrosampling, dosages, and routes of heparin administration) or the method of enzyme assay. For example, the interference of ceruloplasmin was amply documented (Bieganski et al., 1977) when a spectrophotometric assay with *o*-dianisidine was used as the chromogen to determine DAO activity. The "gold standard" is the radiochemical DAO assay with ^{14}C-putrescine as a substrate, first described by Okuyama and Kobayashi (1961) with subsequent modifications (Kusche et al., 1980; Rokkas et al., 1990a, c; Fogel et al., 2005b). On occasion, the expression of enzyme activity other than in standard units (i.e., absorbance, radioactivity) prevented inclusion of the data.

Concerning plasma PHD activity, the optimal dose of heparin and time of blood sampling remain unclear. Different doses of heparin were used and the blood was collected at different intervals following heparin bolus (often 30 or 60 min). Intravenous administration of 15,000 units of heparin led to DAO activity increases within 5 min, with the maximum value at 60 min, persisting to 120 min (D'Agostino et al., 1988b). Similar DAO fluctuations in normal subjects, i.e., 33 units/ml and 30 to 35 units/ml, were observed after intravenous injection of 3000 and 5000 units, respectively (Thompson et al., 1991; Rokkas et al., 1990c); 1000 units of heparin were insufficient to significantly elevate enzyme activity (Thompson et al., 1988b).

Corazza et al. (1988) reported that increasing the heparin dose from 75 to 150 units/kg body mass (ca 5000 to 10000 units) eliminated the overlap between healthy subjects and patients with celiac disease. In some pathological states, for example in patients with UC accompanied by bleeding, lower doses of heparin are advisable (Rokkas et al., 1990c; Thompson et al., 1991). For some patients with hematological malignancies, heparin administration is not recommended (Tsujikawa et al., 1999).

To facilitate the comparison of the results published by various authors, the data were recalculated and DAO activity uniformly expressed in nanomol per minute per gram of protein with standard deviation, based on the information provided by the authors (e.g., definitions of enzyme units, sample sizes, etc.). Measures were standardized according to the following approximation: 1 g of fresh/wet tissue = 100 mg protein and 1 ml of plasma/serum = 70 mg protein.

ACKNOWLEDGMENT

This work was supported by statutory activity UM503-7106-1.

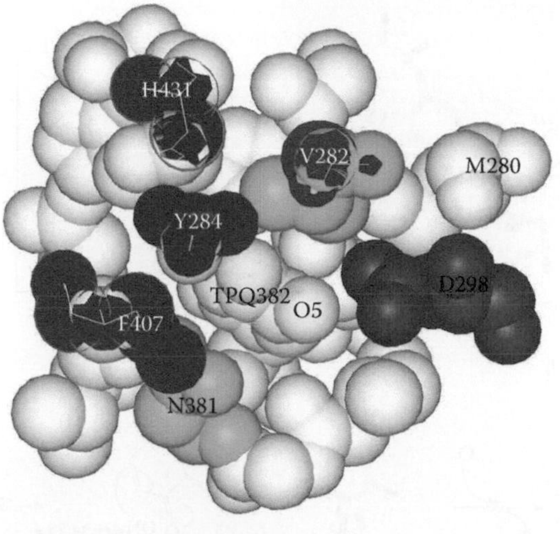

COLOR FIGURE 3.5 Space-filling representation of region surrounding TPQ_{ox} (yellow) in the "active" conformation of wild-type AGAO. V282 and N381 forming a wedge-shape pocket (green). O5 of TPQ is facing D298 (red).

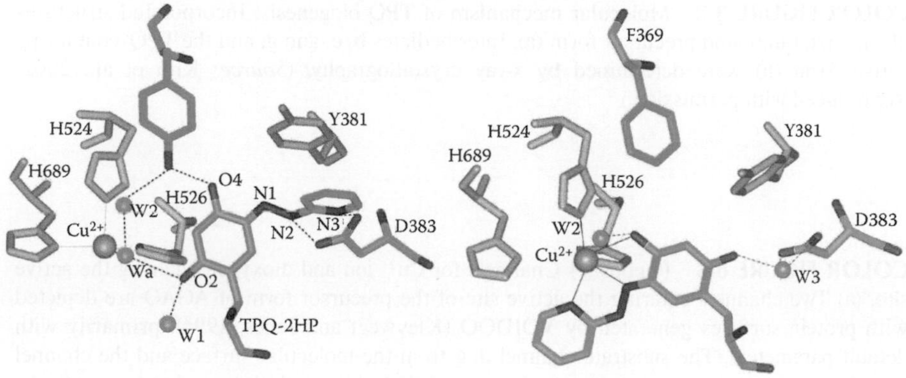

COLOR FIGURE 3.7 (a) Structure of TPQ-2HP adduct in WT ECAO. (b) Structure of TPQ-2HP adduct in Y369F ECAO.

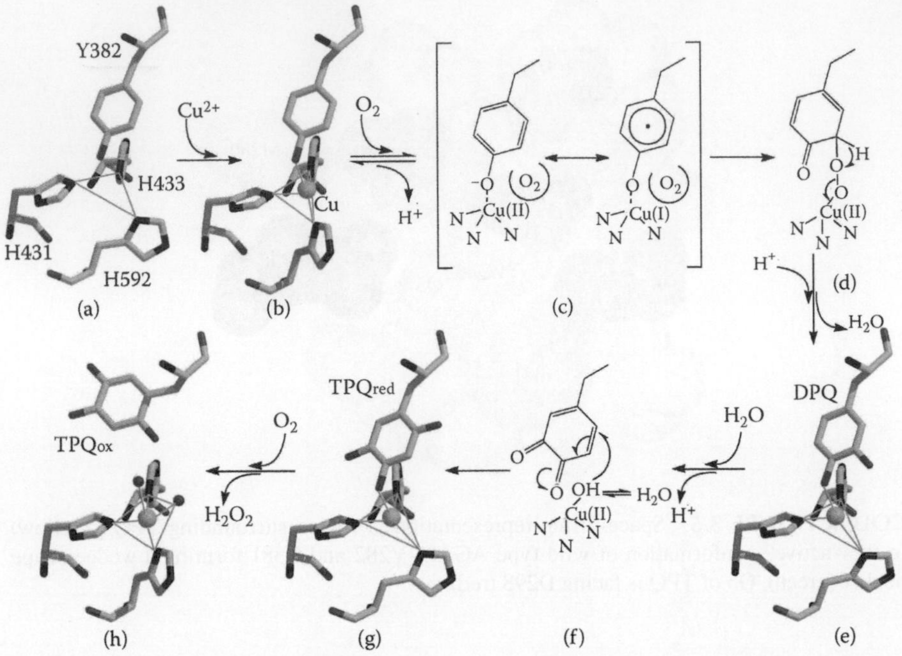

COLOR FIGURE 8.2 Molecular mechanism of TPQ biogenesis. Incorporated structures of the metal-unbound precursor form (a), Intermediates b, e, and g, and the TPQ-containing active form (h) were determined by x-ray crystallography. (*Source:* Kim et al., 2002. Reproduced with permission.)

COLOR FIGURE 8.3 (Opposite) Channels for Cu^{2+} ion and dioxygen entering the active site. (a) Two channels entering the active site of the precursor form of AGAO are depicted with protein surfaces generated by VOIDOO (Kleywegt and Jones, 1994), primarily with default parameters. The substrate channel dug from the molecular surface and the channel associated with the central cavity are drawn with light blue and pink meshes, respectively. The dimeric structure of AGAO was generated by transforming the deposited monomer coordinates (PDB entry code: 1AVK, chain A) into the other subunit according to the crystal symmetry. Although the channels are calculated with the dimer coordinates, only chain A is presented with the surfaces for showing details. The Xe atoms assigned in the Xe-complexed crystal structures of active AGAO, *Pichia pastoris* lysyl oxidase, pea seedling CAO (Duff et al., 2004), and HPAO (Johnson et al., 2007) are overlaid after superimposing each polypeptide chain and are represented by light brown spheres. The funnel-shaped substrate channel is likely used for both substrate amines and the aquated Cu^{2+} ions entering the active site. Conversely, the channel connected to the central cavity formed between the two subunits is one of the two pathways predicted for dioxygen entering into the active site. (b) Enlarged side view of the channels depicted in panel A shown without protein surface. The bold line with an arrowhead represents the predicted pathway of the Cu^{2+} ion and substrate amines. The dotted lines represent the two potential pathways of dioxygen independently proposed from the array of the Xe-binding sites in the hydrophobic core (Johnson et al., 2007) and from the channel connected to the central cavity (Wilce et al., 1997).

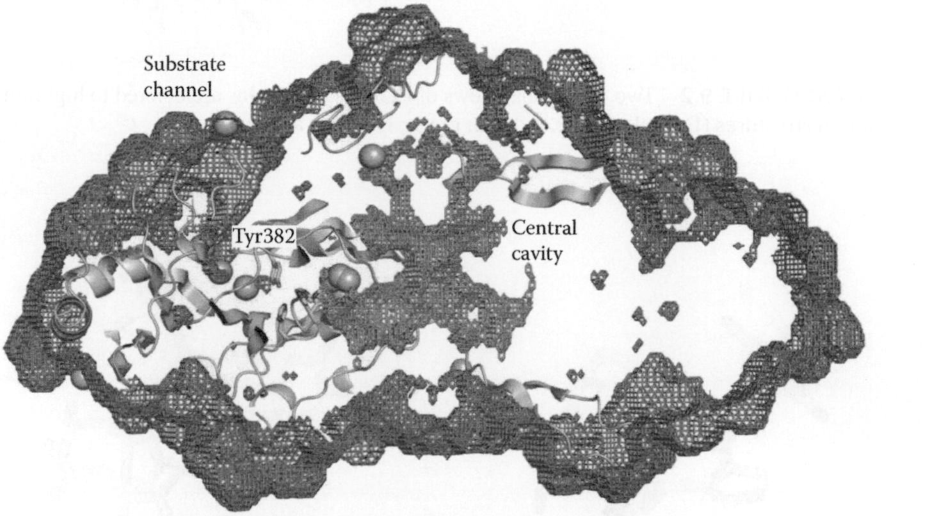
Substrate
channel
Tyr382
Central
cavity

(a)

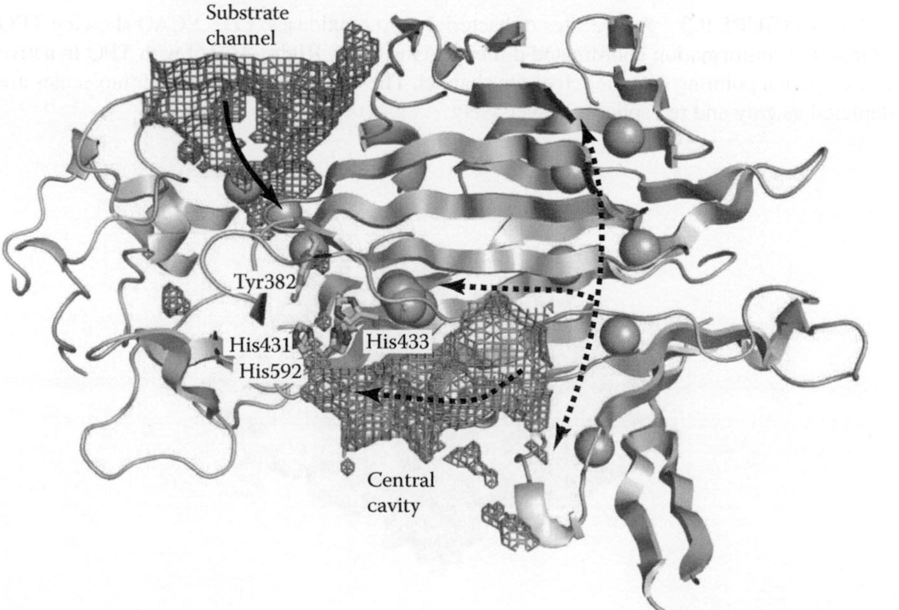
Substrate
channel
Tyr382
His431
His433
His592
Central
cavity

(b)

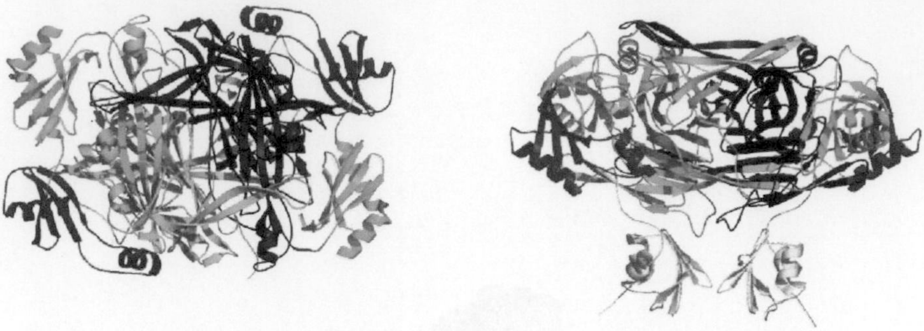

COLOR FIGURE 9.2 Two orthogonal views of ECAO. The chains are colored to highlight domain structures (D1, yellow; D2, red; D3, green; D4, blue and cyan).

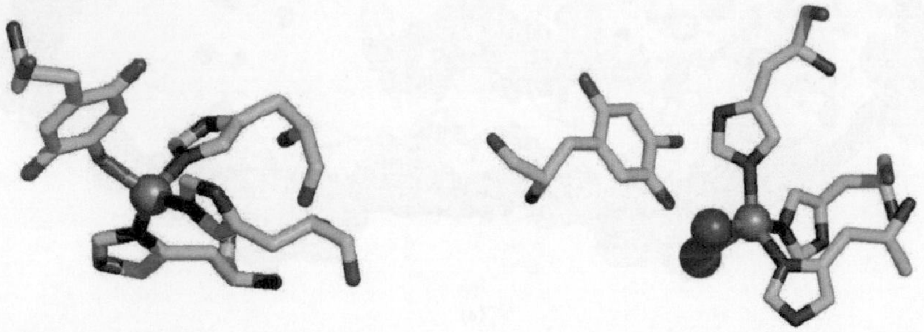

COLOR FIGURE 9.3 Active sites of bacterial amine oxidases. Left: ECAO showing TPQ in inactive conformation coordinated directly to the TPQ. Right: AGAO with TPQ in active conformation pointing into the active site channel. The copper atoms and water molecules are depicted as grey and red spheres, respectively.

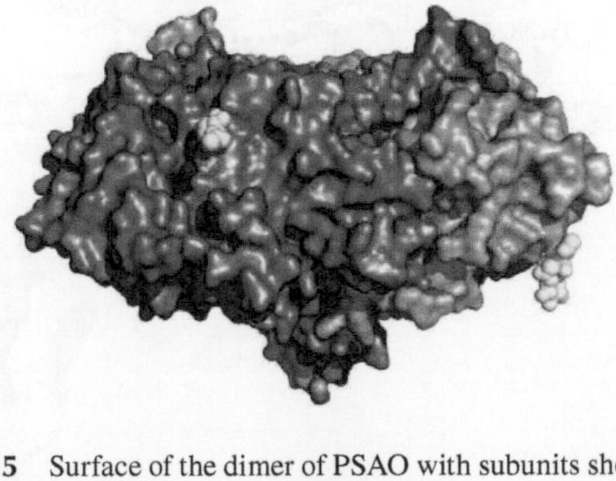

COLOR FIGURE 9.5 Surface of the dimer of PSAO with subunits shown in blue and pink. The attached carbohydrate is represented by yellow space-filling atoms.

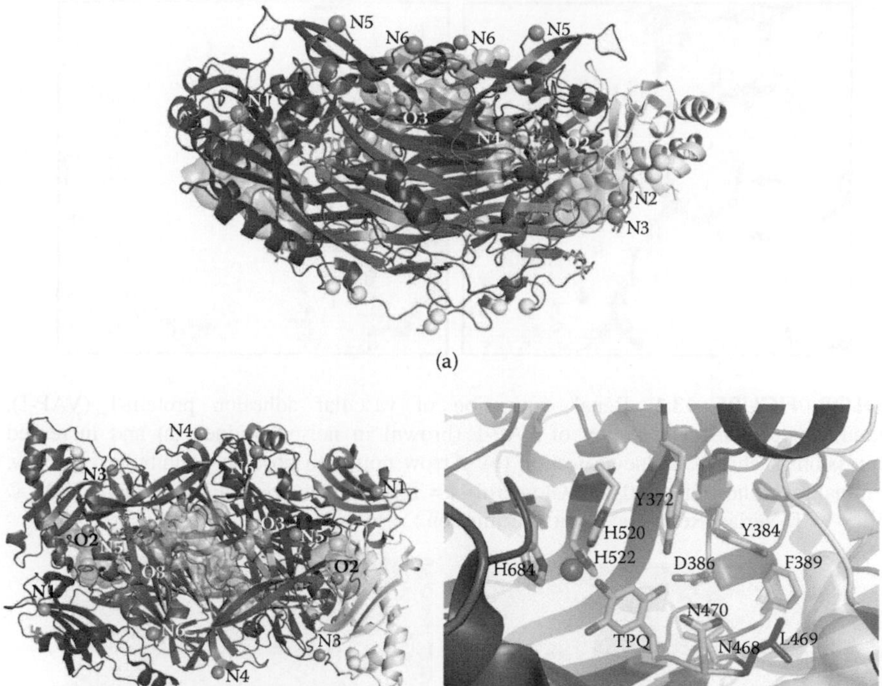

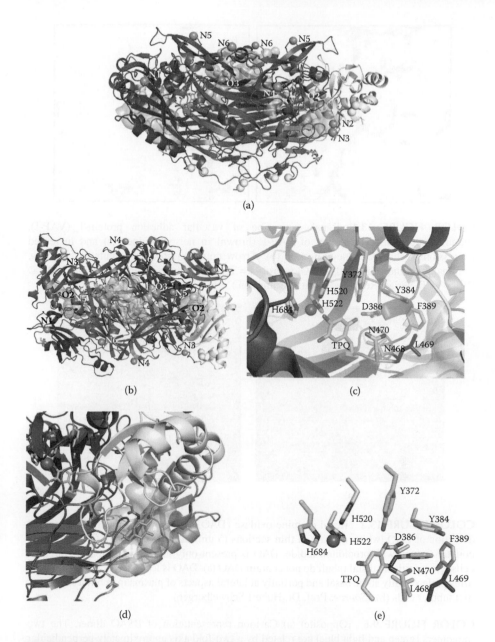

COLOR FIGURE 9.7 Overview of VAP-1 structure. The D2 domains are shown in dark red and cyan, the D3 domains in magenta and light cyan, and the D4 domains in red and blue. The N-glycosylation sites and attached carbohydrates are shown in orange as spheres and as sticks, respectively. The cysteines are shown as yellow spheres. The RGD loop is shown as sticks with yellow carbon atoms. The active site cavity is shown as light violet surface and the central cavity as light pink surface. The active site residues are shown as lines with green carbon atoms and the copper ion as a dark yellow sphere. Leu 469 is shown in red. (a) Side view. (b) View from top toward membrane. (c) Active site. (d) Active site cavity. (e) Active site of 2HP adduct structure. The figures were created with the PyMOL Molecular Graphics System (DeLano Scientific).

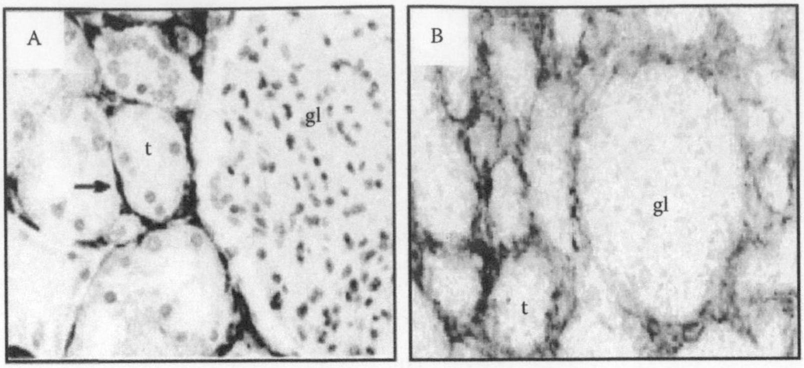

COLOR FIGURE 13.1 Renal expression of vascular adhesion protein-1 (VAP-1). Immunohistochemical detection of VAP-1 (brown) in normal kidney (a) and increased expression in chronic kidney rejection (b). Arrow points to positive peritubular capillary. Original magnification ×200. gl = Glomeruli. t = Tubules. (*Source:* Kurkijarvi et al., 2001. © Wiley-VCH Verlag. Reproduced with permission.)

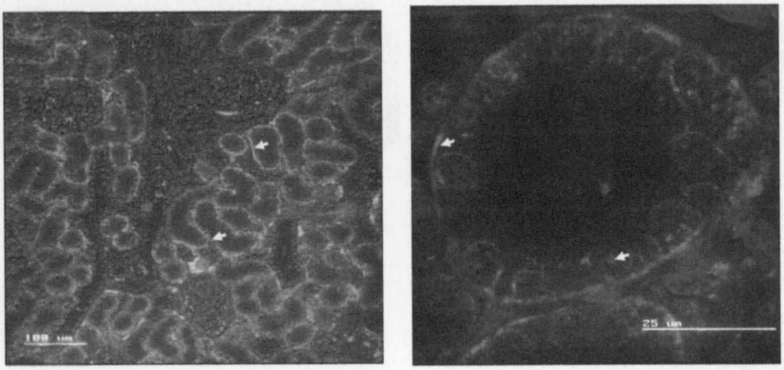

(a) (b)

COLOR FIGURE 13.2 Renal diamine oxidase (DAO) expression. Immunohistochemical detection of DAO (white, arrows) in thin sections (5 μm) of porcine kidney. Cell nuclei are counter-stained with propidium iodide. DAO is present only in proximal tubular epithelial cells; glomeruli and distal tubuli do not contain DAO (a). DAO is localized in vesicular structures preferentially at the basal and partially at lateral aspects of plasma membranes of proximal tubular cells (b). (*Source:* Prof. Dr. Hubert Schwelberger.)

COLOR FIGURE 9.6 (Opposite) (a) Cartoon representation of BSAO dimer. The two monomers (green and light blue) are related by a two-fold axis approximately perpendicular to the plane of the paper. Copper ions are shown as red spheres, secondary cations as yellow spheres, bound sugars as yellow and magenta sticks. (b) Space-filling model of the BSAO dimer seen in the same view as (a). Two arms embracing each other that form one monomer are clearly visible. (c) Internal cavity and access pathways. The volume of the central cavity between monomers is represented in blue, the three small channels and the large mouth connecting the central cavity to the external solvent in orange. Protein atoms are shaded in grey. TPQ (yellow) and the copper ions (green) are represented as CPK models. (From Lunelli et al., 2005. With permission.). (d) BSAO active site. Cu(II) (red) is coordinated by three histidine residues (519, 521 and 683, green) and by a water molecule (light blue). TPQ in the off-Cu position, along with the chain trace of the enzyme, is shown in yellow. (e) Clonidine (green) inhibitor bound to the active site of BSAO (yellow). The phenyl ring of clonidine is stacked with respect to TPQ 470 and Tyr 472 (light blue).

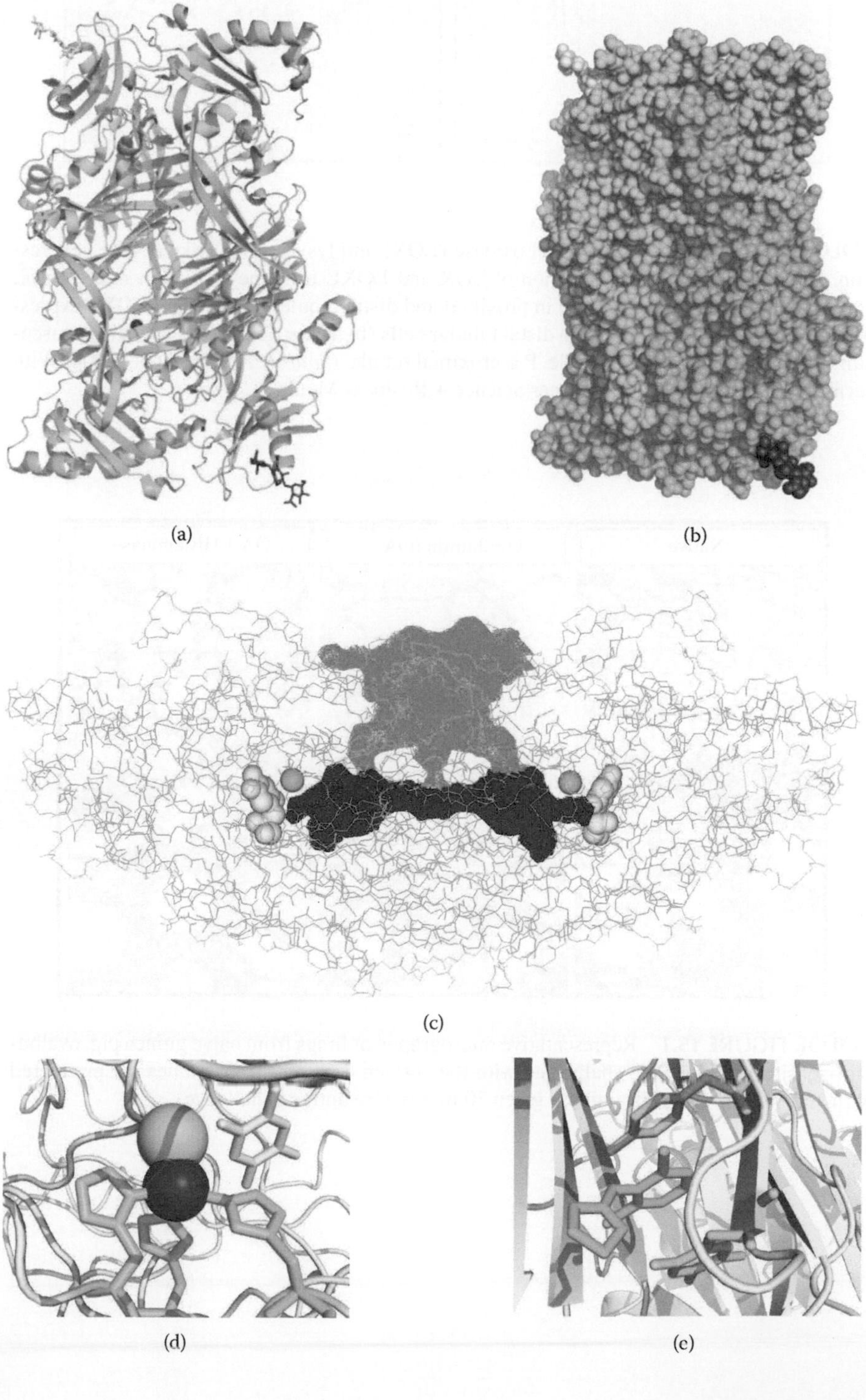
(a)
(b)
(c)
(d)
(e)

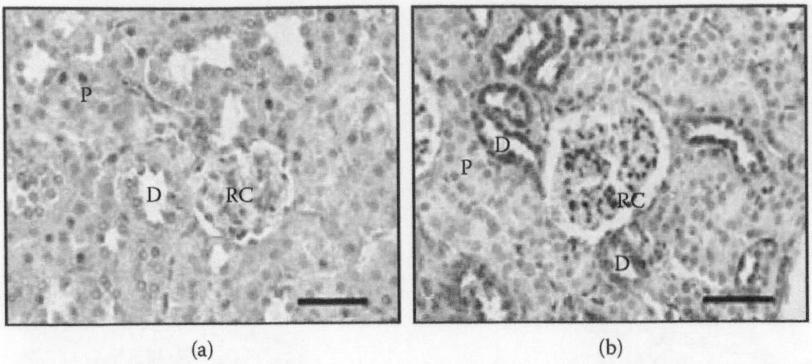

COLOR FIGURE 13.3 Renal lysyl oxidase (LOX) and lysyl oxidase-like (LOXL) expression. Immunohistochemical detection of LOX and LOXL in kidneys of adult mice (black, arrows). LOX is weakly expressed in proximal and distal tubules (a). Stronger LOXL expression is almost exclusively found in distal tubular cells (b). Bar = 50 μm. RC = renal corpusculum (glomerulus). D = distal tubule. P = proximal tubule. (*Source:* Hayashi et al., 2004. With permission from author and Springer Science + Business Media.)

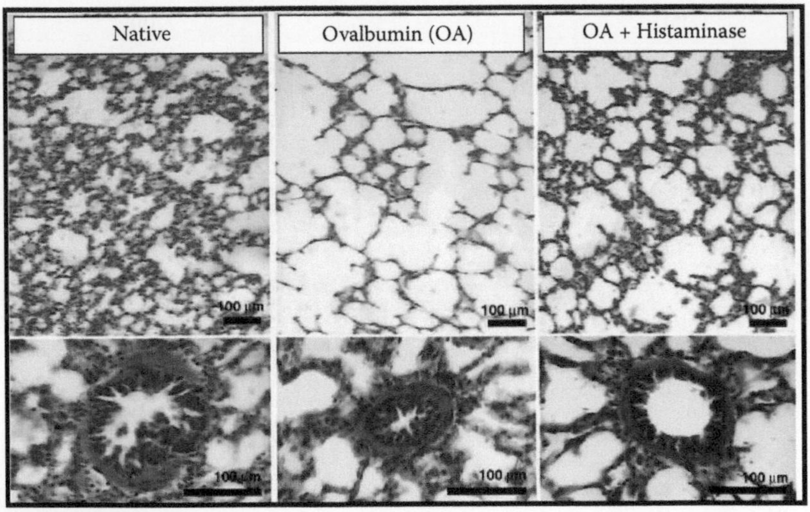

COLOR FIGURE 15.1 Representative micrographs of lungs from naive guinea pig, ovalbumin-sensitized guinea pig challenged with the antigen, and sensitized guinea pig pretreated with aerosolized plant histaminase given 30 min before antigen challenge.

12 Copper Amine Oxidases in Adipose Tissue-Related Disorders

Christian Carpéné

CONTENTS

12.1 INTRODUCTION

The number of obese individuals continues to increase worldwide and obesity is now considered a global epidemic. It was estimated in 2005 that approximately 1.6 billon adults were overweight and more than 400 million adults were obese. At present, about 15% of the world's population is overweight or obese, according to the International Association for the Study of Obesity (website: http://www.iaso.org). Obesity is recognized as a major risk factor for type 2 diabetes, cardiovascular diseases, nonalcoholic fatty liver diseases, and certain types of cancers. The term *adipose-related diseases* as used in this chapter will refer only to obesity and diabetes since the roles of amine oxidases, especially those containing copper (CAOs), in the onset or evolution of cardiovascular disease and cancer are explored in other chapters.

Insulin resistance (namely the deficiency of response to insulin, leading to very blunted glucose uptake into cardiac or skeletal muscles and adipose tissues) is a metabolic disturbance that firmly links excessive white adipose tissue (WAT) with obesity-related complications. Although skeletal muscles quantitatively constitute the major consumers of glucose in the body, adipose tissue plays a qualitative role in glucose handling that is becoming more evident. The association between expanded adipose depots and insulin resistance is likely a complex but tight relationship since numerous clinical investigations indicate that weight gain correlates closely with decreasing insulin sensitivity. In fact, the risk of type 2 diabetes raises progressively with increasing body mass index (BMI, calculated as ratio of weight (kg) on the square of height (m), with obesity defined as BMI > 30). However, the association is reversible, and many human and animal studies have reported that weight loss results in increased sensitivity to insulin and improved glucose tolerance. This chapter will therefore focus on the putative roles of amine oxidase (AO) activities in adipose tissue and their variations observed under diverse obese and diabetic conditions.

12.2 ROLE OF ADIPOSE TISSUE IN OBESITY AND DIABETES

WAT is the body's main depot for energy storage and mobilization. It stores energy in the form of triglycerides (lipogenesis) and supplies it when needed in the form of fatty acids (lipolysis). These processes are exquisitely regulated by integrated neural and endocrine mechanisms. The reference lipogenic hormone is insulin, while adrenaline is known for its β-adrenergic lipolytic capacities. Moreover, WAT is a complex tissue with important endocrine functions since it secretes signalling molecules that impact on multiple target organs.

The best-known product secreted by fat cells and regulating food intake and energy homeostasis is leptin. It has been clearly evidenced that rodents lacking functional leptin signalling are massively obese, as is the case for ob/ob or db/db mice and fatty Zucker rats. Leptin is not the sole adipose-derived product involved in an endocrine loop; WAT secretes numerous factors, from lipids such as fatty acids to peptides or proteins such as the so-called adipokines. However, adipose tissue is not exclusively constituted by adipocytes storing fat in their large lipid droplets; it also contains fibroblasts, endothelial cells, macrophages, lymphocytes, and many other cell types that constitute the stroma–vascular fraction (SVF).

It therefore must be considered that not all the cytokines secreted by WAT are strictly produced by the adipocytes. Many of these adipokines have been recently demonstrated to influence insulin sensitivity both in the WAT and also in many other target organs. Adiponectin is the most potent adipokine able to improve glucose utilization, while resistin should be considered an example of an adipose-derived factor impairing insulin sensitivity. In addition, the influence of these adipokines is crucial on what is currently called the "low-grade inflammation state" of WAT, and vice versa. Inflammation in WAT consists of macrophages and lymphocytes surrounding the adipocytes and exerts deleterious effects on insulin sensitivity. Therefore, no single unique factor is entirely responsible for the link of obesity and insulin resistance, and the interplay between the metabolic and inflammatory pathways in WAT is far from totally explored. This "novel" immuno-endocrine aspect of WAT interferes with its classical metabolic function and adds complexity, but it is useful for further delineating the mechanisms of adipose-related diseases.

The various effects of fatty acids illustrate the intricacies of the immune, endocrine, and metabolic functions of WAT. An excess of circulating free fatty acids (FFAs) was known to cause metabolic disturbances, especially in liver and in skeletal muscle, and to favor hyperglycemia long before the discovery of adipokines. It is conceivable that insulin resistance results from long-term elevation of FFAs because of a competition in most cell types between FFAs and glucose related to oxidative metabolism. Additionally, FFAs alter hepatic glucose output and lipid and lipoprotein handling. FFAs also activate macrophages within adipose tissue and increase inflammation markers. These deleterious actions of FFAs may explain why several antilipolytic drugs have been proposed to treat obesity-associated complications such as metabolic syndrome.

However, these FFA properties appear paradoxical when considering that weight loss, which is associated with a decrease in insulin resistance and in adipose tissue inflammation, is a situation in which adipose tissue lipolysis and subsequent FFA release are stimulated. This discrepancy can be explained by different patterns of adipokines released by adipose tissue, depending on its physiological status (managing excess nutriments or redistributing energy on demand). Indeed, the production of factors with autocrine, paracrine, and endocrine actions is as much regulated as FFA release but the regulating factors and situations have not been completely clarified.

In this context, the CAOs have multiple roles to play in the metabolic, endocrine, and inflammatory aspects of WAT physiopathology, since:

1. Hydrogen peroxide, one end product of AO activity, has been described to mimic insulin effects on adipocytes.
2. The product of the AOC3 gene, namely SSAO/VAP-1 (semicarbazide-sensitive amine oxidase/vascular adhesion protein-1), is highly expressed in adipocytes.
3. Adipose tissue has been proposed as a source of soluble plasma amine oxidase activity.
4. SSAO/VAP-1 is involved in leukocyte extravasation to the sites of inflammation, while low grade inflammation is considered to occur in WAT during obesity.

All these issues will be discussed in this chapter.

12.3 INSULIN-LIKE EFFECTS OF HYDROGEN PEROXIDE AND AMINE OXIDASE SUBSTRATES

12.3.1 Pharmacological Actions of Hydrogen Peroxide

Hydrogen peroxide (H_2O_2) has long been known to mimic a number of insulin actions in several cell types (Table 12.1). In pioneering studies, H_2O_2 seemed to act at an early stage of insulin signalling, as it was found to increase insulin receptor phosphorylation

TABLE 12.1
Some Insulin-Like Effects of H_2O_2

Effect	Cell type	Reference
↑ Glucose transport	Rat brown adipocytes	Czech et al., 1974
	Rat cardiomyocytes	Morin et al., 2002
	Human fat cells	Morin et al., 2001
↑ Glucose oxidation	Rat adipocytes	Czech et al., 1974
		Mukherjee, 1980
↑ Glucose incorporation into lipids	Rat adipocytes	May and de Haën, 1979
		Mukherjee, 1980
↑ Pyruvate dehydrogenase activity	Rat adipocytes	May and de Haën, 1979
↓ Hormone-stimulated lipolysis	Rat adipocytes	Mukherjee, 1980
↓ cAMP production	Rat adipocytes	Mukherjee, 1980
↑ Insulin receptor kinase activity	Rat adipocytes	Hayes and Lockwood, 1987
↑ PI3-K and PKB activities	3T3 mouse fibroblasts	Van der Kaay et al., 1999

Notes: ↑ = increase/ ↓ = decrease; PI3-K = phosphoinositide 3-kinase; PKB = protein kinase B/Akt.

and kinase activity (Hayes and Lockwood, 1987). This activation leads to enhanced tyrosine phosphorylation of intracellular proteins, which is the underlying mechanism by which H_2O_2 exerts its insulin-like effects. However, subsequent studies demonstrated that millimolar concentrations of H_2O_2 enhanced tyrosine phosphorylation of intracellular targets primarily via the inhibition of specific phospho-tyrosine phosphatases (PTPs; Heffetz et al., 1992) that are key regulators of the reversible tyrosine phosphorylations occurring in the insulin signalling pathway. The redox-sensitive cysteine residues present in the catalytic sites of various phosphatases are now recognized targets of hydrogen peroxide, their oxidation leading to enzyme inhibition.

Mukherjee's group (1980) was the first to identify NADPH-oxidase in the plasma membranes of rat adipocytes that generated H_2O_2 in response to insulin. Therefore, H_2O_2 must be considered as both an insulin-mimetic exogenous agent and an intracellular second messenger to insulin action, capable of endogenous production by the stimulated cells.

Kather's group later confirmed the existence of a plasma membrane-bound NADPH-oxidase in mature human adipocytes and in murine 3T3 pre-adipose cells. The membrane-bound NADPH-oxidase is activated upon stimulation of insulin receptor via a coupling to heterotrimeric G proteins and is also regulated by numerous hormones and growth factors (Krieger-Brauer et al., 1997). Despite its unquestionable biological function, it cannot be denied that H_2O_2, as a reactive oxygen species (ROS), may cause oxidative stress. However, its deleterious effects vary, depending on cell type, within a large range of LD_{50} values and with different modes of cell death.

Adipocytes appear peculiarly resistant, since the demonstration on cultured 3T3-L1 adipocytes that a 48-hr exposure to 10 *mM* H_2O_2 reduces cell viability by only 50%. Conversely, a growing body of literature associates high levels of ROS to insulin-resistant states. Using an *in vitro* H_2O_2-generating system, Bashan and coworkers observed insulin-resistance in cultured 3T3-L1 adipocytes resulting from prolonged exposure to oxidative stress (Rudich et al., 1997). However, all these data were obtained with extra-physiological doses of hydrogen peroxide or through use of a long-term and artificial H_2O_2-generating system that has no conceivable physiological counterpart. Although the hydrogen peroxide generated during amine oxidase-dependent oxidation of endogenous or dietary amines probably never reaches such high levels, whether the experimental addition of exogenous substrates of monoamine oxidases (MAOs) or semicarbazide-sensitive amine oxidases (SSAOs) mimics insulin action has been questioned.

The effects of amine oxidase substrates were first studied on the glucose transport activity of adipocytes in the presence and absence of vanadate. A simple explanation for studying the influence of this transition metal is that vanadium is well known to facilitate glucose transport. Moreover, the chemical interaction between H_2O_2 and vanadium rapidly results in peroxidation of the metal ion and subsequent formation of pervanadate. This compound was found to mimic several insulin effects with a much higher potency than vanadate or H_2O_2 alone. In fact, the mechanism of inhibition of PTPs differs among vanadate, H_2O_2, and pervanadate. The latter is an irreversible inhibitor of phosphatases since it involves oxidation of a critical cysteine in the catalytic sites of PTPs.

12.3.2 Insulin-Like Actions of Amine Oxidase Substrates in the Presence of Vanadate

The first two papers reporting the insulin-like action of tyramine (Marti et al., 1998) and benzylamine (Enrique-Tarancon et al., 1998) in rat adipocytes clearly showed that SSAO or MAO activity was able, once activated by its substrate, to generate enough H_2O_2 to interact with the vanadate added to the incubation medium, and therefore promote glucose transporter translocation and glucose uptake. This synergistic action of vanadate and the hydrogen peroxide formed during amine oxidation by amine oxidases present in rat adipocytes is shown for benzylamine (a typical SSAO substrate) in Figure 12.1. Diverse insulin-like actions were then reported for a combination of 0.1 to 1 *mM* SSAO substrates and 0.1 *mM* vanadate (Table 12.2) while these agents lacked clear effects when tested separately.

Thus, the synergism of vanadate and H_2O_2 was reproduced with vanadate and SSAO substrates. Most of these *in vitro* aspects of insulin mimicry were blocked by pretreatment with semicarbazide, diverse CAO inhibitors, or antioxidants. In addition, the other products of amine oxidation (such as benzaldehyde and ammonium in the case of benzylamine) were totally without effect on glucose uptake into rat adipocytes, even when tested at millimolar doses and/or in the presence of vanadate (Iglesias-Osma et al., 2005).

The synergism of benzylamine and vanadate is also observed with vanadyl or tungstate. Most importantly, benzylamine was found to be active *in vivo*, improving the glucose tolerance of conscious and anaesthetized rats, when administered intraperitoneally in combination with a dose of vanadate that would be ineffective alone. The acute antihyperglycemic effect of benzylamine plus vanadate was

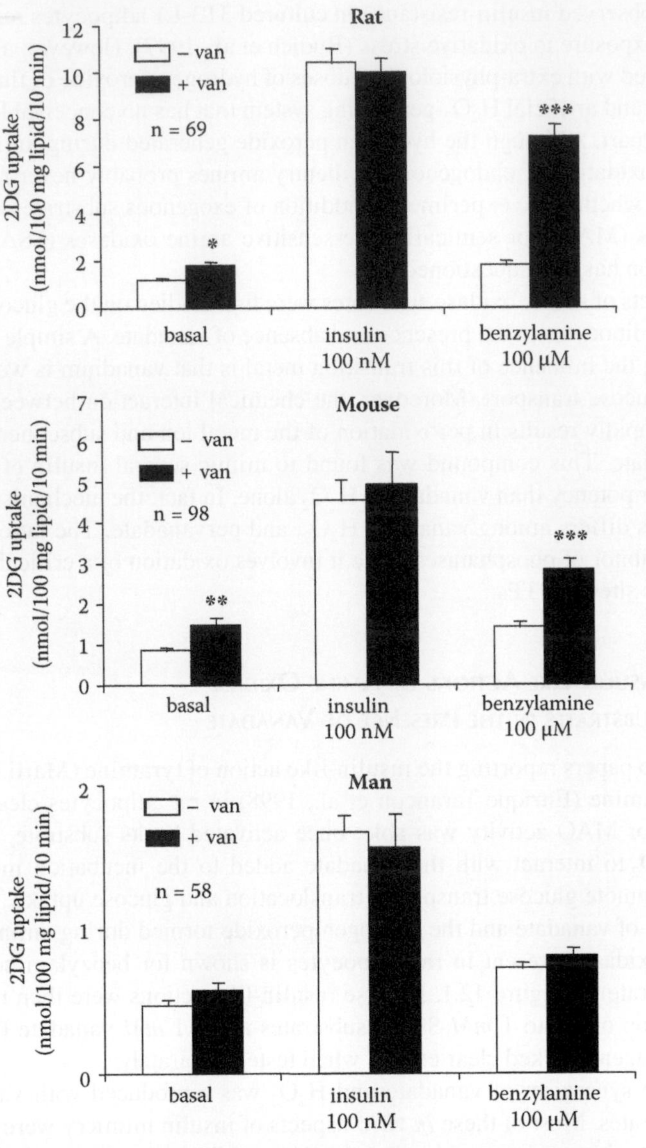

FIGURE 12.1 Influence of vanadate on stimulation of glucose uptake by insulin and benzylamine. Rat (top), mouse (middle), and human (bottom) adipocytes were incubated 45 min without (basal) or with 100 *nM* insulin or 100 *μM* benzylamine in the absence (white columns, –van) or the presence (black columns, +van) of 0.1 *mM* vanadate. 2-Deoxyglucose (2-DG) uptake was then assayed over a 10-min period and expressed on a per-100-mg-cell-lipid basis. The results from control cases of successive independent experiments were compiled in order to show mean values ± SEM of 69, 98, and 58 observations in rats, mice, and humans, respectively. Note the change of Y-axis scale among species. Benzylamine alone significantly activated 2DG-uptake at $p<0.02$ in rats and $p<0.001$ in mice and humans. Differences from corresponding condition without vanadate: * $p<0.02$; ** $p<0.01$; *** $p<0.0001$.

TABLE 12.2
***In vitro* and *in vivo* Insulin-Like Effects of SSAO Substrates plus Vanadate**

Insulin-Like Action	Model	AO Substrate + Vanadate	Reference
Glucose transport stimulation	Rat adipocyte	TYR	Marti et al., 1998
	Rat adipocyte	BZA	Enrique-Tarancón et al., 1998
	Rat adipocyte	MA, O	Enrique-Tarancón et al., 2000
	Mouse adipocyte	BZA, TYR	Yu et al., 2004
	Mouse adipocyte	MA	Yu et al., 2004; Iglesias-Osma et al., 2005
	Murine 3T3 pre-adipocyte	MA	Mercier et al., 2003
Glucose incorporation into lipids	Rat adipocyte	BZA	Carpéné et al., 2006
Lipolysis inhibition	Rat adipocyte	BZA, TYR	Visentin et al., 2003
IRS phosphorylation PI3-K activation GLUT translocation	Murine 3T3 pre-adipocyte	BZA	Enrique-Tarancón et al., 2000
Improvement of glucose tolerance	Normoglycemic rat	BZA, TYR	Marti et al., 2001; Morin et al., 2002
	Type 1 diabetic rat	BZA	Marti et al., 2001
	Type 2 diabetic rat	BZA	Abella et al., 2003
	Diabetic rat or mouse	O	Garcia-Vicente et al., 2007

Notes: BZA = benzylamine; MA = methylamine; TYR = tyramine; O = other amines; IRS = insulin receptor substrates; PI3-K = phosphoinositide 3-kinase; GLUT = glucose transporter.

observed despite the diabetic status of the rodents subjected to glucose challenge: normoglycemic, insulin-deficient (after streptozotocin treatment), or insulin-resistant (Goto-Kakizaki rats or db/db mice; see Table 12.2).

12.3.3 Insulin-Like Actions of Amine Oxidase Substrates in the Absence of Vanadate

An increasing number of benzylamine effects were observed even in the absence of vanadium when studying insulin-like responses in models other than glucose uptake into rat adipocytes. In human adipocytes, the insulin stimulation of glucose transport (which is not as impressive as in rats, Figure 12.1) was partially reproduced by 0.1 to 1 *mM* benzylamine in a manner that was not potentiated by vanadate, whatever the duration or dose of the transition metal added to the medium of isolated fat cells (Morin et al., 2001). This difference appeared to be species-specific since the synergism between vanadate and AO substrates in mouse adipocytes was intermediate to that found in rat and absent in humans (Figure 12.1). However, it did not occur in murine pre-adipocyte lineages (3T3-F442A or 3T3-L1 cell lines; Fontana et al., 2001).

Moreover, the long-term adipogenic effect of benzylamine was not potentiated by vanadate in these latter models. To make short the list of vanadate-dependent and -independent actions of AO substrates, it must be emphasized that lipolysis inhibition, another insulin-like action, was observed with methylamine and benzylamine in human fat cells without need for vanadate (Table 12.3). Indeed, the comparative approach using mature fat cells and cultured pre-adipocytes from different species led to the assessment that vanadate is not necessary to reveal insulin-like effects of amines, although it is sufficient to strongly potentiate them (Figure 12.2).

Conversely, we have no definitive way to improve the maximal response to insulin with benzylamine regarding glucose transport activation or lipolysis inhibition. The *in vitro* responses to amines and to insulin were simply additive when both were tested at submaximal concentrations. As an exception, only the maximal adipogenic effect of insulin on differentiating 3T3-F442A pre-adipocytes was improved by the long-term presence of amines (Fontana et al., 2001). To summarize, it is clear that any amine must be oxidized by AO to generate insulin-like effects (lists in Tables 12.2 and 12.3 are not exhaustive). Hence, insulin mimicry of amines is not limited to fat cells since hydrogen peroxide is also able to activate hexose uptake in skeletal and cardiac muscles (Table 12.1). Tyramine, a substrate of both MAO and SSAO in rodents, is active in these anatomical locations and *in vivo* (Morin et al., 2002).

TABLE 12.3
Short- and Long-Term Insulin-Like Effects of SSAO Substrates

Insulin-Like Action	Model	Substrate	Reference
Glucose transport stimulation	Human adipocyte	BZA, MA	Morin et al., 2001
	Rodent adipocyte	BZA	Iglesias-Osma et al., 2005 Yu et al., 2004
	Murine 3T3 pre-adipocyte	BZA, TYR	Fontana et al., 2001
Lipolysis inhibition	Human adipocyte	BZA, MA	Morin et al., 2001
	Rabbit adipocyte	BZA	Iglesias-Osma et al., 2005
	Mouse adipocyte	BZA, MA	Iglesias-Osma et al., 2005
	Murine 3T3 pre-adipocyte	BZA, TYR	Carpéné et al., 2006
PKB phosphorylation	Murine 3T3 pre-adipocyte	BZA, TYR	Carpéné et al., 2006
Adipogenesis	Murine 3T3 pre-adipocyte	BZA, TYR	Fontana et al., 2001
	Murine 3T3 pre-adipocyte	MA	Mercier et al., 2001
	Human pre-adipocyte	BZA	Bour et al., 2007a
Improvement of hyperglycemia	Rabbit (acute)	BZA, MA	Iglesias-Osma et al., 2005
	Diabetic rat (chronic)	BZA	Soltesz et al., 2007
	Diabetic mouse (acute and chronic)	BZA	Iffiú-Soltész et al., 2007
	SSAO-transgenic mouse (chronic)	MA	Stolen et al., 2004a

Notes: BZA = benzylamine; MA = methylamine; TYR = tyramine; PKB = protein kinase B/Akt. Parentheses indicate acute or chronic *in vivo* administration.

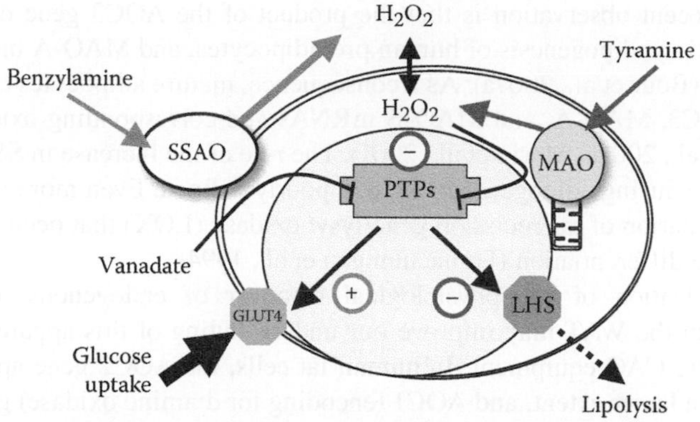

FIGURE 12.2 Hydrogen peroxide-mediated effects of AO substrates in adipose cell; GLUT4 = glucose transporter, insulin-sensitive; HSL= hormone-sensitive lipase; MAO = monoamine oxidase; SSAO = semicarbazide-sensitive amine oxidase.

AO substrates exhibited insulin-like actions even when responses were studied *in vivo*. In conscious rabbits, a sole infusion of benzylamine at 7 μmol/kg (without vanadate) before a glucose challenge was sufficient to lower the hyperglycemic response in a manner sensitive to semicarbazide, reproduced by methylamine, and independent of stimulation of insulin secretion (Iglesias-Osma et al., 2005). Moreover, oral and repeated administration of SSAO substrates to diabetic rodents was successful in improving glucose tolerance and may constitute a novel approach to treat insulin-resistant states (Table 12.3).

Of course, when administered chronically with vanadate, benzylamine more effectively lowers the elevated glucose levels of diverse diabetic rodents than when given alone (Table 12.2). Thus, a combination of AO substrates with low doses of vanadium may improve the bioavailability of the transition metal and create a greater difference between its therapeutic and toxic doses (Mukherjee et al., 2004). The bottom line is to define whether AO substrates can promote *in vivo* insulin-like effects if orally administered to diabetic patients. Can this be achieved at doses lacking adverse effects? Can vanadate be useful for potentiating such putative therapeutic properties without increasing the risk–benefit ratio? These two major concerns must be clarified and whether AO activation can influence adipokine production must be further investigated.

12.4 COPPER-CONTAINING AMINE OXIDASE EXPRESSION IN ADIPOSE TISSUE

12.4.1 Increased MAO and SSAO Expression during Adipocyte Differentiation

An increase of SSAO during adipocyte differentiation in rodents was described as early as 1990 by Raimondi and co-workers. They investigated cells isolated from the stroma–vascular fraction of rat adipose tissue and cultured under adipogenic conditions (Raimondi et al., 1990). A later study detected an impressive increase of SSAO mRNA, protein, and activity during adipocyte differentiation in the murine lineages 3T3-F442A and 3T3-L1 (Moldes et al., 1999).

A more recent observation is that the product of the AOC3 gene dramatically increases during adipogenesis of human pre-adipocytes, and MAO-A increases to a lesser extent (Bour et al., 2007a). As a consequence, mature adipocytes express high levels of AOC3, MAO-A, and MAO-B mRNAs and corresponding oxidase activities (Bour et al., 2007a; Morin et al., 2001). The role of the increase in SSAO/VAP-1 activity found during adipogenesis remains poorly defined. Even more intriguing is the downregulation of ras recission gene/lysyl oxidase (LOX) that occurs early during adipocyte differentiation (Dimaculangan et al., 1994).

A quantification of the physiological turnover of endogenous amines and polyamines in the WAT may improve our understanding of this apparently redundant adipocyte CAO equipment. In human fat cells, the AOC2 gene appears to be expressed to a lesser extent, and AOC1 (encoding for diamine oxidase) gene expression is practically undetectable (Bour et al., 2007a). To date, a relative order of richness in the different AOs may be proposed for human fat cells: AOC3 > MAO-A > MAO-B >> AOC2 > LOX > AOC1 (Figure 12.3).

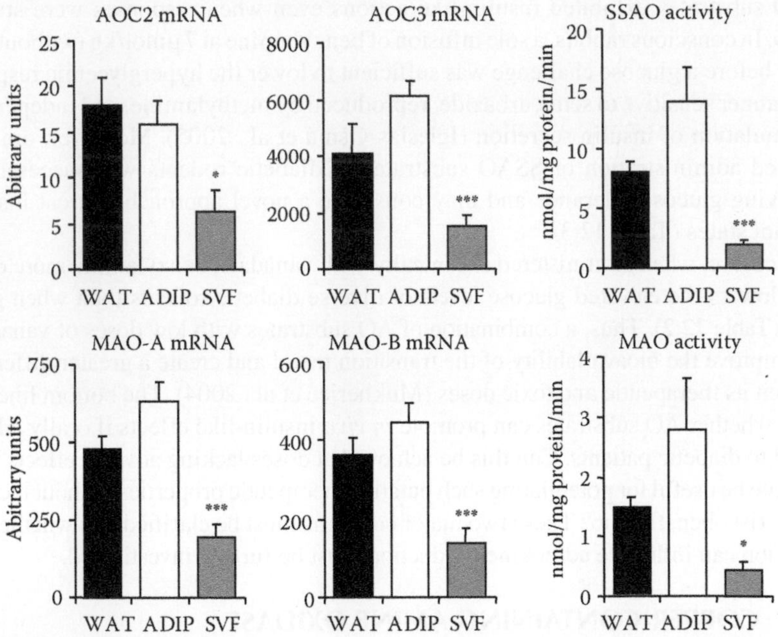

FIGURE 12.3 mRNA abundance of AOC2, AOC3, MAO-A, and MAO-B in stroma–vascular and adipocyte fractions of human subcutaneous adipose depot. Comparison with SSAO and MAO activity. mRNA levels found in whole adipose tissue homogenates (WAT) and corresponding adipocyte (ADIP) or stroma–vascular fractions (SVF) are expressed as arbitrary units, using RT- and 18S mRNA as references. SSAO-dependent oxidation of 0.5 *mM* benzylamine and MAO-dependent oxidation of 0.5 *mM* tyramine are expressed as nmoles of amine oxidized per milligram protein per minute. Mean ± SEM of four to seven determinations. Differences from adipocytes: * $p<0.05$; *** $p<0.001$. (Redrawn from Bour et al., 2007a. *Biochimie* 89: 916. With permission.)

The predominance of AO expression in adipocytes can be visualized in a gene expression database that shows the anatomical distributions of the above genes. It was designed by the Genomics Institute of the Novartis Foundation and is available at www.symatlas.gnf.org.

In human adipose tissue, it is clear that fat cells exhibit higher levels of SSAO/VAP-1 and MAO than any other cells belonging to the stroma–vascular fraction (Figure 12.3). This observation confirms previous findings made on rat brown adipose tissue (Barrand and Fox, 1984). The very high level of SSAO expression in adipocytes was confirmed by independent approaches studying protein expression profiles in different subcellular compartments. SSAO has been detected in plasma membranes, glucose transporter-containing vesicles, caveolae (Souto et al., 2003), and diverse microvesicles of adipocytes (Morris et al., 1997). Nevertheless, the targeted invalidation of the AOC3 gene leads to SSAO-deficient mice that apparently possess functional adipocytes. Insulin responsiveness appears unaltered, and only the response to SSAO substrates is abolished in AOC3-KO mice regarding glucose uptake (Bour et al., 2007b). Unexpected disturbances in cholesterol metabolism and adipokine production are under current investigation (Carpéné et al., unpublished observations) and may help unravel the novel role of AO in adipocyte physiology.

12.4.2 Obesity and Changes in CAO Expression

The soluble form of SSAO/VAP-1 circulating in the plasma has been extensively studied in diverse human pathologies, while reports of amine oxidases present in adipose tissue remain rare. Although plasma amine oxidase is lower in human plasma than in animals (Schwelberger and Feurle, 2008), its easier accessibility led to a substantial number of clinical reports of various pathologies (Boomsma et al., 2003). While plasma SSAO clearly increases in diabetic conditions, as reviewed elsewhere (Yu et al., 2003b; Obata, 2006), its association with the obese phenotype is less consistent.

Therefore, the correlation found between plasma SSAO activity and BMI in type 2 diabetic patients (Mészáros et al., 1999b) may be the consequence of a parallel increase in insulin resistance and adiposity. Elevations of SSAO activity were found in nondiabetic obese patients. Such elevation has thus been proposed as a cardiovascular risk factor (Weiss et al., 2003). However, the association of plasma SSAO and BMI was not confirmed by other studies conducted with nondiabetic (Dullaart et al., 2006) and young obese (Visentin et al., 2004) subjects. Moreover, plasma CAO elevation represents only a 20% increase when compared to healthy control levels.

Numerous other plasma factors predict atherothrombotic, cardiovascular, or type 2 diabetic complications and show a broader range of increase in obesity, e.g., C-reactive protein, tumor necrosis factor-α (TNF-α), interleukin-6, leptin, and insulin. When morbidly obese patients (BMI 38.8) received vertical banded gastroplasty and lost weight after the surgery (BMI 30.8), significant improvements were noted in several phenotypes associated with metabolic syndrome (reduced blood pressure, increased glucose tolerance, reduced plasma insulin, etc), but no significant reduction of mean plasma SSAO/VAP-1 levels. Only a weak tendency to correlate individual changes in plasma SSAO/VAP-1 with decreases of anthropometric measurements of adiposity suggested that obesity and circulating SSAO/VAP-1 may be associated (Li et al., 2005).

A more substantial increase in plasma CAO than those reported in the above studies should be expected in obese subjects, related to the dramatic increase in TNFα in obesity and *in vitro* TNF-α stimulation of SSAO/VAP-1 shedding from adipocytes (Abella et al., 2004). Hence, obesity is not a condition associated with a clear-cut increase in plasma CAO even though adipose tissue has been proposed as a source of soluble SSAO, essentially in diabetic states (Stolen et al., 2004b).

Since plasma AO is involved in the oxidation of circulating methylamine and its subsequent deleterious effects on vasculature (Yu et al., 2003b), attention must be paid to these two circulating parameters, especially when obesity is complicated by cardiovascular injury. Blood methylamine has been proposed to vary around 1 μM, but plasma SSAO activity remains at least 1000-fold lower than the membrane-bound form of WAT (see below). Thus, perivascular fat may constitute a source of vasculotoxic aldehydes, if any, and its richness in SSAO/VAP-1 deserves to be studied.

In subcutaneous adipose tissue, no change in SSAO/VAP-1 activity was found when comparing young obese subjects with age-matched controls. However, in WAT, SSAO oxidized more than 7 nmol benzylamine/mg/min, while plasma SSAO activity barely reached 0.003 nmol benzylamine oxidized/mg protein/min (equivalent to ~20 nmol/mL/hr; Visentin et al., 2004). If we consider WAT masses in obese subjects, the adipose tissue-bound SSAO appears quantitatively more substantial than that circulating in blood. Therefore, we must ask what endogenous substrates reach the adipose tissue-bound SSAO/VAP-1, and whether they change in obesity.

A significant reduction of adipose MAO activity was observed in subcutaneous WAT of young obese subjects while blood MAO was hardly detectable (Visentin et al., 2004). The functions and variations of AOs in WAT are still unknown and need further investigation to delineate their multifunctional roles in the physiopathology of obesity. Since visceral intra-abdominal adipose depots are especially involved in the serious complications of massively obese patients, such information about these tissues is worth obtaining.

Our limited knowledge about the regulation of CAOs in human obesity is slightly enhanced by animal studies showing that SSAO/VAP-1 activity is always elevated in adipose tissue whatever the obese model considered. Indeed, two distinct causes of increase coexist: one is upregulated expression of genes encoding for AOs and the other merely depends on the adipose tissue extension that characterizes obesity.

As a consequence of its hypertrophied mass, the adipose tissue of an obese individual contains much more SSAO/VAP-1 activity than tissue of a lean control, even without an increase in enzyme expression at cellular level. Although obvious, this quantitative aspect must be considered for a tissue that can increase its mass 10-fold in obesity. There are no equivalent organs except growing tumors and placenta. An obesity model of increased SSAO/VAP-1 expression is required before we can focus on another aspect of AO research: proportional increases in expanding adipose tissue.

MAO and SSAO activities have been compared in tissue biopsies from dogs fed a high-fat diet (HFD) in a longitudinal study aiming at describing the cardiovascular and metabolic parameters at the onset of obesity (Wanecq et al., 2006). After 9 weeks of high-fat feeding, increases were noted in both MAO and SSAO activities in the omentum, an intra-abdominal fat depot, when expressed as nmoles of substrate oxidized per milligram of protein per minute (Figure 12.4). Such increased maximal velocity of amine oxidation may result from an increased abundance of

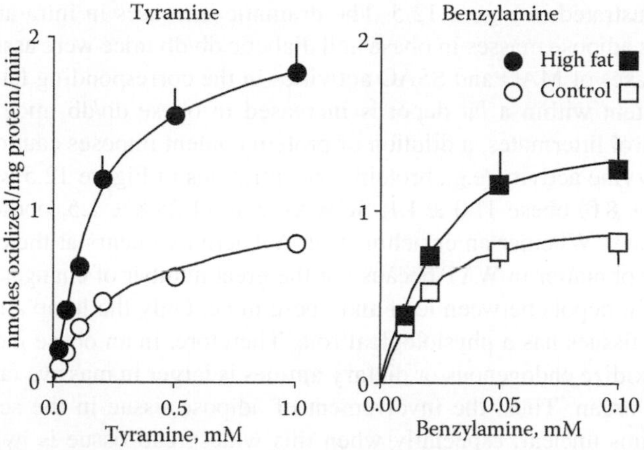

FIGURE 12.4 Dose-dependent amine oxidation in omental adipose tissue from control and high-fat fed dogs. Tyramine or benzylamine oxidation was conducted 30 min after pre-incubation without inhibitor and with the indicated substrate dose. MAO activity (sensitive to 0.1 *mM* pargyline) accounted for 32 to 45% of tyramine oxidation (circles). SSAO activity (sensitive to 1 *mM* semicarbazide) accounted for 92 to 98% of benzylamine oxidation (squares) in both control (open symbols) and obese dogs (closed symbols). Results are expressed as nanomoles amine oxidized per milligram protein per minute. Mean ± SEM of five experiments. (*Source:* Wanecq et al., 2006. *J Physiol Biochem* 62: 113. With permission.)

SSAO/VAP-1 protein. This was confirmed by the Northern blot detection of SSAO mRNAs using a cDNA probe for the human AOC3 gene. The twofold increase in SSAO mRNAs found in omental adipose tissue from HFD dogs likely indicates upregulated SSAO expression.

No changes of MAO and SSAO activities were found in other tissues such as liver, aorta, and surprisingly, subcutaneous WAT, indicating that the observed variations in omental WAT were depot-specific. Conversely, the plasma AO activity decreased from 1.6 to 1.0 nmoles of benzylamine oxidized/mg protein/hr in HFD dogs (again, plasma activity was lower than the adipose tissue-bound form; cf. Figure 12.4). This increase in SSAO expression in omental WAT fits with previous reports proposing SSAO as an adipogenic marker (Moldes et al., 1999). Nonetheless, this model of nutritional obesity does not seem predictive of the disturbances observed in human obesity.

Transgenic mice were used to compare two strains: one that resists fat by controlling its body weight gain and another transgenic line that is obesity-prone. This allowed differentiation of the effect of HFD from the effect of fat mass enlargement. As expected, the former mice resisted the HFD while the latter became obese. No significant changes in SSAO or MAO activities were found in the visceral or subcutaneous WAT of both strains after HFD when expressing activity per milligram protein. However, when considering the overall capacity contained in the fat depots, increases in MAO and SSAO activity were found only in the enlarged WAT of HFD-induced obese mice (Visentin et al., 2005a). Thus, fat store enlargement is accompanied by an overall increase of adipose AO activity, even in the absence of enzyme upregulation.

This is illustrated in Figure 12.5. The dramatic increases in intra-abdominal or subcutaneous adipose masses in obese and diabetic db/db mice were associated with similar increases of MAO and SSAO activities in the corresponding fat pads. Since the lipid content within a fat depot is increased in obese db/db mice when compared to control littermates, a dilution of protein content imposes caution when one compares enzyme activity (e.g., protein concentrations in Figure 12.5 were INWAT control 23.8 ± 8.0, obese 11.0 ± 1.1; SCWAT control 28.8 ± 2.5, obese 11.5 ± 2.7 mg/g wet tissue). Whether an enrichment in AO activity occurs at the cellular level almost does not matter in WAT because of the great number of changes in the compositions of fat depots between lean and obese mice. Only the lump amount of AO restrained in tissues has a physiological role. Therefore, in an obese individual, the capacity to oxidize endogenous or dietary amines is larger in massive fat stores than in any other organ. Thus, the involvement of adipose tissue in the scavenging of amines remains unclear, especially when this widespread tissue is hypertrophied.

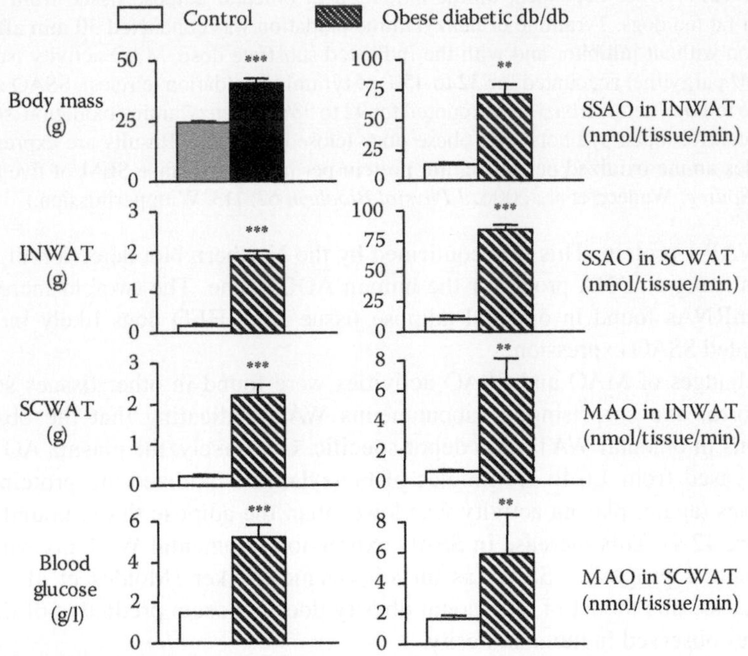

FIGURE 12.5 Increased amine oxidase content in fat stores of diabetic and obese db/db mice. Certain biological parameters of 3-month old db/db mice (hatched columns) and their littermates db/+ (white columns) are reported at left. The intra-abdominal and subcutaneous white adipose tissues (INWAT and SCWAT) were weighed and stored at −80°C. SSAO and MAO activities were determined on thawed samples by measuring the oxidation of 0.1 mM benzylamine sensitive to semicarbazide and that of 0.5 mM tyramine sensitive to pargyline, respectively. Amine oxidase activities (right) are expressed as nanomoles of amine oxidized per total fat depot per minute. Mean ± SEM of four determinations. Differences from db/+ (control) mice: ** $p<0.01$; *** $p<0.001$. No significant difference found between corresponding conditions in INWAT and SCWAT.

Whether this proportional increase in amine degradation leads to an adaptation may be studied in Zucker obese rats since decreases in SSAO expression in their WAT were reported (Moldes et al., 1999). Whether this decreased adipose SSAO expression compensates for the massive WAT extension remains speculative since no parallel decrease in adipose MAO activity was reported. Alternatively, the diminished adipose SSAO activities in obese rats may result from TNF-α-induced downregulation, as evidenced in pre-adipose cell cultures (Mercier et al., 2003).

12.4.3 Pharmacological Interventions on Copper-Containing Amine Oxidases in Obesity

Although our knowledge of AO regulation in obesity remains limited, several experiments have been performed to block MAO and SSAO in obese models. Testing SSAO/VAP-1 inhibitors in obesity involves at least two rationales:

1. Methylamine and benzylamine are known to exert central hypophagic effects that are dependent on neuronal potassium channels and increase when their degradation by SSAO is prevented by inhibitors such as semicarbazide or aminoguanidine (Cioni et al., 2006). As a consequence, sustained SSAO inhibition may reduce food intake and treat obesity.
2. AO substrates activate glucose transport and inhibit lipolysis in isolated fat cells when added at submillimolar doses (Enrique-Tarancón et al., 2000; Visentin et al., 2003). Prolonged AO blockade would cause decreased lipogenesis and increased lipolysis, and may limit WAT enlargement only if sufficient alimentary amines spontaneously reach WAT in obese and hyperphagic models and exert insulin mimicry.

Most treatments of obese rodents performed to date with AO inhibitors have resulted in weight losses that may be explained partially the working hypotheses cited above. Reductions of weight gain were observed in obese rats treated with aminoguanidine (Prévot et al., 2007). Semicarbazide alone (Mercier et al., 2007) or in combination with pargyline (Carpéné et al., 2007) also reduced body weight gain and limited fat deposition in rats (Figure 12.6). Again, chronic administration of the SSAO inhibitor, (E)-2-(4-fluorophenethyl)-3-fluoroallylamine, limited body weight gains of obese mice (Yu et al., 2004).

Unfortunately, SSAO inhibitors cause moderate and long-lasting hyperglycemia (Yu et al., 2004). Conversely, they reduce multiple signs of atherogenesis (Yu et al., 2003b). Various concerns about their prolonged use have been raised: compensatory increases of MAO activity (Holt and Callingham, 1995), vasculotoxicity, and especially arterial elasticity alteration (Langford et al., 1999 and 2002; Mercier et al., 2007). Whether these adverse effects may be limited with novel AO inhibitors remains to be seen. Avoiding inhibition of LOX appears to be an encouraging challenge. Finally, we still have no indication of the reversion of low-grade inflammation sites of WAT after AO inhibitor treatment although such compounds have proven beneficial in treating experimental inflammation (Yu et al., 2006; Noda et al., 2008). Further pharmacological screening and cardiovascular, behavioral, and endocrino-metabolic investigations are needed before AO inhibitors are used to mitigate obesity.

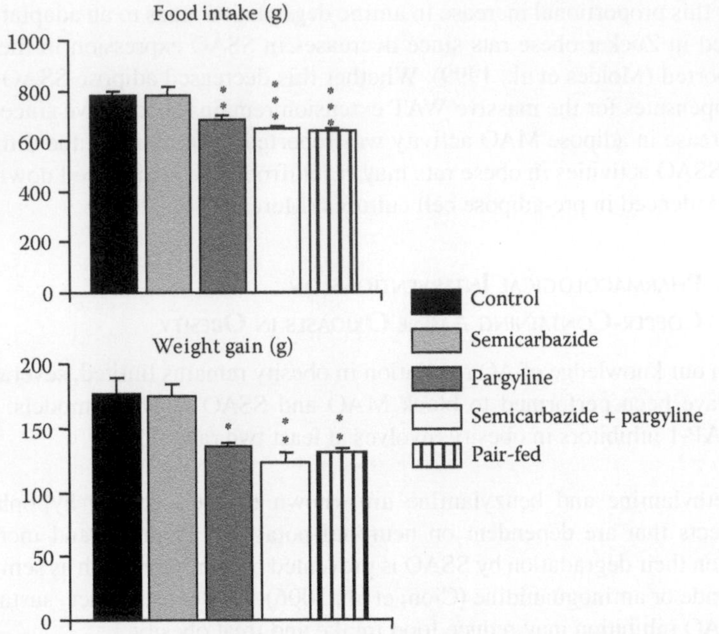

FIGURE 12.6 Effect of 4-week treatment with amine oxidase inhibitors on food intake and weight gain in male obese Zucker rats. Mean ± SEM, five rats per group: controls (dark columns), treated with semicarbazide (300 μmol/kg/day, shaded), pargyline (60 μmol/kg/day, grey) or both (P + S, white). Pair-fed rats (hatched) received the same daily amount of chow as P + S-treated group. Differences from controls: * $p<0.05$; ** $p<0.01$; *** $p<0.001$.

12.5 CAOS AND DIABETES

Plasma AO increases are known to occur in diabetes and are reviewed elsewhere (Boomsma et al., 2003; Obata, 2006) and will not be discussed in this chapter. Instead, it is necessary to mention that increased plasma SSAO is not hazardous on its own. It is the excessive deamination of methylamine and aminoacetone that increases the risk of endothelial injury. Actually, the combination of increased enzyme and substrate supply constitutes the risk factor. Diabetes mellitus involves an increase in plasma AO and also increases of methylamine and aminoacetone levels. As a consequence, formaldehyde, methylglyoxal, protein cross-linkage, and advanced glycation end products are elevated in type 1 and type 2 diabetes and participate in vascular injury, as reviewed by Yu and associates (2003b).

In addition to the reports of increased plasma SSAO activities in diabetic patients (Yu et al., 2003b), increases of soluble SSAO in the plasma of rats and mice rendered diabetic by streptozotocin injection (Göktürk et al., 2004; Visentin et al., 2005b; Somfai et al., 2006) are widely recognized. Conversely, reports of adipose tissue-bound SSAO in streptozotocin-treated rodents are somewhat conflicting (Conforti et al., 1995; Nordquist et al., 2002; Göktürk et al., 2004; Visentin et al., 2005b). Decreases of SSAO activity seem to occur in WAT, at least as a consequence of the dramatic reduction of intra-abdominal and subcutaneous fat pads induced by streptozotocin treatment in rats and mice. Nevertheless, the level of amine oxidation is not totally abolished in WAT, since treatment with benzylamine and vanadate tends to normalize the hyperglycemia and drinking consumption and also favors adipose tissue deposition (Marti et al., 2001). This latter observation and additional in vitro data (Fontana et al., 2001; Mercier et al., 2001; Visentin et al., 2005b) definitely show that SSAO substrates can activate lipogenesis in adipocytes even in the absence of insulin. In keeping with this finding, the benzylamine and tyramine SSAO substrates can decrease hyperglycemia in type 1 diabetic rodents (Marti et al., 2001; Visentin et al., 2005b; Garcia-Vicente et al., 2007; Soltesz et al., 2007). Conversely, SSAO inhibitors, proposed to reduce diabetic vascular complications by preventing plasma methylamine oxidation (Yu et al., 2003b), have never been proven antihyperglycemic (Yu et al., 2004; Visentin et al., 2005b).

A dilemma arises when considering AO-mediated amine oxidation instrumental for treating diabetes. On one hand, it appears beneficial to stimulate SSAO or MAO activity in adipocytes to produce hydrogen peroxide and take advantage of the insulin-mimicking properties described above. On the other hand, the activation of plasma or endothelial SSAO/VAP-1 in the vessels is hazardous because it increases local production of cytotoxic ROS, aldehydes, and ammonium that contribute to the cardiovascular and renal complications of diabetes (Yu et al., 2003b).

The chronic administration of methylamine in mice overexpressing SSAO/VAP-1 in endothelial cells, is a perfect example of this dual aspect of CAO-mediated oxidation of exogenous amines (Stolen et al., 2004a). Such treatment improves glucose tolerance but increases vascular injury. To reduce the risk-to-benefit ratio, it must be proven definitively that the oxidation of methylamine or aminoacetone alone is deleterious for vessels. The formaldehyde generated during methylamine oxidation and the methylglyoxal formed by aminoacetone oxidation have no equivalents in terms of toxicity. Benzaldehyde (from benzylamine oxidation) is likely less cytotoxic than the two former reactive products. Therefore, it appears that methylamine oxidation that surely must be reduced in diabetes must be limited by a mechanism other than SSAO/VAP-1 inhibition alone, since among all the pharmacological inhibitors known to date, no single molecule acts selectively on the soluble enzyme without inhibiting tissue-bound forms. As described above, chronic inhibition of SSAO/VAP-1 may alter arterial stiffness in arterial beds and limit putative insulin-like actions of endogenous or exogenous substrates in adipose depots. Limiting exaggerated inflammatory responses in diverse inflammation models (Yu et al., 2006; Noda et al., 2008) is one of the rare beneficial effects of pharmacological SSAO inhibitors reported to date. This is due to the blockade of leukocyte adhesion by endothelial SSAO/VAP-1 (Jalkanen et al., 2007).

An alternative to enzyme inhibition is to chronically administer an exogenous AO substrate that will compete for methylamine oxidation in blood. Ideally, the exogenous substrate will generate products less reactive than formaldehyde or methylglyoxal and therefore limit vascular injury and at the same time be oxidized in the WAT and promote glucose utilization via hydrogen peroxide formation. One problem is that the SSAO/VAP-1 activity of skeletal muscles is less than that of WAT; as a result, SSAO substrates will facilitate fat deposition and body weight gain, as is the case for other antidiabetic drugs such as thiazolidine diones. This is even more problematic in the case of type 2 diabetes—generally improved by weight loss.

The use of a combination of vanadium plus amine alleviates this concern since the pervanadate formed in WAT can act at a distance in skeletal muscles, making them participate in the increase in peripheral glucose utilization (Abella et al., 2003). As for mitigating obesity with AO inhibitors, further evidence supporting the lack of serious adverse effects of AO substrates is required in addition to the studies of treatments of insulin-resistant diseases with these substrates performed to date. Because of the accumulating functions of CAOs in diverse anatomic locations, the influence of prolonged administration of AO substrates on endocrine-metabolic, hepatic, cardiovascular, renal, inflammatory, and behavioral areas must be determined. These issues and the possible oral use of benzylamine and less toxic SSAO substrates should be the topics of ongoing research.

12.6 CONCLUDING REMARKS

Research of the past decade has increased our understanding of the role adipose tissue plays in obesity, diabetes, and cardiovascular diseases. Adipose tissues do not merely serve as sinks for excess calories; they are highly active metabolic and endocrine organs. Adipocytes secrete a variety of adipokines and express high levels of SSAO/VAP-1. They also exert MAO-A activity that has attracted little attention. The genes encoding MAO-B and CAO such as diamine oxidase and lysyl oxidase appear to be less expressed.

Whatever their levels of expression, the roles of all forms of AOs have not been completely determined. The levels and turnover mechanisms of endogenous AO substrates within adipose tissues are poorly defined. Mounting evidence suggests that adipose tissue dysfunction in obesity plays a prominent role in the progression of insulin resistance. However, the changes in AO expression and activity in obesity and insulin resistance related to adipose tissue are not well defined. While SSAO/VAP-1 is not required for normal insulin responsiveness in adipocytes, its activation by exogenous substrates may promote glucose utilization under both in vitro and in vivo conditions. This ability has been used to successfully treat experimental diabetes with benzylamine, but future experiments must verify whether oxidative stress, weight gain, and vascular damage occur as side effects. Conversely, pharmacological inhibitors of SSAO appear to limit fat deposition and inflammation, seem ineffective in lowering hyperglycemia, and exert deleterious effects on arterial walls. Circumventing the arterial wall effect will require selective SSAO/VAP-1 inhibitors that do not interfere with LOX and consideration of the large proportion of adipose SSAO/VAP-1 in the body due to its elevated expression level and levels of fat stores.

ACKNOWLEDGMENTS

The author would like to thank members of the plastic surgery department of the Rangueil Hospital in Toulouse, France, for providing adipose samples, and of the Zootechnie de l'IFR31 Rangueil for animal care.

13 Copper Amine Oxidases in Adhesion Molecules in Renal Pathology and in Alzheimer's Disease and VAP-1 in Leukocyte Migration

Peter Boor, Mercedes Unzeta, Marko Salmi, and Sirpa Jalkanen

CONTENTS

13.1 COPPER AMINE OXIDASE AND KIDNEYS

Of the known human copper-containing amine oxidases (CAOs), three are involved in renal physiology and pathology: vascular adhesion protein-1 (VAP-1), diamine oxidase (DAO), and lysyl oxidase (LOX). Furthermore, the activity of plasma semicarbazide-sensitive amine oxidase (SSAO) is altered in renal patients.

13.1.1 VAP-1

Similarly to other organs, expression of VAP-1 in human kidneys is localized to endothelial cells, pericytes, and smooth muscle cells of larger vessels. VAP-1 is found only in peritubular and periglomerular endothelium, but not in highly specialized glomerular endothelial cells (Figure 13.1; Salmi et al., 1993; Kurkijarvi et al., 2001). In developing human kidneys, VAP-1 expression is first observed in vascular smooth muscle cells at embryonic week 13. Ten weeks later, the first peritubular capillaries show VAP-1-positive staining (Salmi and Jalkanen, 2006). Glomerular mesangial cells and podocytes as well as tubular cells are VAP-1-negative in both developing and adult kidneys (Kurkijarvi et al., 2001; Salmi and Jalkanen, 2006).

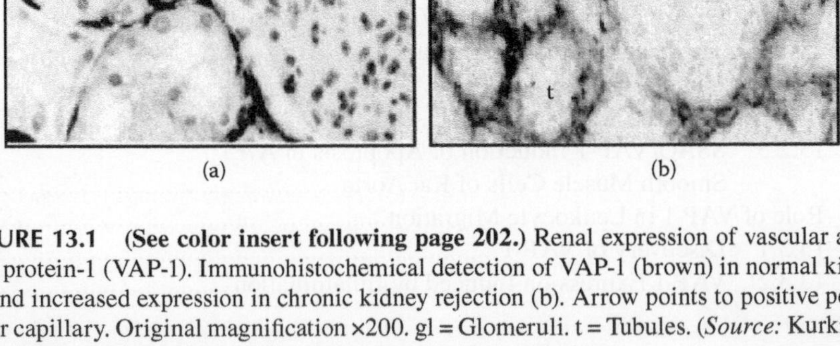

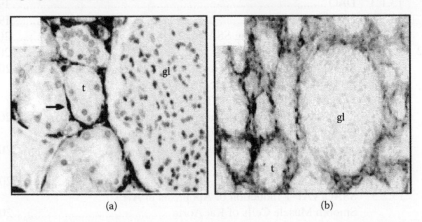

FIGURE 13.1 **(See color insert following page 202.)** Renal expression of vascular adhesion protein-1 (VAP-1). Immunohistochemical detection of VAP-1 (brown) in normal kidney (a) and increased expression in chronic kidney rejection (b). Arrow points to positive peritubular capillary. Original magnification ×200. gl = Glomeruli. t = Tubules. (*Source:* Kurkijarvi et al., 2001. © Wiley-VCH Verlag. Reproduced with permission.)

Wegener's granulomatosis is a severe systemic autoimmune disease characterized by necrotizing granulomatosis of the respiratory tract, necrotizing glomerulonephritis, and small-vessel vasculitis (Day and Savage, 2001; Morgan et al., 2006). The cause is unknown, but endothelial cells are the primary targets of injury, at least partially via anti-endothelial cell auto-antibodies (Morgan et al., 2006). Kidneys of patients with active Wegener's granulomatosis display strong expression of VAP-1 in peritubular and periglomerular microvasculature. They also have higher numbers of circulating endothelial cells positive for VAP-1 and MICA (MHC class I-related chain A) compared to patients in remission. These circulating endothelial cells have a pro-inflammatory phenotype and impair the capacity of endothelial progenitor cells to repair endothelium. Their numbers correlate with the inflammation marker C-reactive protein as well as with the degree of organ damage (Holmen et al. 2005 and 2007).

Inflammatory endothelial damage is a hallmark of a number of renal diseases (Couser, 1998). After *in vitro* stimulation with anti-endothelial cell auto-antibodies from patients with Wegener's granulomatosis, human renal microvascular endothelial cells acquire proinflammatory phenotypes, e.g., the expression of chemokines and VAP-1 is upregulated. While the proinflammatory effects of anti-endothelial antibodies may be blocked by specific inhibitors of the signaling cascade stress-activated protein kinase/c-Jun NH_2-terminal kinase (SAPK/JNK), VAP-1 overexpression cannot (Holmen et al., 2007).

Glomerulosclerosis is an unspecific pathology observed in many progressive glomerular diseases (Ma and Fogo, 2007). Transgenic mice overexpressing human VAP-1 in endothelial cells challenged with atherogenic diets developed glomerulosclerosis. Considering the lack of glomerular VAP-1, this glomerulosclerosis is most likely due to VAP-1-induced hypertension and its proatherogenic effects (Stolen et al., 2004a; Chapters 10 and 11, this volume). A high fat diet induced glomerular hypertrophy and expansion of glomerular parietal cells in wild-type mice was reduced by the transgene expression. Transgenic but not wild-type mice fed with atherogenic diets developed sporadic glomerular cysts (Stolen et al., 2004a). These findings are of interest since parietal cells may contribute to podocyte regeneration and glomerular cysts may indicate disruption of tightly regulated glomerular homeostasis of endothelial and mesangial cells (Eremina et al., 2007; Sagrinati et al., 2006; Liu et al., 2007a; Lowden et al., 1994).

Hypertension is the most important independent risk factor for development and progression of glomerular diseases (Barri, 2006). The hypertensive effect of transgene VAP-1 overexpression and the finding of increased serum VAP-1 in hypertensive patients with diabetes present considerable implications in treatment of all renal patients with elevated blood pressure (Stolen et al., 2004a; Salmi et al., 2002).

To date, kidney transplantation is the treatment of choice for end-stage renal disease. Despite immense improvement in managing transplant patients, acute and chronic rejections are major causes for graft loss (Chapman et al., 2005). In renal biopsies of patients with chronic renal rejection, strong immunohistochemical VAP-1 positivity was found on peritubular capillaries and on inflamed high endothelial venule-like vessels. Chronic kidney rejection is characterized by massive leukocyte infiltrates that are also VAP-1-positive. In both normal and rejected human renal tissue, VAP-1 is expressed as a 170- to 180-kDa homodimeric glycoprotein able to mediate leukocyte binding (Kurkijarvi et al., 2001).

13.1.2 Circulating SSAO

In mice, serum SSAO activity is derived solely from sVAP-1 (Stolen et al., 2004b). Whether this is the case in humans as well remains unclear. Human sVAP-1 circulates as a homodimeric 180-kDa molecule, is not age- or sex-dependent, and exhibits low intra- and interindividual variability in healthy humans (Kurkijarvi et al., 1998; Chapter 5, this volume). Serum VAP-1 is higher in patients before kidney transplantation and declines to nearly normal levels shortly thereafter (up to 40 days). Pretransplant serum VAP-1 concentrations were slightly but insignificantly higher in seven patients with acute rejection compared to those without rejection (n = 10; Kurkijarvi et al., 2001).

In contrast, plasma SSAO enzyme activity in chronic renal failure patients on hemodialysis (n = 152) was 19% lower if compared to healthy controls (n = 107; Boomsma et al., 2005a). In a much smaller patient cohort, SSAO activity in plasma of hemodialyzed patients was lower by approximately one-third compared to healthy controls. Hemodialysis sessions did not impact SSAO activity. Interestingly, plasma SSAO activity was only slightly (by 17%) and insignificantly lower in patients on peritoneal dialysis and normal in nondialyzed patients with chronic renal failure (Nemcsik et al., 2007). In the same study a different method for measurement of plasma SSAO enzyme activity (production of H_2O_2 by SSAO) revealed a contradictory result. Hemodialyzed patients revealed 66% higher enzyme activity than normal controls. How can these results be explained? Renal failure (uremia) is characterized by retention of a large number of small solutes including the SSAO substrates such as methylamine (Baba et al., 1984; Yu and Dyck, 1998) that are normally cleared by the kidneys. High concentration of natural substrate can cause a decrease in enzyme activity assessed by a method based on transformation of added substrate, but not by measuring the amount of product of the enzyme reaction (Nemcsik et al., 2007). It is possible that circulating SSAO is higher in uremic patients compared to healthy subjects. This difference may be masked in a classical substrate-based measurement of SSAO activity due to natural substrate accumulation in uremia. This explanation better fits the current concept of SSAO as a marker of inflammation and organ damage, since uremia is characterized by strong activation of the immune system and toxic (uremic) organ damage.

13.1.3 DAO

The activity of DAO (amiloride-binding protein-1, histaminase) in kidneys and serum was first described decades ago, but its precise function in renal physiology and pathology remains enigmatic. In mammals, including humans, one of the highest DAO activity levels occurs in the kidneys. Conversely, DAO is nearly undetectable in rodent kidneys (Schwelberger, 2004). Renal DAO was identified as a protein that binds the amiloride diuretic in pig kidneys (Barbry et al., 1990) and exerts amiloride-sensitive diamine oxidase activity (Novotny et al., 1994). Renal DAO is localized intracellularly at the basal and partially at the basolateral sides of proximal tubular cells (Figure 13.2) (Schwelberger et al., 1998 and 1999). In pigs, DAO is released after stimulation with heparin, the best described DAO stimulant, and only a minor fraction with a half-life of about an hour appears in circulation (Schwelberger, 2007).

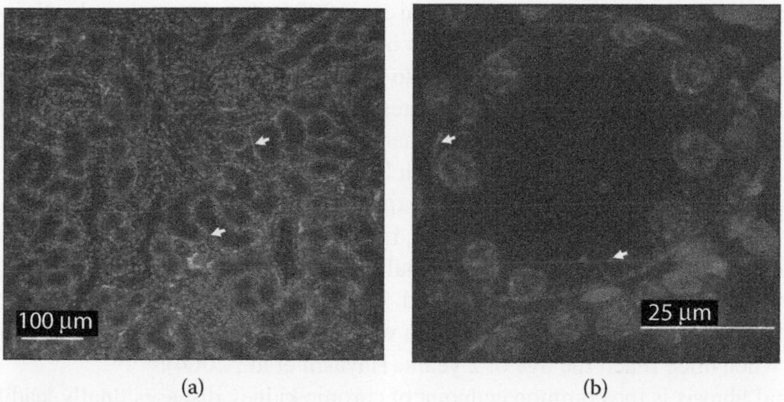

FIGURE 13.2 (See color insert following page 202.) Renal diamine oxidase (DAO) expression. Immunohistochemical detection of DAO (white, arrows) in thin sections (5 μm) of porcine kidney. Cell nuclei are counter-stained with propidium iodide. DAO is present only in proximal tubular epithelial cells; glomeruli and distal tubuli do not contain DAO (a). DAO is localized in vesicular structures preferentially at the basal and partially at lateral aspects of plasma membranes of proximal tubular cells (b). (*Source:* Prof. Dr. Hubert Schwelberger.)

Compared to healthy individuals, heparin administration to uremic patients induces only limited DAO release into plasma (Gang et al., 1976a; Stein et al., 1994). Renal DAO activity is upregulated via activation of β_2 adrenergic receptors in normal and hypertrophic kidneys (Desiderio et al., 1982 and 1985) or via ethanol (Sessa et al., 1984).

Data concerning DAO activity in chronic renal failure are inconsistent. Compared to normal healthy subjects, DAO activity in uremic patients on hemodialysis is elevated in plasma and decreased in urine (Tam et al., 1979; Di Silvestro et al., 1997). Similar findings were described for uremic rats (Kopple et al., 1978). Contrasting results showed elevated urinary DAO in patients with renal failure (Giarnieri et al., 1985) or comparable plasma DAO activities in healthy subjects and uremic patients (Gang et al., 1976a). Different and rather unspecific methods of DAO measurement, very small patient numbers, and recent data describing circulating SSAO do not allow us to draw definite conclusions from these studies.

Histamine promotes renal damage through its effects on glomerular and arteriolar endothelium (Gill et al., 1991). DAO represents an important catabolic pathway of histamine, mainly in glomeruli (Abboud, 1983). In ischemia–reperfusion injury of the kidneys, intravenous infusion of DAO reduced vascular leakage and renal injury and normalized kidney function. Inhibitors of histamine, diphenylhydramine and ranitidine, were similarly effective, supporting the role of histamine in ischemia–reperfusion injury (Kaneko et al., 1998).

13.1.4 LOX

Normal rat kidneys exhibit high LOX activities (Rucker et al., 1996). In adult mice, immunohistochemistry revealed very weak LOX expression only in distal and partly in proximal tubular cells (Figure 13.3; Hayashi et al., 2004). Tubular epithelial cells indeed appear to be major sources of LOX *in vitro* in various animal models and patients with renal fibrosis (Aydin et al., 2008; Di Donato et al., 1997; Goto et al., 2005; Jansen and Csiszar, 2007; Higgins et al., 2007). At present, nothing is known about the cellular localization and expression of LOX in human kidneys. No data cover the expression of the four lysyl oxidase-like (LOXL1 through LOXL4) CAOs in human kidneys.

In mice, renal LOXL is differentially expressed during development and aging. In fetal mouse kidneys (embryonic day 16.5), LOXL was abundant in collecting tubules, undifferentiated tubules, and mesenchymal cells. In newborn mice, the same localization with weaker expression was found. LOXL in adult mice was confined to the same areas as LOX and LOXL expression was much more prominent. This dynamic is lost when mice reach the age of 2 years (Hayashi et al., 2004).

Renal fibrosis is the common endpoint of chronic kidney diseases finally leading to organ failure. We presently lack effective treatment to halt or reverse kidney fibrosis (Boor et al., 2007). LOX and LOXL may play crucial roles, particularly in the two major pathways of hypoxia and transforming growth factor-β (TGF-β)-induced fibrosis.

In vitro a number of renal cell lines including tubular cells express LOX (Di Donato et al., 1997; Jansen and Csiszar, 2007). LOX expression is induced in tubular cells after stimulation with TGF-β (Di Donato et al., 1997; Goto et al., 2005). Hypoxia induces LOX and LOXL2 expression in a hypoxia-inducible factor-1α (HIF-1α)-dependent manner (Higgins et al., 2007). In tubular cells, hypoxia-mediated LOX upregulation or stable transfection with LOX induces epithelial-to-mesenchymal transition, a key process in the development of renal fibrosis (Higgins et al., 2007; Jansen and Csiszar, 2007). In addition to TGF-β and hypoxia, LOX mRNA is also induced in tubular cells when co-cultured with endothelial cells (Aydin et al., 2008). Inhibition of LOX blocked hypoxia-induced migration of tubular cells in a scratch assay (Higgins et al., 2007).

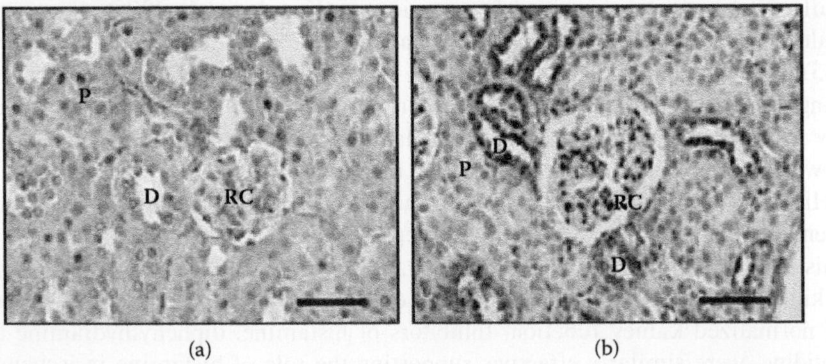

FIGURE 13.3 (See color insert following page 202.) Renal lysyl oxidase (LOX) and lysyl oxidase-like (LOXL) expression. Immunohistochemical detection of LOX and LOXL in kidneys of adult mice (black, arrows). LOX is weakly expressed in proximal and distal tubules (a). Stronger LOXL expression is almost exclusively found in distal tubular cells (b). Bar = 50 μm. RC = renal corpusculum (glomerulus). D = distal tubule. P = proximal tubule. (*Source:* Hayashi et al., 2004. With permission from author and Springer Science + Business Media.)

In early stage of a rat model of toxic progressive nephropathy with renal fibrosis (adriamycin nephropathy), LOX mRNA was upregulated up to threefold in glomeruli, medulla, and whole kidneys. In later stages of adriamycin nephropathy with diffuse fibrosis, LOX mRNA returned to levels found in healthy animals but LOX catalyzed collagen cross links increased (Di Donato et al., 1997). Adenine phosphoribosyltransferase knock-out mice developed severe nephrolithiasis that led to loss of renal function and development of renal fibrosis. Microarray experiments revealed renal LOX as one of the most strongly upregulated genes in the kidneys of these mice (Wang et al., 2000).

Institute of Cancer Research (ICR)-derived glomerulonephritis (ICGN) mice developed hereditary nephrotic syndrome associated with kidney fibrosis. Compared to healthy kidneys from ICR mice, fibrotic kidneys of the ICGN strain exhibited four- and threefold upregulated LOX mRNA and LOX activities, respectively. In healthy ICR mice, weak LOX mRNA expression was localized in renal endothelial cells. Fibrotic kidneys of ICGN mice displayed strong *de novo* LOX expression in tubular cells (Goto et al., 2005). In unilateral ureteral obstruction (UUO), a primary model of tubulointerstitial fibrosis, LOX and LOXL2 were upregulated. Pharmacological inhibition of LOX and LOXL in mice with UUO reduced fibrosis comparably to knock-out of HIF-1α. Increased LOXL2 mRNA was found in microdissected tubulointerstitial compartments from renal biopsies of patients with diabetic nephropathy, IgA nephropathy, or hypertensive nephrosclerosis, compared to healthy living donor kidneys (Higgins et al., 2007).

13.1.5 Compounds Inhibiting SSAO in Nephrology

Some compounds that inhibit SSAO activity (Boomsma et al., 2003) exert nephroprotective effects, e.g., hydralazine, an antihypertensive, protects kidneys via a decrease in blood pressure (Vergona et al., 1987). Aminoguanidine is another compound with nephroprotective properties, but its clinical development was halted due to major side effects (Yu and Zuo, 1997). The extent to which the inhibitory effect of SSAO is responsible for nephroprotection of these compounds is unknown.

The function of SSAO in kidneys is only starting to be elucidated. It appears that SSAO may play an important role in various renal diseases. VAP-1 is closely linked to processes that are crucial in development and progression of kidney pathologies, e.g., inflammation, hypertension, and endothelial changes. However, the lack of intervention studies makes it difficult, if not impossible, to predict the potential of SSAO modulation in treatment of renal diseases. Somewhat conflicting results obtained from studying SSAO enzyme activity in patients with failing kidneys further complicate the matter. The promising potential of LOX as a treatment target in renal fibrosis should be confirmed in specific intervention studies. The role of DAO remains largely unexplained.

13.2 CAOS AND ALZHEIMER'S DISEASE

Alzheimer's disease (AD) is a neurodegenerative disorder of the central nervous system (CNS) characterized by a progressive loss of cognition capacity and memory impairment that lead to dementia. The main physiopathological hallmarks of AD are the intracellular neurofibrillary tangles containing aggregates of hyperphosphorylated tau protein and the senile plaques formed mainly by extracellular amyloid-β protein (Aβ) deposits (Selkoe, 2001) Several studies show that the 42-amino acid amyloid species ($A\beta_{1-42}$) in senile plaque deposits, particularly in oligomeric and fibrillar assembly intermediates, is highly toxic to cells and may be the main cause of neurodegeneration in AD (Catalano et al., 2006; Meredith, 2005; Walsh and Selkoe, 2004). Moreover, experimental data indicate that insolubility of amyloid plaques is caused by an extensive covalent protein cross linking (Smith et al., 1996) and that the growth of $A\beta_{1-42}$ aggregates is accelerated by the formation of advanced glycation end products (AGEs; Loske et al., 2000; Munch et al., 1997; Prabhakaram and Ortwerth, 1994).

Aβ deposits are found in brain parenchyma and in cerebrovascular tissue, inducing cerebral amyloid angiopathy (CAA)—a condition present in most cases of AD, characterized by the deposition of Aβ in the tunica media and adventitia of leptomeningeal vessels and intracortical microvessels, thus inducing the degeneration of vascular smooth muscle and endothelial cells (Vinters et al., 1996). The fact that AD and cerebrovascular diseases share risk factors and are found together in many situations supports the hypothesis of a link between vascular degeneration and AD (and other dementias). It has been suggested that the accumulation of Aβ in vessel walls causes functional deterioration of the blood–brain barrier, which is essential for the correct transport and clearance of Aβ from the parenchyma (Deane et al., 2004; Zlokovic, 2005).

Semicarbazide-sensitive amine oxidase [E.C.1.4.3.6, amine: oxygen oxidoreductase (deaminating) (copper-containing), SSAO], also known as vascular adhesion protein-1 (VAP-1; Salmi and Jalkanen, 1992) constitutes a large family of enzymes present in most mammalian species. SSAO is present in blood plasma as a soluble form and is also associated with cell membranes (Precious and Lyles, 1988; Lyles, 1996). The physiological role of SSAO is still far from clear. SSAO is considered a multifunctional enzyme whose function varies depending on the tissue in which it is expressed (O'Sullivan et al., 2004). SSAO activity stimulates glucose transport in adipocytes (Enrique-Tarancón et al., 1998) and participates in lymphocyte trafficking in endothelial cells where it is induced under inflammatory conditions (Smith et al., 1998). Its primary function appears to be dependent on its amino oxidase activity in most cell types.

SSAO catalyzes the oxidative deamination of primary aromatic and aliphatic amines, generating ammonia, hydrogen peroxide, and the corresponding aldehyde. Aminoacetone and methylamine are considered the physiological SSAO substrates (Precious et al., 1988), and their oxidation generates methylglioxal and formaldehyde, respectively (Dar et al., 1985). The products generated by SSAO have been considered potential risk factors for stress-related angiopathy (Yu et al., 1997 and 2003b; Yu, 2001), due to their capacity to induce lipid peroxidation and oxidative stress.

13.2.1 SSAO/VAP-1 and Neurological Diseases

SSAO/VAP-1 present in endothelial cells, is induced under inflammatory conditions and participates in lymphocyte trafficking (Smith et al., 1998). Inflammatory mechanisms are induced in ischemia, for example, as a manifestation of long-term atherosclerotic disease. These mechanisms are responsible for tissue injury, producing endothelial damage and extravasation of polymorphonuclear leukocytes to the brain parenchyma. Plasma SSAO/VAP-1 levels in this pathological condition have been studied and controversial results were reported. In one study, no statistical differences were found for SSAO/VAP-1 plasma activities between controls and patients with cerebral embolism (Garpenstrand et al., 1999b), suggesting that changes in the circulating form of SSAO do not play any important role in this cerebrovascular event.

Conversely, it was recently reported that acute cerebral ischemia induces SSAO release from the brain vasculature, thus increasing the SSAO/VAP-1 in serum and diminishing the membrane-bound form of vascular SSAO in infarcted brain areas (Airas et al., 2007). The strong and rapid reduction of SSAO/VAP-1-expressing vessels in infarcted areas may be related to the loss of basal lamina and alteration of endothelial structures.

Multiple sclerosis (MS) is a chronic immune disease of the CNS characterized by perivenous inflammatory infiltrates in areas of demyelination and axonal loss. Increases of plasma SSAO/VAP-1 were found in MS patients with inflammatory activities.

Experimental autoimmune encephalitis (EAE) in rodents is a reproducible animal model of human MS. The EAE pathogenesis involves presentation of myelin antigens to T cells in the periphery, migration of activated T cells to the CNS, and development of inflammation and demyelination upon recognition of the same antigens in the CNS. A new SSAO/VAP-1 enzymatic activity inhibitor that is able to reduce disease incidence and severity in this *in vivo* murine model of human MS has been designed and synthesized (O'Rourke et al., 2007; Wang et al., 2006).

13.2.2 SSAO/VAP-1 in Pathological Conditions Related with Alzheimer's Disease and Cerebral Amyloid Angiopathy

The pathogenesis of AD is closely related to vascular disorders (Kalaria, 1999). CAA is present in most cases of Alzheimer's disease (Kalaria and Ballard, 1999). The formaldehyde generated through the catalytic activity of SSAO can generate cross links between proteins and contribute to β-amyloid misfolding and its anomalous deposition in cerebrovascular tissues. It has been hypothesized that SSAO/VAP-1 present in cerebrovascular tissues may be involved in the pathogenesis of Alzheimer's disease and vascular dementia (Yu, 2001).

13.2.3 Overexpression of SSAO/VAP-1 in Cerebrovascular Tissues of AD and CAA Patients

The presence of SSAO/VAP-1 in microvascular tissue and cerebral parenchyma remains controversial (Andree and Clarke, 1981; Dostert et al., 1989; Zuo and Yu, 1994). Biochemical and immunohistochemical approaches indicate that SSAO is present in human microvessels and meningeal vessels and absent in neurons and glia (Castillo et al., 1999). In addition, its histological localization in the tunica media and tunica intima of meningeal membranes has been reported (Castillo et al., 1998).

The enzyme present in human microvessels and meningeal vessels from AD patients has been biochemically characterized. Higher protein expression was detected by Western blotting using polyclonal anti-SSAO antibodies, especially in meningeal vessels (Unzeta et al., 2007).

In addition, the expression of SSAO/VAP-1 was immunohistochemically assessed in postmortem AD samples. Moderate to strong increases in SSAO immunoreactivity were seen in the AD samples between the intima and the muscular layers of arteries containing amyloid deposits (Ferrer et al., 2002). Double immunolabeling for Aβ and SSAO/VAP-1 showed clear co-localization of the abnormal amyloid deposition and increased SSAO expression—the former distributed at the periphery of the vascular SSAO overexpression. Oxidative stress caused by increased Cu/Zn superoxide dismutase 1 (SOD1) immunoreactivity was also observed (Ferrer et al., 2002). Because of the strong link of Aβ, oxidative stress, and neurodegeneration proposed in AD (Behl et al., 1994; Hensley et al., 1995), the upregulation of SSAO observed in human cerebrovascular tissue from postmortem AD patients may contribute through the oxidative stress generated by its own catalytic action to the vascular damage underlying CAA-AD.

13.2.4 Increased Plasma SSAO/VAP-1 in Severely Affected AD Patients

Soluble SSAO/VAP-1 in human plasma is altered in several pathological conditions such as diabetes type 1 and 2 (Boomsma et al., 1995), congestive heart failure (Boomsma et al., 1997), nondiabetic obesity (Mészáros et al., 1999b), atherosclerosis (Gronvall-Nordquist et al., 2001; Karadi et al., 2002), and inflammatory liver disease (Garpenstrand et al., 1999a; Kurkijarvi et al., 2000). The overexpression of SSAO detected in cerebrovascular tissues of CAA-AD patients (Ferrer et al., 2002) correlated with altered SSAO/VAP-1 activity in human plasma. A clear increase of plasma SSAO/VAP-1 activity in moderate-to-severe and severe AD was observed based on NINCDS-ADRA criteria (McKhann et al., 1984), independent of patient age (del Mar-Hernandez et al., 2005; Unzeta et al., 2007).

These results suggest that the SSAO overexpression found in AD brain vessels may be responsible for the increased plasma SSAO activity observed in severe AD patients. This rise may result from increased shedding of SSAO from cell membranes based on the proposal that soluble SSAO is derived from the membrane-bound enzyme (Abella et al., 2004; Stolen et al., 2004a). Since SSAO is an adhesion protein whose expression is induced under inflammatory conditions (Arvilommi et al., 1997), an increase of SSAO activity in advanced AD may be the result of vascular degeneration and inflammation. Overexpression of membrane-bound SSAO by vascular cells and its release into plasma may amplify oxidative stress and contribute to vascular damage in AD.

13.2.5 SSAO/VAP-1 Induction of Apoptosis in A7r5 Smooth Muscle Cells of Rat Aorta

Under pathological conditions in which methylamine levels and plasma SSAO/VAP-1 activity are elevated (Hernandez et al., 2006; Yu and Zuo, 1993), the catalytic action of SSAO may be an important source of toxicity through its products generated such as H_2O_2 and formaldehyde. H_2O_2 is a major reactive oxygen species and the main generator of oxidative stress that is implicated in several diseases. Formaldehyde is a very reactive aliphatic aldehyde that is considered a powerful inflammatory agent (H. d'A-Heck, 1988; Yu and Deng, 1998), and recent reports potentially implicate endogenous aldehydes in β-amyloid misfolding, oligomerization, and fibrillogenesis (Chen et al., 2006).

To investigate whether the soluble SSAO/VAP-1 in plasma contributes to vascular cell damage, *in vitro* toxicity studies were performed to evaluate the cytotoxic effect of soluble SSAO activity from bovine serum on rat aorta A7r5 smooth muscle cells and primary cultures of human aorta. Methylamine and tyramine oxidation by soluble SSAO induced cytotoxicity and apoptosis in rat and human smooth muscle cells prevented by SSAO inhibitors [semicarbazide and MDL-72974A ([E]-2-(4-fluorophenethyl)-3-fluoroallylamine hydrochloride kindly provided by Dr. Peter Yu of the University of Saskatchewan in Canada)] (Hernandez et al., 2006).

The H_2O_2 and formaldehyde products of SSAO/VAP-1 were cytotoxic while ammonia was not. Among the diverse factors capable of inducing oxidative stress, H_2O_2 plays a key role because it is generated by most sources of oxidative stress and can diffuse freely in and out of cells and tissues (Barbouti et al., 2002). It is well known that oxidative stress is an underlying factor contributing to apoptotic processes. However, formaldehyde played the main pro-apoptotic role, detected by chromatin condensation, caspase-3 activation, poly(ADP-ribose)polymerase (PARP), cleavage, and cytochrome c release to cytosol (Hernandez et al., 2006). Moreover, elevated levels of aldehydes are found in Alzheimer's disease and the involvement of endogenous aldehydes in amyloid deposition of AD was recently reported (Chen et al., 2007). Conversely, formaldehyde and free radicals generated from H_2O_2 may contribute synergistically to oxidative stress and vascular damage (Lichszteld, 1979).

The cytotoxic effect of plasma SSAO/VAP-1 activity has been corroborated with the membrane-bound enzyme. The metabolism of methylamine by a SSAO/VAP-1-stable transfected A7r5 smooth muscle cell line induced a dose- and time-dependent cytotoxic effect via the mitochondrial apoptotic pathway. Methylamine oxidation leads to caspase-3 activation, one of the main apoptosis triggers, as previously observed with soluble SSAO (Hernandez et al., 2006). The apoptotic signaling study showed that methylamine treatment of cells expressing SSAO (hSSAO/VAP-1) induced a deregulation of the BAX/Bcl-2 pro/anti-apoptotic protein balance (this ratio represents the pro/anti-apoptotic protein balance), a determinant event of mitochondrial pathway activation.

In the signaling sequence, p53 phosphorylation was induced and a decrease of Bcl-2 levels was observed. This may allow BAX oligomerization and induce mitochondrial permeabilization, in turn activating caspase-9 and caspase-3. The toxicity recovery in the presence of α-pifinitrin, a p53 transcriptional activity inhibitor, suggested a main role of p53 in mediating vascular cell death (Solé et al., 2008). Based on a recent report, SSAO-mediated deamination increased Aβ deposition on blood vessel walls (Jiang et al., 2008).

These results taken together suggest that under pathological conditions such as Alzheimer's disease in which membrane-bound SSAO is overexpressed, its chronically elevated activity may lead to vascular damage through its own catalytic action.

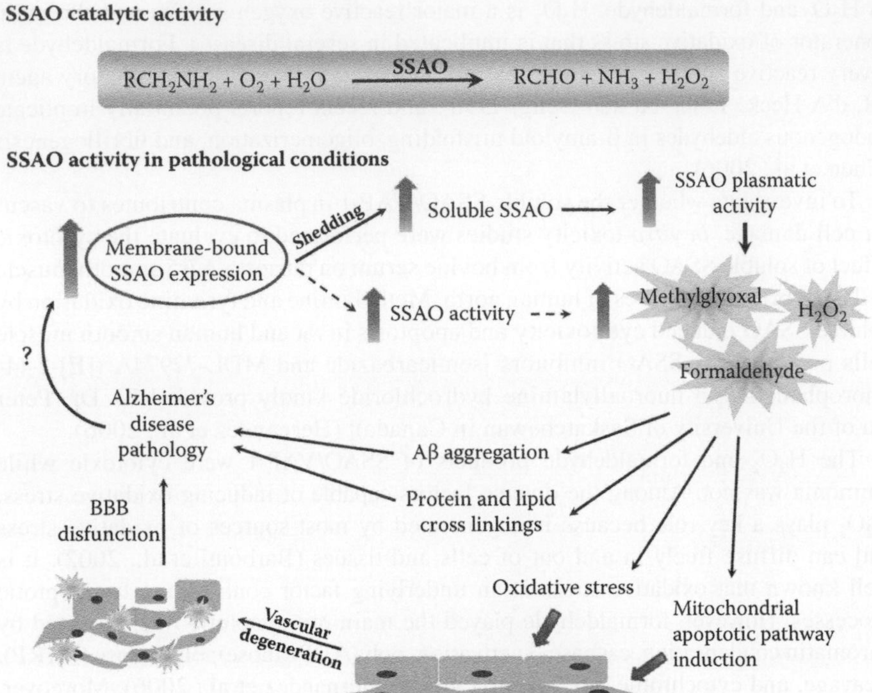

FIGURE 13.4 SSAO activity under pathological conditions.

This vascular damage may be reinforced by elevated plasma SSAO/VAP-1 via an unknown shedding process from the membrane-bound isoform, contributing to cerebrovascular damage underlying this neurodegenerative pathology (Figure 13.4). Further experimental work is required to elucidate the mechanism responsible for SSAO/VAP-1 overexpression in such neurological disorders.

13.3 ROLE OF VAP-1 IN LEUKOCYTE MIGRATION

Continuous lymphocyte recirculation between the blood and lymphatic organs forms the basis for proper functioning of the immune system. Recirculation allows lymphocytes to patrol the body and create an immune response whenever needed to eliminate microbes or kill malignant cells (Butcher and Picker, 1996; von Andrian and Mempel, 2003). Although only lymphocytes possess recirculation capacity, other leukocytes such as granulocytes and monocytes can leave the blood stream and enter sites of inflammation. In fact, leukocyte infiltration to inflammatory sites is a hallmark of inflammation (Luster et al., 2005). Leukocytes leave the blood through a cascade of interactions with vascular endothelial cells mediated by leukocyte receptors and their counter-receptors (ligands) on endothelium (Ley et al., 2007). The four phases of the cascade are (1) tethering and rolling, (2) activation, (3) adhesion, and (4) transmigration.

Transient interactions of selectins and their sialomucin ligands are involved in tethering and rolling (Ley and Kansas, 2004). Activation of integrins is triggered by chemokine binding to their receptors (Kunkel and Butcher, 2002). Activated integrins then bind to their immunoglobulin ligands, resulting in firm adhesion of leukocytes to vessel walls (Luo et al., 2007). Thereafter, the leukocytes invade the endothelium, seeking a proper place to transmigrate. Most leukocytes transmigrate via interendothelial junctions and the rest enter tissues via a transcellular route (Vestweber, 2007; Figure 13.5).

13.3.1 Discovery of VAP-1

By the late 1980s, selectins, sialomucins, chemokines, integrins, and immunoglobulin superfamily members were known but all the characteristics of tissue-specific leukocyte trafficking were still to be revealed. For example, it was obvious that lymphocyte migration from the blood into inflamed synovia of arthritic joints is functionally distinct from lymphocyte migration into peripheral lymph nodes or mucosa-associated lymphatic tissues, but the molecular mechanisms behind the selective interactions of lymphocytes and synovial endothelial cells were unknown (Jalkanen et al., 1986).

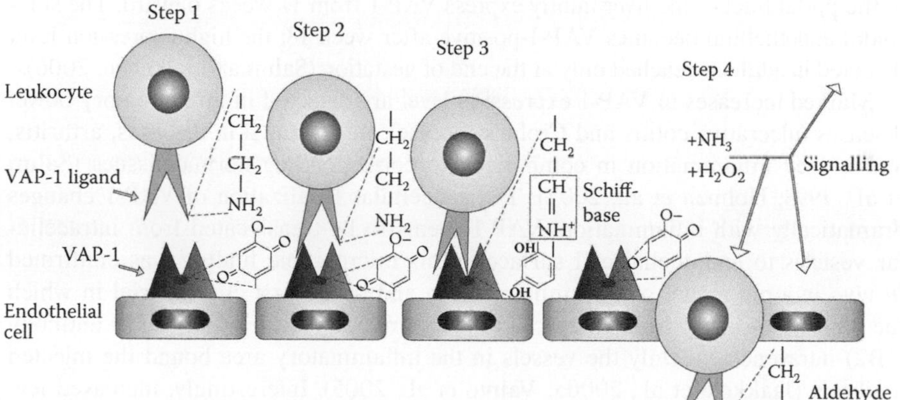

FIGURE 13.5 VAP-1 in leukocyte extravasation cascade. VAP-1 is present on surfaces of endothelial cells and leukocytes express ligands and/or substrates for it. During the initial contacts between leukocytes and endothelial cells (step 2) the leukocyte ligands make contacts with VAP-1 surface epitopes defined by the function-blocking monoclonal antibodies. The amine presented on the same or different leukocyte surface molecule penetrates the enzymatic channel of VAP-1 and the catalytic reaction is initiated (step 3). Through a transient Schiff base formation, the enzymatic reaction leads to covalent binding of the substrate-bearing (leukocyte) and enzyme-bearing (endothelial) cells. During the oxidative phase of the SSAO reaction, ammonium and hydrogen peroxide are released and may exert further adhesion-promoting effects on the vessels and leukocytes (step 4). Details of this putative model have not yet been demonstrated experimentally.

To determine the endothelial cell molecules responsible for lymphocyte trafficking into the synovium, monoclonal antibodies against isolated vessels from inflamed synovial tissues of rheumatoid arthritis patients were prepared. One of the antibodies (1B2) selected in the screening process brightly stained vessels in inflamed synovium. It also inhibited lymphocyte binding to synovial vessels in *in vitro* binding assays. Based on these characteristics, it was named vascular adhesion protein (VAP)-1 (Salmi and Jalkanen, 1992).

13.3.2 VAP-1 Expression Induced by Inflammation

The monoclonal anti-VAP-1 antibody was used to screen normal human tissues and samples from various inflammatory lesions (Salmi and Jalkanen, 1992; Salmi et al., 1993). It quickly became evident that VAP-1 is widely distributed throughout the body. Vessels in most tissues showed VAP-1 positivity. On closer examination, the positivity often localized to smooth muscle cells or pericytes surrounding the endothelium. Moreover, in normal flat endothelial vessels, VAP-1 seemed to localize preferentially within intracellular vesicles. Besides pericytes and smooth muscle cells, the anti-VAP-1 antibody also stained follicular dendritic cells and adipocytes.

During human embryonic development, vascular expression of VAP-1 is detected relatively early and its expression undergoes significant changes in liver, intestinal lamina propria, heart, and kidneys. For example, the central vein and smooth muscle cells in the portal tract of the liver faintly express VAP-1 from 11 weeks onward. The sinusoidal endothelium becomes VAP-1-positive after week 18; the high expression level detected in adults is reached only at the end of gestation (Salmi and Jalkanen, 2006).

Marked increases in VAP-1 expression level are detected in inflammatory bowel diseases (ulcerative colitis and Crohn's disease), in several skin diseases, arthritis, and kidney inflammation in comparison to corresponding normal tissues (Salmi et al., 1993; Holmen et al., 2007). The subcellular localization of VAP-1 changes dramatically with inflammation: VAP-1 seems to be translocated from intracellular vesicles to endothelial cell surfaces. This microscopic finding was confirmed *in vivo* in a pig model of skin inflammation and in the first clinical trial in which the patients suffering from allergic skin inflammation received anti-VAP-1 antibody (1B2) intravenously. Only the vessels in the inflammatory area bound the injected antibody (Jaakkola et al., 2000a; Vainio et al., 2005). Interestingly, increased levels of soluble VAP-1, mostly of endothelial origin, appear in certain inflammatory conditions (Boomsma et al., 2003). Soluble VAP-1 enhances lymphocyte binding to endothelial cells (Kurkijärvi et al., 1998; Stolen et al., 2004a). The current scenario is that it binds to lymphocyte surface molecules, triggering the adhesion strengthening machinery of the lymphocytes.

High levels of VAP-1 are also detected in vessels of certain tumors such as squamocellular carcinomas of head and neck and primary hepatocellular carcinomas (Irjala et al., 2001; Yoong et al., 1998). Interestingly, vessels in hepatic metastases of colorectal carcinomas remain VAP-1-negative (Yoong et al., 1998).

In recent years, monoclonal antibodies recognizing mouse and rat VAP-1 have been produced. Those antibodies and cross-reactive antibodies against human VAP-1 were used to study the expression and function of VAP-1 in mice, rats, dogs, pigs, and rabbits.

Similar inflammation-induced upregulation of VAP-1 as demonstrated in humans was detected in several experimental settings including inflammation of the skin, joints, pancreas (diabetes), and liver in these animals (Jaakkola et al., 2000a; Bono et al., 1999; Martelius et al., 2000; Bonder et al., 2005). However, the expression pattern of VAP-1 reveals species-specific differences. For example, while sinusoidal endothelial cells in human liver are highly VAP-1-positive, they remain VAP-1-negative or express very little in mouse liver. Moreover, the intertubular vessels of human kidneys express high levels of VAP-1; they are negative in murine kidneys (Bono et al., 1999). Also the substrate specificities of murine and human VAP-1 vary. Human VAP-1 effectively deaminates only methylamine and benzylamine; the mouse counterpart also deaminates tyramine and tryptamine (Bono et al., 1999). These differences indicate that experimental results from different species should not be generalized.

13.3.3 VAP-1 Mediation of Leukocyte Binding to Vasculature

Several types of *in vitro* assays have been used to study the role of VAP-1 in mediating leukocyte–endothelial cell interactions. In the widely used *in vitro* frozen section adhesion assay, leukocytes are incubated on freshly cut frozen sections of different organs in rotatory conditions. During incubation, the leukocytes specifically bind to vessels in the section. This assay was used to test the functionality of VAP-1 in fetal-type vessels and inflammation sites. Binding of cord blood lymphocytes to fetal-type vessels in lymph nodes and mucosa-associated lymphatic tissues was inhibited by anti-VAP-1 antibody, suggesting that VAP-1 may play a role in lymphocyte trafficking during human ontogeny (Salmi and Jalkanen, 2006).

Similarly, anti-VAP-1 antibodies effectively blocked lymphocyte binding to vessels at different sites of inflammation in adult tissues. This clearly demonstrated that VAP-1 was fully functional and supported lymphocyte binding to inflamed vessels in synovium observed in preliminary screening and also in the gut and skin (Salmi and Jalkanen, 1992; Salmi et al., 1993; Arvilommi et al., 1996). Moreover, tumor vessels were able to bind effector lymphocytes such as tumor infiltrating lymphocytes in a VAP-1-dependent fashion in *in vitro* binding assays, suggesting a role for VAP-1 in extravasation of tumor infiltrating lymphocytes *in vivo* and thus in tumor immunity (Irjala et al., 2001; Yoong et al., 1998).

Chamber and capillary flow assays allow analysis of the different steps of leukocyte–endothelial cell interaction. These assays were used to measure VAP-1 dependence in leukocyte interactions with sinusoidal endothelial cells of liver, human umbilical vein endothelial cells infected with adenoviral construct encoding VAP-1, and primary rabbit heart endothelial cells (Salmi et al., 2001; Lalor et al., 2002; Koskinen et al., 2004). The results of these analyses indicate that VAP-1 may be involved in both lymphocyte and granulocyte rolling, adhesion, and transmigration steps during leukocyte–endothelial cell interaction.

13.3.4 VAP-1 Enzymatic Activity Regulation of Leukocyte Trafficking

When the cDNA encoding the VAP-1 was eventually cloned and sequenced, its enzymatic function as a semicarbazide-sensitive amine oxidase became evident (Smith et al., 1998). Involvement of an ectoenzyme in leukocyte trafficking opened a new era. Extensive investigations have been required to determine whether the enzymatic function of VAP-1 is indeed important in leukocyte–endothelial cell interaction and at what stage during the multi-step adhesion cascade it exerts its function (Salmi and Jalkanen, 2005).

Using primary endothelial cells from rabbit heart, it was shown that inhibition of enzymatic activity significantly reduced the number of rolling and adhering lymphocytes in flow chamber assays. In contrast, the soluble reaction products (aldehydes and hydrogen peroxide) appeared unnecessary for VAP-1-dependent rolling and adhesion (Salmi et al., 2001). VAP-1 inhibitors also decreased lymphocyte adhesion to and transmigration through hepatic sinusoidal endothelial cells (Lalor et al., 2002). Binding of peripheral mononuclear cells to transfectants expressing human VAP-1 may be inhibited by SSAO inhibitors but not by MAO inhibitors (Salter-Cid et al., 2005).

As the inhibitors may potentially have broader target specificity than VAP-1, the role of VAP-1 in the leukocyte extavasation cascade was confirmed using human umbilical vein endothelial cells expressing either wild-type or enzymatically inactive VAP-1 carrying a single point mutation. Unlike enzymatically inactive VAP-1, the wild-type increased leukocyte binding to human umbilical endothelial cells and this was inhibitable by enzyme inhibitors of VAP-1. These studies clearly demonstrate that enzymatic activity of VAP-1 is important in leukocyte rolling and also in the transmigration step of leukocyte extravasation (Koskinen et al., 2004).

Enzymatic activity of VAP-1 is not, however, only important in mediating binding of leukocyte and endothelium as the end products of the activity. Hydrogen peroxide, ammonium, and aldehyde are potent inflammatory mediators and can efficiently modify the inflammatory microenvironment. These end products of the enzymatic reaction activate transcription nuclear factor-κ-B, upregulate expression of endothelial adhesion molecules (P-selectin, E-selectin, vascular cell adhesion molecule-1, and intercellular adhesion molecule-1), and increase secretion of CXCL8 chemokine (Lalor et al., 2007; Jalkanen et al., 2007). Induction of these molecules led to enhanced leukocyte adhesion to vasculature and is likely to be important for recruitment of leukocytes to sites of inflammation.

13.3.5 VAP-1 Regulation of Leukocyte Migration *in vivo*

The physiological role of VAP-1 in the leukocyte extravasation cascade has been shown in multiple models. The intravital videomicroscopy used in these studies allows real-time analyses of leukocyte–endothelial interactions in living animals. Early work was limited to the use of anti-human VAP-1 antibodies that cross reacted with rabbit VAP-1 (Salmi et al., 1997). When human lymphocytes isolated from inflamed tonsils were administered intravenously to rabbits, the leukocyte–endothelial cell interactions were reduced in the presence of an anti-VAP-1 monoclonal antibody.

The role of VAP-1 in leukocyte extravasation was studied in a more physiological setting with inflamed microvasculatures of rabbits (Tohka et al., 2001). Treatment of the rabbits with anti-human VAP-1 antibodies significantly reduced the rolling, firm adhesion, and transmigration of granulocytes in this model. Notably, the most dramatic effect on rolling was a significant 35% increase of rolling velocity, implying that anti-VAP-1 antibodies inhibit the formation of normal contacts between leukocytes and endothelial cells required for slow rolling. Blockade of VAP-1 also caused a marked 45% reduction in the number of firmly adherent cells.

Development of function-blocking anti-mouse VAP-1 antibodies later confirmed the *in vivo* role of VAP-1 in leukocyte recruitment. Intravenous injection of anti-mouse VAP-1 antibodies rapidly decreased the ability of leukocytes to roll normally in inflamed cremaster vessels (Stolen et al., 2005). VAP-1 appears to play a particularly important role in leukocyte migration in liver (Bonder et al., 2005). Intravital analyses of leukocyte interaction with sinusoidal liver endothelium showed that the binding of Th2 type CD4 T helper cells is almost entirely dependent on VAP-1. Th1 type T cells and granulocytes interacted with liver microvessels relatively independent of VAP-1. Thus, neutralization of VAP-1 with anti-VAP-1 antibodies clearly shows that this ectoenzyme is involved in the multi-step extravasation cascade *in vivo*. Moreover, at least in certain vascular beds, VAP-1 may control emigration of distinct leukocyte subsets.

SSAO enzyme inhibitors have also been shown to interfere with the leukocyte extravasation cascade *in vivo*. Treatment of rats with LJP1207, a selective SSAO inhibitor, resulted in diminished leukocyte–endothelial contacts in a stroke model (Xu et al., 2005). Adhesion and transmigration were both dramatically reduced when the catalytic activity of VAP-1 was inhibited. Remarkably, even if the administration of the SSAO inhibitors was postponed (6 hr after onset of reperfusion), the adhesion cascade was impaired. These experiments strongly suggest that the catalytic activity of VAP-1 is important for its adhesive function in the multi-step extravasation cascade.

The ultimate proof for the *in vivo* function of VAP-1 in leukocyte migration came from studies using VAP-1-deficient mice (Stolen et al., 2005). Generation of this mouse line was instrumental to allow experimentation free of the potentially nonspecific effects of exogenously administered antibodies and enzyme inhibitors. VAP-1-deficient mice were apparently normal under basal conditions (Table 13.1). They bred normally and showed no obvious macroscopic or microscopic aberrations. However, intravital videomicroscopy revealed that lymphocytes rolled faster in the high endothelial venules of Peyer's patches and fewer cells were firmly adherent in the absence of VAP-1.

The altered lymphocyte–endothelial contacts translated into mild defects in short-term homing experiments. Lymphocyte homing into mesenteric lymph nodes and spleen diminished by ~20% in mice lacking VAP-1 when compared to wild-type controls. However, lymphocyte homing to Peyer's patches or peripheral lymph nodes was not statistically significantly altered. These data imply that VAP-1 may contribute to physiologic constitutive patrolling of lymphocytes between the blood and certain lymphatic organs.

Analyses of the extravasation cascade in VAP-1-deficient mice under inflammatory conditions provided the most unambiguous evidence for the involvement of VAP-1 in leukocyte migration (Stolen et al., 2005). In the absence of VAP-1, the rolling velocity of granulocytes in mildly inflamed vessels was five times higher than that in control mice. Moreover, ~50% fewer firmly adherent cells were present in these mice, and the number of transmigrated cells was reduced by 70%.

TABLE 13.1
VAP-1-Deficient Mice

Parameter	Phenotype
Breeding	Normal
Gross morphology	Normal
Leukocyte counts	
Blood	Normal
Lymphatic organs	Normal in PLN, MLN, and spleen
	Reduced PP lymphocytes in young mice
Serum Ig	Normal IgG, IgM
	Slightly reduced IgA
Vascular barrier function	Normal
Adhesion cascade	Impaired
	Faster rolling
	Decreased firm adhesion
	Decreased transmigration
Lymphocyte homing	Slightly impaired in MLN and spleen
Inflammatory response	No change: Yersinia
	Mildly decreased
	Coxsackie B
	Staphylococcus aureus
	Decreased
	TNF-induced peritonitis
	Anti-collagen antibody-induced arthritis
	Oral OVA immunization

Notes: PP = Peyer's patches; OVA = ovalbumin; PLN = peripheral lymph node; MLN = mesenteric lymph node.

VAP-1-deficient mice also demonstrated that the effects of anti-VAP-1 monoclonal antibodies on leukocyte traffic are specific, inasmuch the antibodies had no effect in mice lacking VAP-1. Collectively, the use of VAP-1 gene targeted mice confirmed the novel concept that the multi-step extravasation cascade is modulated by endothelial ectoenzymes *in vivo*.

13.3.6 Attenuation of Inflammatory Reactions *in Vivo* via VAP-1 Inhibition

Based on the adhesive role of VAP-1 in leukocyte binding to endothelium, dissection of its contribution to inflammatory disease pathogenesis has been an important goal of VAP-1 research. The anti-adhesive potential of VAP-1 targeting was addressed initially with the rabbit model (Tohka et al., 2001). Chemical peritonitis was induced in rabbits, they were treated with anti-VAP-1 or control antibodies, and the number of granulocytes emigrating from the blood into the inflamed peritoneal cavity was counted. The analyses showed that anti-VAP-1 antibody treatment diminished the number of extravasated cells by 70%. The efficacy of anti-VAP-1 antibodies in alleviating inflammation was later confirmed in several models (Bonder et al., 2005; Merinen et al., 2005; Martelius et al., 2004). The anti-VAP-1 antibodies appear to attenuate migration of multiple leukocyte subtypes including neutrophils, monocytes, and lymphocytes to sites of acute and chronic inflammation (Table 13.2).

After elucidation of the role SSAO activity in controlling leukocyte extravasation cascade, a remarkable number of *in vivo* studies of the enzyme inhibitors in various settings were conducted. In an early experiment, BTT-2052, a small molecule indane hydrazino alcohol, was used to prevent leukocyte migration to paws in

TABLE 13.2
VAP-1 Inhibition by Antibodies and SSAO Inhibitors Alleviates Inflammation *in vivo*

Antibody Treatment

Model	Major Effect	Species	Reference
Peritonitis			
PP + IL-1[a]	Leukocyte infiltration 70% ↓	Rabbit	Tohka et al., 2001
PP + IL-1	Granulocyte infiltration 50% ↓	Mouse	Merinen et al., 2005
Liver Inflammation			
Allograft	Lymphocytes in allografts 60% ↓	Rat	Martelius et al., 2004
ConA[b]	ALT[c] level 90% ↓	Mouse	Bonder et al., 2005
Insulitis			
NOD[d]	Incidence of diabetes 50% ↓	Mouse	Merinen et al., 2005
Skin Inflammation			
CCL21[e]–induced	Monocyte infiltration 60% ↓	Mouse	Merinen et al., 2005

Inhibitor Treatment

Mode	Major Effect	Species	Inhibitor	Reference
Arthritis				
Adjuvant	Clinical inflammation score 40% ↓	Rat	BTT-2052[f]	Marttila-Ichihara et al., 2006
mAb[g]–induced	Clinical inflammation score 80% ↓	Mouse	BTT-2052	Marttila-Ichihara et al., 2006

(Continued)

TABLE 13.2 (*Continued*)

	Inhibitor Treatment			
Mode	**Major Effect**	**Species**	**Inhibitor**	**Reference**
Lung Inflammation				
LPS[h] inhalation	Leukocytes in BAL[i] lavage 40% ↓	Mouse	MDL-72974[j]	Yu et al., 2006
LPS inhalation	Leukocytes in BAL lavage 50% ↓	Rat	LJP1586[k]	O'Rourke et al., 2008
Ischemic reperfusion	H_2O_2 production 25% ↓	Rat	Semi-carbazide	Ucar et al., 2005
Skin Inflammation				
Carrageenan	Paw edema 70% ↓	Rat	LJP1207	Salter-Cid et al., 2005
Carrageenan	Leukocyte infiltration 50% ↓	Rat	BTT-2052	Koskinen et al., 2004
Carrageenan	Leukocyte infiltration 60% ↓	Mouse	LJP1586	O'Rourke et al., 2008
Colitis				
Oxazolone-induced	Survival 50% ↑	Mouse	LJP1207[l]	Salter-Cid et al., 2005
Endotoxemia				
LPS-induced	Mortality 50% ↓	Mouse	LJP1207	Salter-Cid et al., 2005
Ischemia–Reperfusion Injury				
Stroke	Neurological ability 80% ↑	Rat	LJP1207	Xu et al., 2005
EAE[m]				
Prophylactic	Clinical severity score 25% ↓	Mouse	LJP1207	Wang et al., 2006
Flare	Clinical severity score 60% ↓	Mouse	LJP1207	O'Rourke A et al., 2007
Remission	Clinical severity score 60% ↓	Mouse	LJP1207	O'Rourke A et al., 2007
Uveitis	Retinal leukocyte transmigration 70% ↓	Rat	U-V002	Noda et al., 2008

Notes: [a] Proteose peptone + interleukin-1; [b] Concanavalin A; [c] Alanine aminotransferase; [d] Nonobese diabetic; [e] CCL21chemokine; [f] (1S,2S)-2-(1-methylhydrazino)-1-indanol; [g] Anti-collagen type II antibody cocktail; [h] Lipopolysaccharide; [i] Bronchoalveolar lavage; [j] (E)-2-(4-fluorophenethyl)-fluoroallylamine; [k] Z-3-fluoro-2-(4-methoxybenzyl)allylamine; [l] N′-(2-phenylallyl)hydrazine; [m] Experimental autoimmune encephalomyelitis.

response to skin irritation (Koskinen et al., 2004). The compound efficiently blocked SSAO activity and reduced leukocyte infiltration in a dose-dependent manner. Since then several groups have successfully used SSAO inhibitors with different chemical structures to alleviate inflammation in acute and chronic diseases (Table 13.2).

One very important set of experiments was carried out by Linnik's group (O'Rourke et al, 2007). They used relapsing–remitting experimental autoimmune encephalomyelitis (EAE), a widely used animal model of crippling human multiple sclerosis. LJP1207, an SSAO inhibitor, effectively reduced the clinical severity and incidence of the disease when administered prophylactically during the induction of the disease (Wang et al., 2006). Remarkably, it also profoundly suppressed the disease when the initiation of therapy was postponed until the peak of initial flare (O'Rourke et al., 2007).

The LJP1207 also prevented further relapses when the start of treatment was delayed until the first spontaneous remission. Thus, prophylactic SSAO inhibition not only alleviated the development of inflammatory reactions, but SSAO blockade was also effective in situations in which therapy was started when full-blown inflammatory manifestations were already present. SSAO inhibitors have also been effective in alleviating more acute fatal inflammatory responses, for example, in sepsis and colitis models, and was accompanied by attenuation of the cytokine storm characteristic to these diseases (Salter-Cid et al., 2005). These data collectively show that inhibition of the catalytic activity of SSAO/VAP-1 can profoundly alleviate the courses of multiple inflammatory reactions.

The use of various chemicals to inhibit SSAO activity raises the question of the specificity of the effects. A few inhibitors have undergone extensive selectivity analyses and conferred strong selectivity against SSAO when compared to MAO-A and MAO-B (O'Rourke et al., 2007 and 2008; Marttila-Ichihara et al., 2006). They have also been shown not to affect a wide panel of kinases, phosphatases, and cell surface receptors. Certain compounds exhibited activity against other SSAO family members (copper containing amine oxidases 1 and 2, AOC1 and AOC2; Marttila-Ichihara et al., 2006), but most have not been tested against these close relatives of VAP-1. The exclusion of certain off-target effects naturally does not provide definitive proof of specificity. Therefore, testing the inhibitors in VAP-1-deficient mice with different disease models is needed to provide ultimate verification of the specificities of observed beneficial effects.

Use of VAP-1-deficient mice also verified the anti-inflammatory potential of therapies aimed to neutralize VAP-1 activity. Mice lacking VAP-1 show reduced clinical and histopathological inflammation in peritonitis, pancreatitis, arthritis, and mucosal immune responses when compared to wild-type mice (Stolen et al., 2005, Marttila-Ichihara et al., 2006; Koskinen et al., 2007). See Table 13.1. The antimicrobial responses against *Staphylococcus aureus* skin inflammation, *Yersinia enterocolitica* gastroenteritis, and Coxsackie B4 virus-triggered pancreatitis were normal or only very mildly affected by the absence of VAP-1. This suggests that therapeutic neutralization of VAP-1 will not lead to generalized immunosuppression, although specific responses against certain microbes may of course be affected.

13.3.7 Dualistic Nature of VAP-1 Function in Leukocyte Migration

Mechanistic studies performed *in vitro* and *in vivo* point to a dualistic function of VAP-1 in leukocyte migration. The current working hypothesis suggests that leukocytes initially bind to endothelial VAP-1 using receptors for the anti-VAP-1 antibody-defined surface epitopes of VAP-1 (Salmi and Jalkanen, 2005). The molecular nature of these putative leukocyte receptors remains to be determined. This interaction and activities of other adhesion molecule pairs bring leukocytes and endothelial cells into close contact. This allows penetration of SSAO substrates present on the surfaces of leukocytes into the enzymatic channel of VAP-1.

When the catalytic reaction is triggered a covalent but transient Schiff base formed between the enzyme and substrate temporarily brings the two cell types together. Spontaneous cleavage of the Schiff base then allows leukocytes to continue their migration through vessel walls. The exact nature of SSAO substrate action on leukocyte surfaces also remains to be unraveled. However, free NH_2 groups on peptide N-termini and certain amino acid side chains can modulate SSAO activity (Salmi et al., 2001; Yegutkin et al., 2004). Various amino sugars have also been proposed as VAP-1 substrates (O'Sullivan et al., 2003a).

The ammonium and hydrogen peroxide released during the catalytic SSAO reaction may also contribute to leukocyte migration. The produced hydrogen peroxide produced can induce the expression of adhesion molecules (such as P- and E-selectins) and chemokines (such as MCP-1 and IL-8) *in vitro* and *in vivo* (Salter-Cid et al., 2005; Lalor et al., 2007; Jalkanen et al., 2007). Moreover, SSAO inhibition led to diminished P-selectin expression in an *in vivo* model of eye inflammation (Noda et al., 2008). This indicates possible cross talk between VAP-1 and traditional adhesion molecules. The modulation of the microenvironment by the end products of VAP-1-driven oxidation may prime the entire leukocyte extravasation cascade.

This working model may explain why both anti-VAP-1 antibodies and SSAO inhibitors block leukocyte extravasation. It also explains the observation that anti-VAP-1 monoclonal antibodies and SSAO inhibitors do not exhibit synergistic or additive effects in *in vitro* models (Salmi et al., 2001; Lalor et al., 2002; Koskinen et al., 2004). It is important to note that anti-VAP-1 antibodies do not inhibit SSAO activity and that SSAO inhibitors do not affect the availability of the epitopes of anti-VAP-1 monoclonal antibodies (Bonder et al., 2005; Koskinen et al., 2004). Similarly, the impaired leukocyte transmigration through endothelial cell layers expressing catalytically dead VAP-1 constructs is compatible with the proposed model (Koskinen et al., 2004). Our current experiments using gene-modified mice carrying point mutations that render VAP-1 catalytically inactive will hopefully provide definitive proof for the current dualistic model.

13.3.8 VAP-1 Antibodies in Clinical Trials

The original mouse anti-human VAP-1 antibody (clone 1B2) was administered to patients in phase I clinical trials. It bound to VAP-1 on inflamed endothelial cells and showed no adverse effects (Vainio et al., 2005). Chimeric (Kirton et al., 2005) and fully humanized anti-VAP-1 antibodies were then developed. A fully human anti-human VAP-1 monoclonal antibody has already entered phase I clinical trials. It will be interesting to see whether these novel biological reagents exhibit safety and efficacy in blocking inflammation.

Use of anti-VAP-1 antibodies, SSAO inhibitors, and experiments with VAP-1-deficient mice indicate that this unique ectoenzyme regulates leukocyte migration under physiological and pathological conditions. In animal models, multiple inflammatory diseases were successfully alleviated by inhibiting the function of VAP-1 by antibodies or small molecule inhibitors. Testing of the clinical potential of VAP-1 inhibition may provide new armamentarium in the fight against inflammatory reactions.

14 Inhibitors of Copper Amine Oxidases
Past, Present, and Future

Marek Šebela and Lawrence M. Sayre

CONTENTS

14.1 SHORT HISTORICAL OVERVIEW OF COPPER AMINE OXIDASE INHIBITOR STUDIES

Historically, two key points have guided the development of inhibitors of CAOs, namely the two prosthetic groups of these enzymes: the organic active carbonyl cofactor responsible for oxidative deamination of primary amine substrates and the active site copper (cupric ion). Indeed, most inhibitors of the enzymes act at one or the other of these two active-site entities. It has long been known that the enzymes are inhibited by carbonyl reagents such as hydrazine and semicarbazide.

Zeller (1940), who first demonstrated that pig kidney amine oxidase (PKAO) was inhibited by cyanide, attributed this inhibition to the presence of a carbonyl compound in the prosthetic group of the enzyme. During the 1960s, many CAOs were purified to homogeneity from plant and mammalian sources. Evidence strongly suggested that the carbonyl component of the plasma enzyme may belong to pyridoxal phosphate (Blaschko and Buffoni, 1965). Thus, it was assumed that all amine oxidases inhibited by carbonyl reagents contained this moiety.

Mann (1961) worked with a pure pea seedling amine oxidase (PSAO) to obtain clear spectroscopic and kinetic evidence of a carbonyl compound and a copper complex at the active site of the enzyme. In ensuing research, various copper-chelating agents were characterized as inhibitors of the pea enzyme (Hill and Mann, 1962; Bardsley et al., 1974b). During the 1960s and 1970s, Macholán and coworkers described numerous substrate analogs that functioned as powerful competitive inhibitors of PSAO and PKAO (Macholán 1969 and 1974; Macholán et al., 1967).

Highly specific mechanism-based inhibitors of CAOs have been continuously designed and evaluated since the 1970s. Their designs were stimulated by the discovery of topaquinone (TPQ) as a ubiquitous cofactor of CAOs (Janes et al., 1990). The compounds were tested with mammalian, microbial, and plant CAOs (Peč and Frébort, 1992b; Shepard et al., 2002; Jeon et al., 2003a, b; Lamplot et al., 2004; O'Connell et al., 2004). Effective inhibitors of so-called semicarbazide-sensitive amine oxidases (SSAOs, a term originally used for CAOs associated with the plasma membranes of vascular and nonvascular smooth muscle cells, adipocytes, and other tissues and now also includes soluble enzymes) are of great current interest because of their desired applications as therapeutic agents (Wang et al., 2006). In fact, it will be highly desirable to obtain inhibitors with notable selectivity toward SSAO over mitochondrial flavoprotein MAO-A and MAO-B.

14.2 IRREVERSIBLE INHIBITION BY CARBONYL REAGENTS

Irreversible inhibitors are generally specific for a group or class of enzymes since they specifically alter the active sites of their targets and do not destroy entire protein structures (Segel, 1993). As a consequence, the inhibition occurs via the formation of a covalent bond between the inhibitor and an essential functional group of the enzyme that belongs to the cofactor or to an important amino acid residue.

Two types of irreversible inhibitors have been developed for the CAOs. One is a range of hydrazines and hydrazine derivatives that act by bimolecular derivatization of the topaquinone C5 carbonyl group. The other type is a range of mechanism-based inactivators that act by taking advantage of the substrate activity of the enzyme, minimally involving substrate Schiff base formation and α-carbon proton abstraction by the conserved general base. The latter type will be discussed below.

Regarding carbonyl reagents, the most effective irreversible inhibitors of amine oxidases from bovine serum (Buffoni and Ignesti, 1975; Morpurgo et al., 1992), pig kidney (Crabbe et al., 1975; Finazzi-Agrò et al., 1977), *Aspergillus niger* (Yamada et al., 1965; Frébort et al., 1994), and enzymes from pea and lentil seedlings (Mann, 1955; Padiglia et al., 1992) are arylhydrazines (phenylhydrazine, 4-nitrophenylhydrazine) and hydrazides of carbamic and thiocarbamic acids (semicarbazide, thiosemicarbazide, and phenylsemicarbazide; Figure 14.1). Interestingly, hydrazine, benzylhydrazine, and alkylhydrazines induce inactivation of plasma/serum amine oxidase. The activity recovers over time due to turnover-like chemical evolution of the inactive cofactor hydrazones in equilibrium with their azo tautomers to forms that can revert to the active quinone cofactor (Morpurgo et al., 1989; Lee et al., 2001b).

Plant and mammalian CAOs are also irreversibly inhibited by hydrazides of carboxylic and amino acids (Morpurgo et al., 1988; Peč et al., 1992a; Figure 14.1). Aminoguanidine (Crabbe et al., 1975; Tamura et al., 1989; Padiglia et al., 1999b), hydroxylamine (Yamada et al., 1965; Padiglia et al., 1992), and 2-hydrazinopyridine (De Matteis et al., 1999; Saysell et al., 2000) have also been described as irreversible inhibitors of mammalian, microbial, and plant CAOs that interact with the cofactor (Figure 14.1).

Phenylhydrazines produce colored adducts with the reactive carbonyl group of the TPQ cofactor that provide strong absorption in the visible spectrum. For example, the yellow phenylhydrazone derivative of lentil seedling amine oxidase (LSAO) is characterized by a molar absorption coefficient at 445 nm of 64,000 $M^{-1}cm^{-1}$ (Padiglia et al., 1992). For that reason, the color reaction was frequently used for spectrophotometric titrations of the cofactor and determining the number of functional active sites. Controversial results reported yields of only one (Lindström and Petersson, 1978) or three (Buffoni and Ignesti, 1975) prosthetic groups per mole of holoenzyme. However, the subsequent results confirm the presence of one active cofactor in each subunit of dimer (Padiglia et al., 1992), in accordance with the published crystal structures of amine oxidases (see Chapter 9).

The apparently conflicting observations of active site titration may be attributed to differences in experimental conditions such as incubation times and purity of enzyme preparations (Lindström and Petersson, 1978; Padiglia et al., 1992). Half-site reactivity was also observed, probably caused by structural changes in the dimer induced upon binding of the reagent to the cofactor in one enzyme subunit (Morpurgo et al., 1992). Finally, the lower stoichiometry observed may relate to a partial defective formation of the cofactor from its tyrosine precursor (Steinebach et al., 1996). Pyridoxamine is a different nonhydrazine type of carbonyl reagent that inhibits rabbit SSAO (Buffoni et al., 1989; Figure 14.1). It may act by forming a resonance-stabilized product Schiff base with TPQ.

14.3 COMPETITIVE INHIBITORS

The classical approach to developing competitive inhibitors of diamine oxidases (DAOs) is based on the assumption that the enzyme recognizes its substrates through binding of the two diamine ammonium termini. Compounds that contain two positively charged but nonmetabolizable groups connected through a tether of appropriate length should bind to the enzyme as do normal substrates and thereby serve as competitive inhibitors. Bardsley and coworkers published a series of papers in the 1970s showing that *bis*-isothiouronium, -guanidinium, -dimethylsulfonium, and -trimethylammonium compounds were effective competitive inhibitors of diamine oxidase from human placenta and from pig and human kidneys (Bardsley and Ashford, 1972; Crabbe and Bardsley, 1974a, b; Shindler and Bardsley, 1976). More recently, various *bis*-amidine drugs such as the aromatic diamidines—stilbamidine, phenamidine, propamidine (Figure 14.1), pentamidine, and berenil (diminazene aceturate)—as well as the cyclohexanetetranol-based *bis*-guanidino streptidine were all shown to be effective reversible inhibitors of mammalian DAOs (Balaña-Fouce et al., 1986; Karvonen, 1987; Cubría et al., 1993), although kinetics suggest noncompetitive behaviors for some of these compounds with PKAO (Cubría et al., 1991). A related study revealed that guanabenz, a mono-amidino drug, inhibits pig plasma benzylamine oxidase (Banchelli et al., 1986). Arcaine (1,4-diguanidinobutane) and methylglyoxal *bis*(guanylhydrazone) are competitive inhibitors of amine oxidases (Figure 14.1; Padiglia et al., 1998b and 1999b).

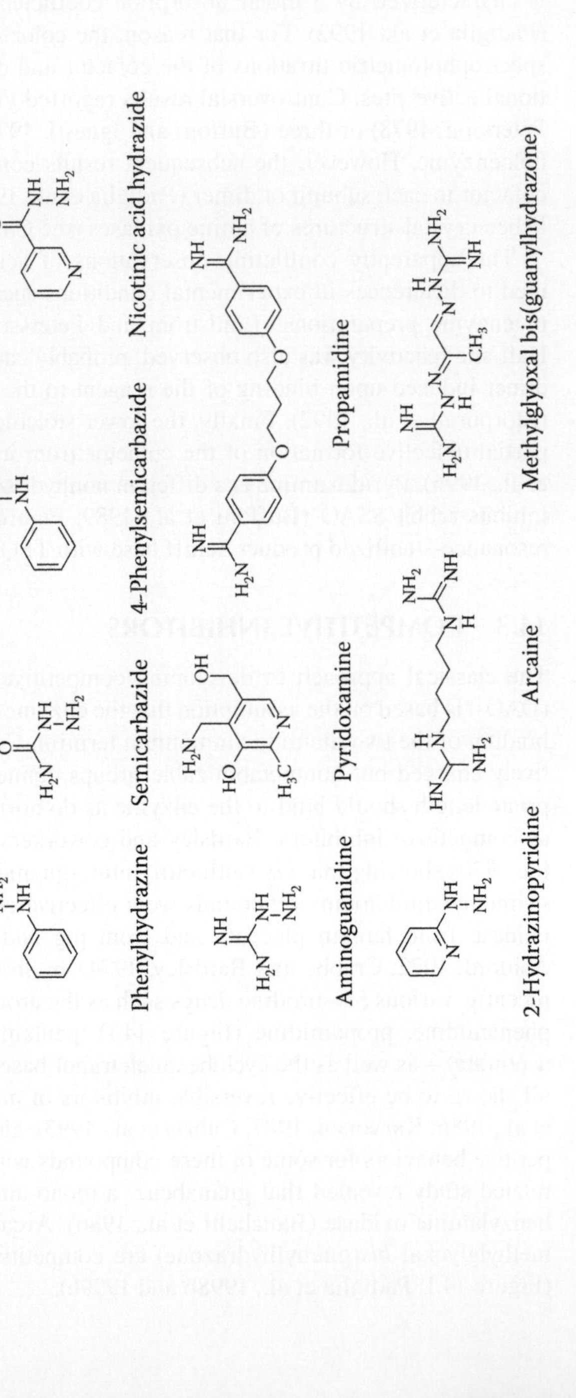
Phenylhydrazine
Semicarbazide
4-Phenylsemicarbazide
Nicotinic acid hydrazide
Aminoguanidine
Pyridoxamine
Propamidine
2-Hydrazinopyridine
Arcaine
Methylglyoxal bis(guanylhydrazone)

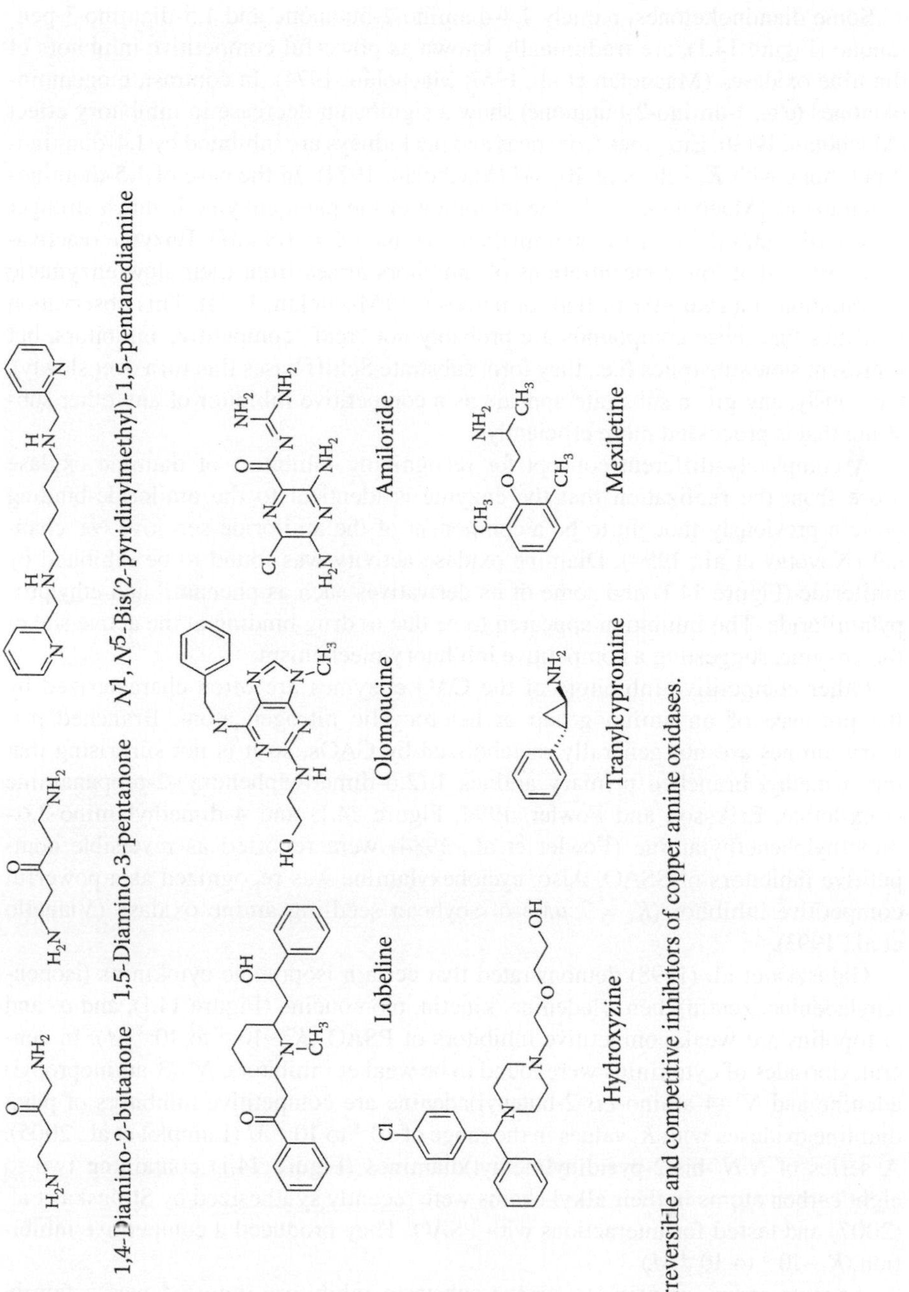

FIGURE 14.1 Irreversible and competitive inhibitors of copper amine oxidases.

Some diaminoketones, namely 1,4-diamino-2-butanone and 1,5-diamino-3-pentanone (Figure 14.1), are traditionally known as powerful competitive inhibitors of diamine oxidases (Macholán et al., 1967; Macholán, 1974). In contrast, monoaminoketones (e.g., 1-amino-2-butanone) show a significant decrease in inhibitory effect (Macholán, 1969). Enzymes from peas and pig kidneys are inhibited by 1,4-diamino-2-butanone with K_i values of 10^{-7} *M* (Macholán, 1974). In the case of 1,5-diamino-3-pentanone (Macholán, 1974), the inhibition of the plant enzyme is much stronger ($K_i = 0.015\ \mu M$) than for the mammalian enzyme ($K_i = 1.8\ \mu M$). Enzyme reactivation observed at low concentrations of inhibitors arises from their slow enzymatic degradation at a rate ~1% of that for putrescine (Macholán, 1974). This observation indicates that these compounds are probably not "real" competitive inhibitors, but represent slow substrates (i.e., they form substrate Schiff bases that turn over slowly). Obviously, any given substrate appears as a competitive inhibitor of any other substrate that is processed more efficiently.

A completely different concept for recognizing inhibitors of diamine oxidase arose from the realization that the enzyme is identical to the amiloride-binding protein previously thought to be a component of the amiloride-sensitive Na^+ channel (Novotny et al., 1994). Diamine oxidase activity was found to be inhibited by amiloride (Figure 14.1) and some of its derivatives such as phenamil and ethylpropylamiloride. The inhibition appeared to be due to drug binding at the active site of the enzyme, suggesting a competitive inhibitory mechanism.

Other competitive inhibitors of the CAO enzymes are often characterized by the presence of an amino group or heterocyclic nitrogen atom. Branched primary amines are not generally metabolized by CAOs, so it is not surprising that the α-methyl branched primary amines 1-(2,6-dimethylphenoxy)-2-propanamine (mexiletine; Eriksson and Fowler, 1994; Figure 14.1) and 4-dimethylamino-2,α-dimethylphenethylamine (Fowler et al., 1984) were reported as reversible competitive inhibitors of SSAO. Also, cyclohexylamine was recognized as a powerful competitive inhibitor ($K_i = 7\ \mu M$) of soybean seedling amine oxidase (Vianello et al., 1993).

Galuszka et al. (1998) demonstrated that certain isoprenoic cytokinins (isopentenyladenine, zeatin), benzyladenine, kinetin, olomoucine (Figure 14.1), and *o*- and *m*-topolins are weak competitive inhibitors of PSAO (K_i ~10^{-4} to 10^{-3} *M*). In general, ribosides of cytokinins were found to be weaker inhibitors. N^6-(3-aminopropyl) adenine and N^6-(4-amino-*cis*-2-butenyl)adenine are competitive inhibitors of plant diamine oxidases with K_i values in the range of 10^{-5} to 10^{-4} *M* (Lamplot et al., 2005). A series of *N,N′*-bis(2-pyridinylmethyl)diamines (Figure 14.1) containing two to eight carbon atoms in their alkyl chains were recently synthesized by Stránská et al. (2007) and tested for interactions with PSAO. They produced a competitive inhibition (K_i ~10^{-5} to 10^{-4} *M*).

Another series of primary amine substrate inhibitors reported was a family of pyridoxamine-like 4-aminomethylpyridines with alkoxy- alkylthio-, or alkylamino groups at positions 3 and 5 (Bertini et al., 2005). Some displayed reversible inhibitory selectivity for benzylamine oxidase and SSAO with respect to diamine oxidase, lysyl oxidase, and MAO. Others behaved as reversible nonselective inhibitors of various CAOs including PSAO and *Hansenula polymorpha* amine oxidase.

PSAO is competitively inhibited by some groups of alkaloids. Cinchona alkaloids (except for quinine) and lobeline, a *Lobelia inflata* alkaloid (Figure 14.1) inhibit the enzyme competitively with K_i values of 10^{-4} to 10^{-3} *M*. Interestingly, PKAO is inhibited by the cinchona alkaloids noncompetitively (Peč and Macholán, 1976).

Other piperidine alkaloids from *L. inflata* (e.g., lobelanine and lobelanidine) showed weak competitive inhibition effects (Peč, 1985). Sedamine alkaloids provided inhibition constants in the range of 10^{-5} to 10^{-3} *M* (Adámková et al., 2001). *trans*-2-Phenylcyclopropylamine, i.e., tranylcypromine (Figure 14.1), has been described as a powerful competitive inhibitor of *Escherichia coli* amine oxidase (ECAO; Saysell et al., 2002). The compound similarly inhibits PSAO and PKAO with a K_i of 10^{-4} *M* (Peč and Šimůnková, 1988). PKAO is competitively inhibited (K_i ~10^{-5} to 10^{-4} *M*) by the diuretic, antihypertensive, and anticoagulant drugs, amiloride, clonidine, and gabexate mesylate, respectively (Federico et al., 1997). Hydroxyzine (Figure 14.1) is a relatively potent competitive inhibitor of bovine plasma amine oxidase (BSAO) but inhibits only weakly the membrane-bound form of the enzyme from bovine lung (O'Sullivan et al., 2006). Tetraphenylphosphonium chloride and its various analogs including tetraphenylarsonium chloride were found to act as powerful competitive inhibitors of BSAO metabolism of polyamine substrates (Di Paolo et al., 2004). Molecular wires based on Ru(II) or Re(I) complexes (Figure 14.2), which strongly and competitively inhibit microbial amine oxidase, will be discussed below.

14.4 COPPER CHELATING AGENTS AND OTHER NONCOMPETITIVE AND UNCOMPETITIVE INHIBITORS

Copper chelating agents have been long known as characteristic noncompetitive inhibitors of CAOs. By definition, noncompetitive inhibition is a kind of reversible inhibition. Inhibitor binding to an enzyme reduces its activity but does not affect the binding of substrate. As a result, the extent of inhibition depends only on the concentration of the inhibitor (Segel, 1993).

Copper chelators target active site copper that is then eliminated from the catalytic reaction. The chelators displace water from its equatorial coordination sites in the enzyme-bound copper complex (Padiglia et al., 1998b). CAOs are inhibited ($K_i = 10^{-5}$ *M*), for example, by 1,10-phenanthroline; 2,2′-bipyridyl; diethyldithiocarbamate (DDC) and 8-hydroxyquinoline (Hill and Mann, 1962; Bardsley et al., 1974b; Figure 14.2). The latter two compounds provide copper complexes that are not sufficiently soluble in aqueous solution (Bardsley et al., 1974b). The DDC–copper complex easily precipitates and therefore has been used as a reagent to obtain copper-depleted enzymes after high-speed centrifugation. The copper-free enzyme can be reconstituted by adding cupric ions (Hill and Mann, 1964). Relatively slow formation of complexes between enzyme-bound copper and the chelating agents requires longer pre-incubation (10 to 20 min) to achieve full inhibition (Peč and Frébort, 1992a).

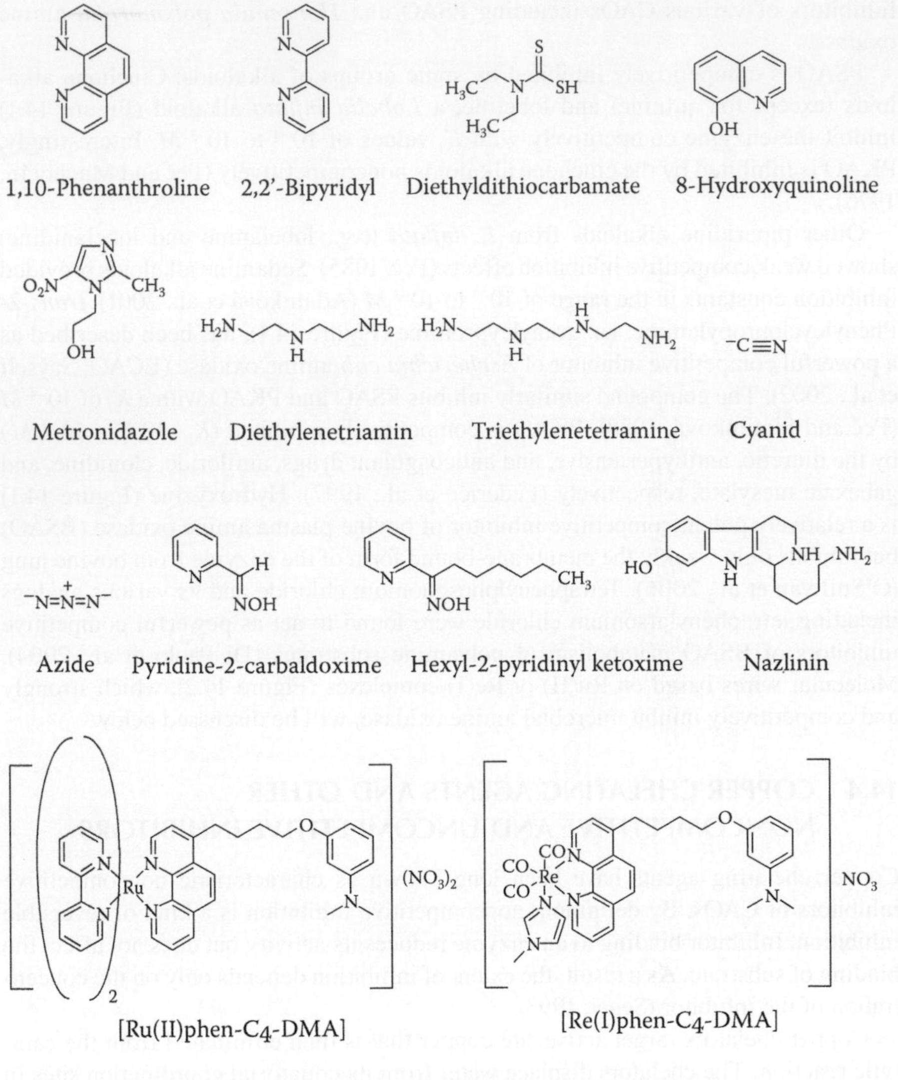

FIGURE 14.2 Noncompetitive and uncompetitive inhibitors of copper amine oxidases plus Ru(II) and Re(I) molecular wires that inhibit the enzymes competitively.

Cyanide (Figure 14.2) is a strong inhibitor of CAOs. Zeller (1940) concluded that animal diamine oxidase is inhibited by cyanide because it acts as a carbonyl reagent; the inhibition of the plant enzyme by cyanide was attributed to the same cause by Kenten and Mann (1952). The discovery of copper at the active site suggested that the inhibition by cyanide may be due, at least in part, to the properties of this compound as a metal reagent (Hill and Mann, 1962). Cyanide has been applied for copper removal from amine oxidases after a previous reduction of Cu(II) to Cu(I) by dithionite (Suzuki et al., 1984). Cyanide binding to

Cu(II), Cu(I), and topaquinone (via cyanohydrin formation) has been implicated as a mechanism for inhibition of amine oxidases by this anion (He et al., 1995b). Recent results indicate that cyanide binds to the Cu(I)–semiquinone catalytic intermediate that is then "deactivated" for oxygen in the cofactor reoxidation step (McGuirl et al., 1997).

The inhibitory effects of other copper ligands such as azide can also be rationalized with regard to the equilibrium between Cu(II)–aminoquinol and Cu(I)–semiquinone active site intermediates (McGuirl et al., 1997). Interestingly, cyanide is an uncompetitive inhibitor toward turnover of amine substrates by PSAO and *Arthrobacter globiformis* amine oxidase (AGAO), meaning that the inhibition arises only when the compound binds to the substrate-reduced form of the enzyme (Shepard et al., 2004). Azide (Figure 14.2) was previously known as the only uncompetitive inhibitor of the pea enzyme (Padiglia et al., 1998b). However, its effects on mammalian, microbial, and plant amine oxidases have recently been re-evaluated in detail (Juda et al., 2006). Inhibition by azide with respect to oxygen seems to be competitive in CAOs (K_i of 10^{-2} *M*), but it is problematic to deconvolute experimentally its effects on the reductive half-reaction (noncompetitive or competitive).

Short polyethylenediamines (Figure 14.2) have been recognized as noncompetitive inhibitors of pea seedling amine oxidase (Peč and Frébort, 1992a). Diethylenetriamine inhibits with a K_i of 7 *μM*. Triethylenetetramine shows even higher inhibition potency characterized by a K_i of 0.6 *μM*. The inhibition mode is partially noncompetitive in this case (Peč and Frébort, 1992a). Such powerful inhibitory effects may arise from an interaction with the active site copper. In contrast, the diaminoether counterparts of the above compounds, i.e., 1,5-diamino-3-oxapentane and 1,8-diamino-3,6-dioxaoctane, respectively, are very good substrates for PSAO (Šebela et al., 2003 and 2007). Note also that 1,6-diamino-3-azahexane has been shown to inhibit the enzyme competitively with a K_i of 12 *μM* (Cragg et al., 1990). Competitive inhibition of PSAO by tetraethylenepentamine (K_i = 20 *μM*) and pentaethylenehexamine (K_i = 148 *μM*) were reported by Vianello et al. (1999).

Among other significant noncompetitive inhibitors of CAOs, pyridine-derived aldoximes and ketoximes (Figure 14.2) with K_i values of 10^{-6} to 10^{-4} *M* toward PSAO and AGAO (Mlíčková et al., 2001) and various heterocyclic compounds (see below) should also be mentioned. 1-(ß-Hydroxyethyl)-2-methyl-5-nitroimidazole (metronidazole, Figure 14.2) noncompetitively inhibits human, rabbit, and rat intestinal diamine oxidases with an inhibition constant of 0.1 *mM* (Befani et al., 1995). As reported by Manna et al. (2002), the enzymes from pig kidney and bovine serum are noncompetitively inhibited by certain 1-acetyl-3,5-diphenyl-4,5-dihydro-(1H)-pyrazoles (K_i ~10^{-6} to 10^{-4} *M*). Imidazole, methylimidazoles, *N*-acetylimidazole, histamine, and *N*-τ-methylhistamine are relatively potent inhibitors of mammalian diamine oxidase but show no influence on plant enzymes. Anserine and carnosine are noncompetitive inhibitors of pig kidney and pea seedling enzymes with K_i values of 10^{-6} *M* (Biegański et al., 1982b). The pig kidney enzyme is also noncompetitively inhibited by the *Nitraria schoberi* alkaloid nazlinin (Figure 14.2) and its 3,4-dihydro derivative (Cheng et al., 1995).

14.5 MECHANISM-BASED CAO INHIBITORS

Mechanism-based inhibitors (MBIs) are unreactive substrate analogs that produce a reactive moiety when processed by the normal catalytic mechanism of an enzyme, leading to irreversible inactivation. This "Trojan horse" concept explains why such agents are often called suicide substrates and/or inhibitors. In some cases, the normal mechanism leads directly to a covalently modified form of the enzyme, and loss of enzyme activity follows a simple pseudo first-order exponential. Some examples for CAOs will be discussed.

In most cases, however, conversion of the substrate analog results in a reactive product that may bind covalently to the cofactor or an active site amino acid residue or may escape from the active site and allow for turnover of the enzyme. In the latter case, if the substrate analog is totally consumed before the enzyme is totally inactivated, the loss of activity over time will reach a plateau. This latter case involves a partitioning that can be quantified by determining rates of product formation and loss of active enzyme or by a partition technique that plots activities at plateau points as a function of the ratio of [I]/[E]. Because MBIs act first by serving as substrates and CAOs are normally considered to have a strict requirement for metabolizing unbranched primary amines, it is not surprising that most MBIs identified to date for CAOs are primary amines. Longu et al. (2005a) recently published a review of various MBIs of plant CAOs.

Historical origins — The first hint that CAOs were sensitive to carbonyl reagents suggested the involvement of a pyridoxal-like active carbonyl organic cofactor and a transamination-like mechanism for conversion of unbranched primary amine to aldehyde. MBIs were sought long before the true nature of the cofactor TPQ became clear. Pioneering work on the mechanism-based inactivation of pyridoxal enzymes (e.g., amino acid transaminases) by Abeles translated into strategies for generation of reactivity arising from transamination of primary amines via the cofactor-derived Schiff base (Abeles and Maycock, 1976). In the 1970s, Abeles and coworkers were the first to describe the MBIs for BSAO in the form of amine analogs of the modified amino acid derivatives they developed for pyridoxal enzymes. These first MBIs included propargylamine, 2-chloroallylamine, and 1,4-diamino-2-butyne (DABY). The same approach was reported by Rando and De Mairena (1974). The principal idea was that α-C deprotonation would result either directly (in the case of β-unsaturation) or upon a 1,2 elimination of HX (in the case of a halo-substituted C=C) in an electrophilic allenyl moiety. Alternatively, transamination would result in an electrophilic α,β-unsaturated imine that upon hydrolysis, would release an α,β-unsaturated aldehyde. Abeles preferred the former explanation; he felt that if the latter consequence were operative, allylamine, a substrate for BSAO, would instead be an inactivator due to its metabolism to acrolein.

Propargylamines — Following up on Abeles' lead, multiple investigators pursued the development of propargylamine- or haloallylamine-based inhibitors and, in the case of the plant and animal *diamine* oxidases, propargylic diamines such as DABY or haloallyl diamines. It might not have been predicted that DABY would act as an inactivator of BSAO, since it might have been expected to act as a selective inhibitor of diamine oxidases. Indeed DABY was shown to inactivate diamine oxidase from

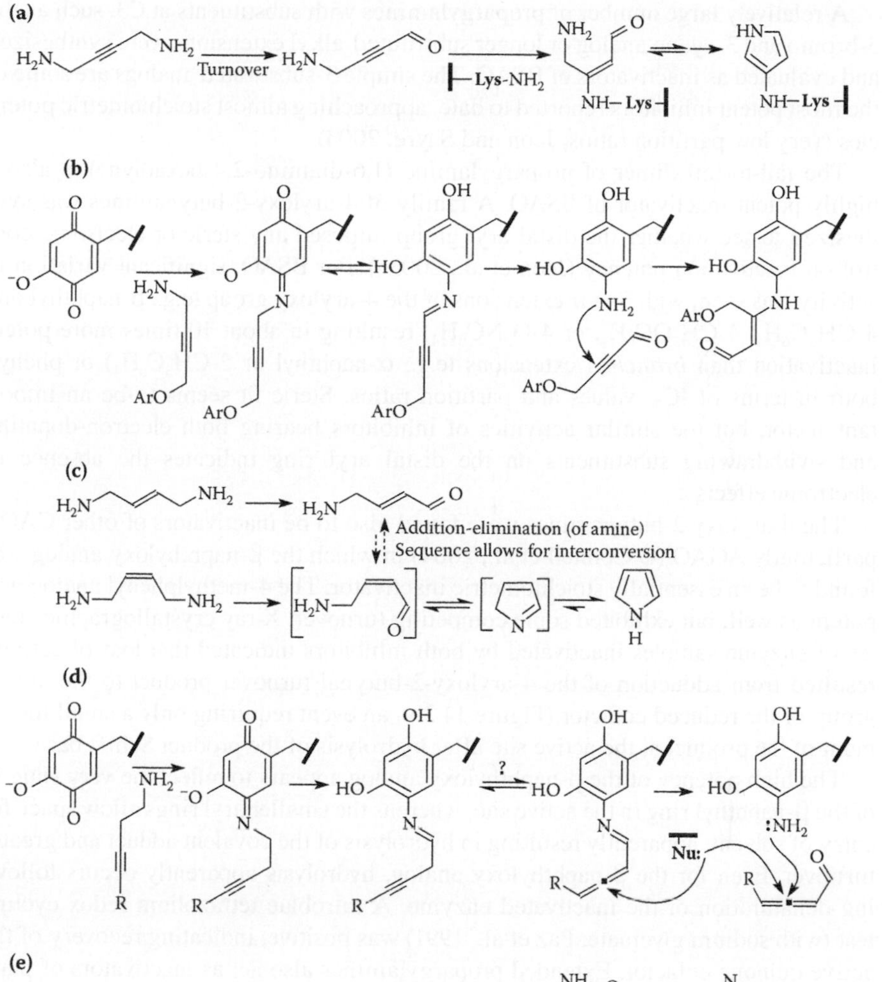

FIGURE 14.3 Some mechanisms of suicide substrate and inhibitor action.

peas (Peč and Frébort, 1992b). The mechanism of inactivation appears to involve catalytic turnover to 4-amino-2-butynal, which undergoes conjugate addition by a Lys residue in or near the active site, to yield an adduct that cyclizes to give a pyrrole (Figure 14.3a). Residues outside the active site may also be modified (Frébort et al., 2000a). DABY is also oxidized by the two CAOs present in *Aspergillus niger* that then suffer time-dependent irreversible inactivation arising from pyrrole formation on an active site Lys (Frébort et al., 1994). The site of modification for one of the two enzymes was shown by proteolytic digestion and mass spectrometry to be Lys 356.

A relatively large number of propargylamines with substituents at C3, such as the 3-bromo and 3-cyano analog or longer substituted-alkyl extensions were synthesized and evaluated as inactivators of BSAO. The simple 3-substituted analogs are some of the most potent inhibitors reported to date, approaching almost stoichiometric potencies (very low partition ratios; Jeon and Sayre, 2003).

The tail-to-tail dimer of propargylamine (1,6-diamino-2,4-hexadiyne) is also a highly potent inactivator of BSAO. A family of 4-aryloxy-2-butynamines was synthesized to see whether the distal aryl group imposed any steric or electronic control on inactivation potency (Jeon et al., 2003b). For BSAO, significant variation in activity was seen, with *linear* extensions of the 4-aryloxy group (e.g., β-naphthyl and 4-$CH_3C_6H_4$, 4-$CH_3OC_6H_4$, or 4-$O_2NC_6H_4$) resulting in about 10 times more potent inactivation than *branched* extensions (e.g., α-naphthyl or 2-$CH_3C_6H_4$) or phenyl, both in terms of IC_{50} values and partition ratios. Steric fit seems to be an important factor, but the similar activities of inhibitors bearing both electron-donating and -withdrawing substituents on the distal aryl ring indicates the absence of electronic effects.

The 4-aryloxy-2-butynamines were found also to be inactivators of other CAOs, particularly AGAO (O'Connell et al., 2004) in which the β-naphthyloxy analog was found to be an essentially stoichiometric inactivator. The 4-methylphenyl analog was potent as well, but exhibited some competing turnover. X-ray crystallographic studies of enzyme samples inactivated by both inhibitors indicated that loss of activity resulted from adduction of the 4-aryloxy-2-butynal turnover product to the amino group of the reduced cofactor (Figure 14.3b), an event requiring only a small movement of the product at the active site after hydrolysis of the product Schiff base.

The high potency of the β-naphthyloxy analog appears to reflect the very snug fit of the β-naphthyl ring in the active site, whereas the smaller aryl rings allow space for entry of solvent, apparently resulting in hydrolysis of the covalent adduct and greater turnover. Even for the β-naphthyloxy analog, hydrolysis apparently occurs following denaturation of the inactivated enzyme. A nitroblue tetrazolium redox cycling test (with sodium glycinate; Paz et al., 1991) was positive, indicating recovery of the active quinone cofactor. Extended propargylamines also act as inactivators of plant CAOs. While N^6-aminoalkyl or -aminoalkenyl derivatives of adenine acted mainly as substrates, N^6-(4-aminobut-2-ynyl)adenine was found to be a powerful inactivator of three different plant CAOs (Lamplot et al., 2005).

Allylamines and haloallylamines — As stated earlier, allylamine is a noninhibitory substrate for plasma amine oxidase. In terms of allylic diamines, 1,4-diamino-2-butene was shown to serve as a substrate for plant and animal diamine oxidases (Macholán et al., 1975; Peč et al., 1991 and 1992b) and can also be used as a convenient assay of enzyme activity since the product enal undergoes cyclocondensation to a pyrrole (which in turn could be quantified in a chromogenic reaction). However, pyrrole formation requires generation of the *cis* enal, suggesting that the substrate 1,4-diamino-2-butene must be either an *E*/*Z* mixture or pure *Z* isomer (Figure 14.3c). Investigation of the individual *E* and *Z* isomers (He et al., 1995a) showed that 1,4-diamino-2(*E*)-butene is a rather potent inactivator of PKAO (the *Z* isomer was a weak inhibitor), apparently because the *trans* enal turnover product cannot immediately cyclize and thus is capable of alkylating a nucleophilic residue in the enzyme active site.

Attachment of a halogen to the C=C of allylamine greatly potentiates its inactivation potency. While Abeles reported modestly high inhibitory potency of 2-chloroallylamine for BSAO, the *cis*- and *trans*-3-chloroallylamines were found to be even more potent than propargylamine (Jeon and Sayre, 2003). These compounds, through an addition–elimination (of chloride) mechanism, may follow an inactivation mechanism that converges with that of propargylamine. The haloallylic diamine, 1,4-diamino-2-chloro-2-butene, was also shown to inhibit BSAO, analogously to DABY (Jeon et al., 2003a) along with AGAO and ECAO, and was an even more potent (almost stoichiometric) inactivator of equine plasma amine oxidase, *Pichia pastoris* lysyl oxidase, and PSAO (Shepard et al., 2002).

Another area of development of haloallylamine inhibitors of CAOs started with Merrell Dow's finding that certain 2-aryl-3-haloallylamines (aryl = substituted phenyl) the company developed to inhibit the flavin-dependent MAOs were also irreversible inhibitors of CAOs. For example, potent inhibition of MAO-B and SSAO occurred with arylfluoroallylamines whereas arylchloroallylamines such as (*E*)-2-phenyl-3-chloroallylamine, although not quite as potent, were selective inhibitors of SSAO (Palfreyman et al., 1986 and 1994; Lyles et al., 1987). Relying on BSAO as a representative CAO, a more recent evaluation of the stereochemical influence of 2-phenyl-3-haloallylamines on selectivity confirmed the previously observed potency rank order of the 3-halo group as F > Cl > Br, and showed that although the (*E*) isomers were reported as more potent inhibitors of rat brain MAO, the (Z) isomers were more potent inactivators of BSAO (Kim et al., 2006). The selectivity preferences (IC_{50} ratios) for both enzymes ranged from 3 to 10.

Homopropargylamines and allenylamines — A new mechanistic concept, with no analog in pyridoxal enzyme research, was the proposal that CAO-mediated transamination of a *homo*propargylamine (1-amino-3-butyne), although not immediately yielding an electrophilic turnover product, could result in subsequent tautomerization to an electrophilic allene, either at the product Schiff base stage or turnover aldehyde stage (Figure 14.3d). Either the allenyl product Schiff base or aldehyde could alkylate an active site amino acid residue or, in the latter case, the reduced cofactor. Homopropargylamine was shown to be a potent irreversible inactivator of BSAO, as was 2,3-butadienylamine allenyl amine that would be metabolized to the same allenyl aldehyde arising from turnover of 1-amino-3-butyne (Qiao et al., 2004). Certain C4-extended homopropargylamines were also shown to be potent inhibitors (Qiao, 2004).

On the basis of the interesting inhibitory activity of the DABY diamine, it was of interest to determine the behaviors of 1,5-diamino-2-pentyne (DAPY), which is propargylic at one end and homopropargylic at the other end, and 1,6-diamino-3-hexyne, which is homopropargylic at both ends. Both were shown to be highly potent inhibitors of BSAO, whereas the *N*-monomethyl and *N,N*-dimethyl derivatives (DAPY has two different ones) were very weak inhibitors (Qiao, 2004). The latter observation has important mechanistic implications that are unexplainable at this time.

DAPY was also shown to be a substrate and inactivator of plant CAOs (Lamplot et al., 2004). Evidence that inactivation may reflect lysine modification by a mechanism analogous to the lysine pyrrolation reported for DABY (*vide supra*), was the isolation from the incubation solution of a cyclocondensation product (a resonance-stabilized 4-amino-2,3-dihydropyridine) resulting from trapping of the turnover product by the amine substrate (Figure 14.3e).

2-Haloalkylamines — According to proven strategies for mechanism-based inactivation of pyridoxal enzymes by the incorporation of a β-halogen into an amino acid (e.g., 3-chloroalanine), incorporation of a β-halogen into a primary amine should result in potential inactivation of CAOs. Indeed, a number of groups reported that 2-bromo- and 2-chloroethylamine act as irreversible inhibitors of various CAOs of plant and animal origins (Medda et al., 1997a; Kinemuchi et al., 2001). Evidence suggests formation of a covalently bound form of the cofactor; 2-bromoethylamine appears to display considerable selectivity for SSAO (Yu et al., 2001; Tabata et al., 2003). In contrast, 3-bromopropylamine is a reversible inhibitor only of CAOs. Two mechanisms may be involved in formation of a stable cofactor derivative (Figure 14.4a). In the first, α-deprotonation can induce β-elimination of halide to generate a vinylamine substrate Schiff base that can aromatize by conjugate addition of the hydroxyl group at C4 to give a cyclic cofactor adduct. Alternatively, the substrate Schiff base can undergo transamination to the product Schiff base of 2-haloacetaldehyde, which is activated for S_N2 displacement of halide by the C4 hydroxyl to give the same final cycloadduct. In the latter mechanism, generation of aldehyde turnover product may be a possible step along the pathway for inactivation, but if so, it must react with the reduced cofactor before it leaves the active site. Because 1,2-diaminoethane is also an irreversible inhibitor of plant CAOs (Medda et al., 1997a), the most likely mechanism would involve formation of a *bis*-Schiff base of the TPQ cofactor prior to proton abstraction which, subsequent to transamination could undergo a second round of oxidation to yield an aromatic pyrazine derivative of the cofactor (Figure 14.4b), as proposed for lysyl oxidase (*vide infra*).

Curiously, *in vivo* studies of the effect of 2-bromoethylamine on SSAO revealed differences in its action and the action of semicarbazide (e.g., how the two inhibitors affect each other when one is given before the other and vice versa), suggesting their interactions at two different sites (Vidrio and Medina, 2005). This observation is inconsistent with both inhibitors forming adducts of the same TPQ moiety.

Aromatization-prone inhibitors — The apparently strict preference of CAOs for metabolizing primary amines (e.g., *N*-methylbenzylamine is not a substrate) and the fact that α-branched primary amines (e.g., α-methylbenzylamine) are also not substrates appears to suggest a steric exclusion control since both types of amines can enter into substrate Schiff base formation with TPQ (a secondary amine would yield an iminium species). Thus, the discovery that the allylic secondary amine 3-pyrroline and especially its 3-aryl derivatives acted as mechanism-based inactivators of BSAO (Lee et al., 1996) was very surprising and implies that these compounds are first accepted as substrates.

Analogs bearing electron-withdrawing 3-aryl groups (e.g., 3-pyridyl-3-pyrrolines) are some of the most potent inhibitors of BSAO known (Zhang et al., 2007).

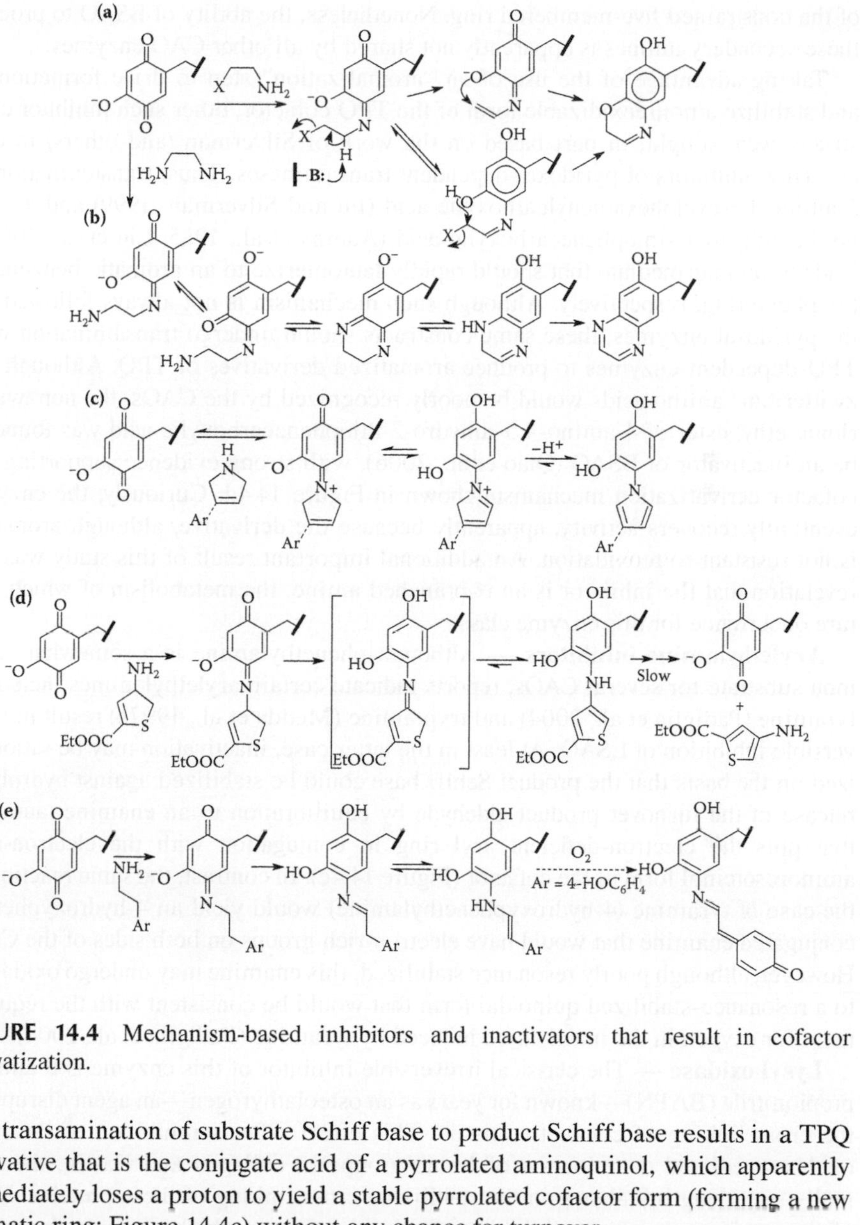

FIGURE 14.4 Mechanism-based inhibitors and inactivators that result in cofactor derivatization.

The transamination of substrate Schiff base to product Schiff base results in a TPQ derivative that is the conjugate acid of a pyrrolated aminoquinol, which apparently immediately loses a proton to yield a stable pyrrolated cofactor form (forming a new aromatic ring; Figure 14.4c) without any chance for turnover.

Accordingly, the kinetics of inactivation are simple pseudo first-order exponentials. Pyrrole formation on the cofactor was confirmed for one of the 3-aryl-3-pyrrolines by LC-MS analysis of the thermolytic digest of the inactivated enzyme (Zhang et al., 2007). The 3-pyrrolines are of particular interest in that they stoichiometrically inactivate the TPQ-dependent enzyme and also serve as pure substrates of the flavin-dependent amine oxidase (Lee et al., 2002). The ability of the 3-pyrrolines to be processed as substrates by the enzyme was proposed to reflect the small steric size of the constrained five-membered ring. Nonetheless, the ability of BSAO to process these secondary amines is apparently not shared by all other CAO enzymes.

Taking advantage of the use of an "aromatization" step to drive formation of and stabilize a nonreoxidizable form of the TPQ cofactor, other such inhibitor constructs were sought, in part based on the work of Silverman (and others) in discovering inhibitors of pyridoxal-dependent transaminases. Thus, transamination of 5-amino-1,3-cyclohexadienylcarboxylic acid (Fu and Silverman, 1999) and 4-amino-4,5-dihydro-2-thiophenecarboxylic acid (Adams et al., 1985; Liu et al., 2007b) leads to an intermediate that should rapidly tautomerize to an aromatic benzene or thiophene ring, respectively. Although such mechanism is not always followed for the pyridoxal enzymes, these same constructs should undergo transamination with TPQ-dependent enzymes to produce aromatized derivatives of TPQ. Although the zwitterionic amino acids would be poorly recognized by the CAOs, the nonzwitterionic ethyl ester of 4-amino-4,5-dihydro-2-thiophenecarboxylic acid was found to be an inactivator of BSAO (Qiao et al., 2006), with strong evidence supporting the cofactor derivatization mechanism shown in Figure 14.4d. Curiously, the enzyme eventually recovers activity, apparently because the derivative, although aromatic, is not resistant to reoxidation. An additional important result of this study was the revelation that the inhibitor is an α-branched amine, the metabolism of which is a rare occurrence for this enzyme class.

Arylethylamine inhibitors — Although phenethylamine is a somewhat common substrate for several CAOs, reports indicate certain arylethylamines including tyramine (Padiglia et al., 2004) and tryptamine (Medda et al., 1997b) result in irreversible inhibition of LSAO. At least in the latter case, inactivation may be rationalized on the basis that the product Schiff base could be stabilized against hydrolytic release of the turnover product aldehyde by equilibration to an enamine tautomer that puts the electron-deficient aryl ring in conjugation with the electron-rich aminoresorcinol form of the cofactor (Figure 14.4e). In contrast, the same reaction in the case of tyramine (4-hydroxyphenethylamine) would yield an 4-hydroxyphenyl-conjugated enamine that would have electron-rich groups on both sides of the C=C. However, although poorly resonance stabilized, this enamine may undergo oxidation to a resonance-stabilized quinoidal form that would be consistent with the requirement for oxygen in the inactivation induced by tyramine (Padiglia et al., 2004).

Lysyl oxidase — The classical irreversible inhibitor of this enzyme is ß-aminopropionitrile (BAPN)—known for years as an osteolathyrogen—an agent disrupting the cross linking of connective tissue. This effect reflects inhibition of the lysyl oxidase-mediated deamination of lysine groups in collagen and elastin, the normally occurring initial step in the stabilizing cross linking of these proteins. Most of the work on developing inhibitors for lysyl oxidase, which contains the modified lysyl tyrosylquinone cofactor, was carried out by Kagan and coworkers, and aimed at developing potential antifibrotic agents. This group was the first to show by a labeling study that BAPN inhibition is accompanied by covalent modification of the enzyme by the entire molecule (Tang et al., 1983). They later found that the electron-withdrawing nitrile moiety of BAPN can be replaced by chlorine, bromine, or the nitro function to yield 2-haloethylamines or 2-nitroethylamine that also act as mechanism-based irreversible inhibitors of the enzyme (Tang et al., 1984), though the distinctive behavior of the 2-haloethylamines from the 2-cyano-/2-nitro ethylamines was noted.

It was later shown that some vicinal diamines are also irreversible inhibitors of the enzyme (Gacheru et al., 1989). Both *cis*-1,2-diaminocyclohexane and ethylenediamine but not *trans*-1,2-diaminocyclohexane were inhibitory, consistent with the proposition that inactivation involves a bifurcated interaction with the quinone cofactor to produce a cyclic adduct (the two amino groups in the *trans* isomer have improper stereochemical disposition). Although the quinone cofactor of lysyl oxidase is different from TPQ, its transamination chemistry is essentially identical, and it appears that the mechanism of inactivation of lysyl oxidase by the 2-haloethylamines and vicinal diamines follows the reactions described above for TPQ enzymes (Figures 14.4a, b). Regarding BAPN, the results of comparative studies indicate a sufficient difference between the sensitivities of lysyl oxidase and the other amine oxidases to this inhibitor to warrant the consideration of BAPN as an antifibrotic agent *in vivo* (Tang et al., 1989). Its mechanism (along with that of 2-nitroethylamine) has not yet been elucidated, but it may reflect simply the formation of a product Schiff base that is stabilized against hydrolysis by tautomerization to resonance-stabilized cyano- or nitro-enamines.

More recently, homocysteine thiolactone, selenohomocysteine lactone (K_I values 8 to 20 μM), and to a lesser extent, homoserine lactone, were found to be competitive, irreversible inhibitors of lysyl oxidase, with evidence provided for a mechanism involving derivatization of the quinone cofactor (Liu et al., 1997). The details, however, have yet to be revealed. Finally, *N*-(5-aminopentyl)aziridine, but not analogs with shorter or longer tethers, was shown to competitively inhibit lysyl oxidase and then induce irreversible inhibition over time (Nagan et al., 1998).

Aziridines — Analogs of diamines such as putrescine and cadaverine bearing aziridinyl groups at one end were shown to be irreversible inactivators of diamine oxidase (Conner et al., 1992). These agents are not classical mechanism-based inhibitors since they already contain reactive aziridinyl groups, and are best classified as active-site-directed agents. Affinity of the inhibitors for the active site appeared to be dependent on alkyl chain length, suggesting that binding promotes the reactivity of the aziridinyl group.

14.6 DESIGNING AND SIGNIFICANCE OF AMINE OXIDASE INHIBITORS

In the past, inhibitors of prokaryotic and eukaryotic CAOs were studied predominantly to uncover the chemical structures and properties of the active site cofactors and characterize the respective mechanisms of reaction (Abeles and Maycock, 1976; Morpurgo et al., 1992; Lee et al., 2001b). During several decades of investigations, numerous compounds were characterized as typical and more or less selective inhibitors of the enzymes.

Great importance is attached to carbonyl reagents, powerful competitive inhibitors, and to highly-selective mechanism-based inhibitors. Semicarbazide, a carbonyl reagent, has been long known as a powerful inhibitor of MAOs associated with plasma membranes and its specificity defined the common semicarbazide-sensitive amine oxidase (SSAO) term. In mammals, two forms of SSAO have been identified: tissue-bound and soluble plasma forms. The latter group appears to result from proteolytic cleavage of the membrane-bound enzyme (Lyles, 1996). SSAO expression at the cell surface is tightly regulated and undergoes significant upregulation at inflammation sites. It has been reported that increases in the levels of plasma and/or membrane SSAO occur in many inflammation-associated diseases including rheumatoid arthritis, inflammatory bowel disease, diabetes, atherosclerosis, chronic heart failure, and perhaps also Alzheimer's disease (Wang et al., 2006). SSAO-mediated adhesion and transendothelial migration of leukocytes through endothelial cell layers is mediated by its enzymatic activity and can be blocked by small molecule inhibitors (Salmi et al., 2001).

The fact that SSAO exerts both enzymatic and adhesion activities and the strong correlation between its upregulation in many inflammatory conditions make it an interesting therapeutic target (Wang et al., 2006) and we anticipate the future significance of the amine oxidase inhibitor groups. Most reported SSAO inhibitors were developed for other therapeutic uses or served as simple chemical reagents with highly reactive structural elements (Kinemuchi et al., 2004; Wang et al., 2006). Most of them lack the combination of high potency and selectivity for SSAO.

The primary aromatic monoamines, each containing a single methyl group on the α-carbon adjacent to the amino group, inhibit SSAO, but they were initially evaluated as inhibitors of the MAO-A and MAO-B flavoprotein enzymes (Kinemuchi et al., 2004). Another example, phenelzine (a carbonyl reagent), is equally potent on SSAO and MAO-A (Andree and Clarke, 1982a). Many haloallylamine SSAO inhibitors also strongly inhibit MAO-B (Lyles et al., 1987). Interestingly, MDL72161, a fluoroallylamine derivative originally designed as an MAO-B inhibitor to treat Parkinson's disease, exhibits a high inhibitory potency on rat aorta SSAO ($IC_{50} = 2.5\ nM$) with no apparent selectivity over MAOs. A series of hydrazine derivatives were synthesized and evaluated as highly potent and selective SSAO inhibitors (Wang et al., 2006). A potent and selective SSAO inhibition was also achieved using hydroxylamine and propargylamine that exhibit little inhibitory effect on MAO activity (Kinemuchi et al., 2004).

The requirement of inhibition selectivity stimulated designing and testing of mechanism-based inhibitors. Two reported synthesis strategies are the incorporation of a halogen or unsaturation at the β-positions of amine substrates. For example, 2-bromoethylamine is a highly selective and potent mechanism-based inhibitor of membrane-bound SSAO without inhibiting MAO activity *in vivo* and *in vitro* (Kinemuchi et al., 2001).

Various β-unsaturated compounds were successfully tested in reactions with plant diamine oxidases (Longu et al., 2005a) and the enzyme from bovine plasma (Jeon et al., 2003a). Another and revolutionary approach to CAO inhibitor design was recently described by Contakes et al. (2005). A series of highly potent channel-blocking inhibitors of AGAO were synthesized. Molecular wires comprising a Ru(II)- or Re(I)-complex head group, an aromatic tail group, and an alkane linker reversibly inhibited the enzyme and exhibited K_i values between 6 μM and 37 nM (Figure 14.2). The crystal structure of a Ru-wire–AGAO conjugate clearly demonstrated that the wire occupied the active site channel tracing the path of substrate from solvent to active site. Because active site channel residues vary substantially among CAOs, carefully designed metallowires may function as highly selective inhibitors. Based on findings with different head groups, tail-group substituents, and linkers, the authors predict that selective inhibitors with picomolar affinities will be available in the near future.

ACKNOWLEDGMENTS

M.S. greatly appreciates financial support of the Ministry of Education, Youth and Sports of the Czech Republic (Grant MSM6198959216). L.M.S. acknowledges support from the National Institutes of Health (GM48812) and the American Diabetes Association (1-06-RA-117).

15 Pharmacological Applications of Copper Amine Oxidases

Emanuela Masini and Laura Raimondi

CONTENTS

15.1 PHARMACOLOGICAL APPLICATIONS

15.1.1 IMPLICATIONS OF CONTROLLING BZAO ACTIVITY IN DIABETES: THE PRESENT

The significance of increased plasma levels of BzAO occurring in diabetes patients is far from clear. Increased plasma BzAO levels are correlated with glycated hemoglobin (Boomsma et al., 1995 and 2005a; Xu et al., 2005) Moreover circulating BzAO has also been proposed as an independent prognostic marker for mortality in heart failure (Boomsma et al., 2000b), suggesting a relationship of plasma enzyme activity, the duration of diabetes, and its cardiovascular complications.

It is currently accepted that most cardiovascular pathologies have a common base in endothelial dysfunction characterized by metabolic (insulin-resistance and oxidative stress) and functional (altered vasodilatation) aspects. Plasma BzAO may be indicative of the metabolic aspects of endothelial dysfunction and, more generally, of changes in glucose homeostasis. Because subclinical, asymptomatic, endothelial dysfunction is hard to detect, an understanding of BzAO timing in the circulation is crucial to investigating its prognostic value. Whether drug therapies (e.g., statins and renin angiotensin system blockers) for improving endothelial dysfunction exert beneficial effects by controlling (reducing?) BzAO plasma activity is still unknown. Another unknown factor is whether reduced plasma BzAO levels are among the beneficial effects of these drugs or whether strategies to reduce the onset of type 2 diabetes in at-risk populations may also prevent increases in plasma levels of BzAO.

If BzAO plasma levels serve as markers of diabetes, they may also increase in patients treated with diabetogenic drugs such as β-blockers, diuretic thiazides, and corticosteroids. Again, epidemiological data are lacking.

Different perspectives may be predicted for tissue-bound SSAO highly expressed in insulin-sensitive cell types. In adipocytes, membrane-bound SSAO-dependent substrate deamination via local production of hydrogen peroxide (H_2O_2) has the potential to exert insulin-like effects, even in the absence of insulin, on glucose uptake (Zorzano et al., 2003), adipose differentiation, and lipolysis (Carpéné et al., 2001). However, even if the production of H_2O_2 is not a peculiarity of membrane-bound SSAO (many other enzyme activities can produce H_2O_2 including DAO) the particular localization of the peroxide produced by membrane-bound SSAO seems to facilitate insulin receptor substrate-1 phosphorylation (Zorzano et al. 2003), thus exerting local beneficial insulin-sensitizing effects.

On the other hand, membrane-bound SSAO activity spreads toxic aldehydes in the microenvironment that may initiate a deleterious cycle involving protein and DNA cross linkage related to angiotoxicity (Yu and Zuo, 1997), a typical consequence of hyperglycemia and an index of diabetes complications. However, as a whole, the benefit of reducing membrane-bound SSAO activity prevails over the local insulin-like effects of the peroxide produced (Stolen et al., 2004a). These authors demonstrated a more favorable role of membrane-bound SSAO inhibitors over substrates in reducing the severity of diabetes-related cardiovascular complications with atherosclerotic bases. In particular, in adipocytes and vascular smooth muscle cells, SSAO substrates generate a favorable microenvironment to remove local insulin resistance but remain neutral related to the control of plasma glycemia. Conversely, SSAO inhibitors protect against angiotoxicity. The addition of aminoguanidine (pigimaline), a guanidine-like compound, to common antidiabetic treatments, has been used for several years to alleviate the protein aging associated with diabetes (Abdel-Rahman and Bolton, 2002; Hou et al., 1998; Yu and Zuo, 1997; Friedman, 1995; Cameron and Cotter, 1993) and the extent of glycated hemoglobin (Yu and Zuo, 1997). Aminoguanidine treatment has a neutral effect on glycemia, even though inhibition of SSAO activity implies increasing tissue levels of enzyme substrates. Aminoguanidine is not selective for SSAO but inhibits the DAO and the NO synthases (Ansar and Ansari, 2006).

15.1.2 Future Diabetes Treatments

Pharmacological treatment of insulin resistance represents a first line approach to reduce diabetes-related complications. In the complex picture of future insulin-sensitizing agents, the antidiabetic activity of vanadium compounds has recently garnered immense interest as an oral therapy. Vanadium compounds lowered glycemia and normalized plasma lipid profiles in an animal model of diabetes (Yamazaki et al., 2005; Kordowiak et al., 2005; Mukherjee et al., 2004; Abella et al., 2003), exerting insulin-like effects downstream of the insulin receptor and probably inhibiting phosphatase activities. Concerns about their safety profiles presently limit clinical applications.

In recent years, major efforts have focused on developing new chemical formulations to minimize the side effects of metal drugs including vanadium esters. A novel combination of amine substrates for membrane-bound SSAO and low concentrations of vanadium esters has been proposed to increase the production of peroxovanadate, the active insulin-mimetic compound of pharmacological interest (Yraola et al., 2007a).

15.1.3 SSAO Activity and Neurodegenerative Diseases: Evidence for Future Pharmacological Implications?

Cerebral amyloid angiopathy characterized by the deposition of β-amyloid in brain vessels, inducing the degeneration of vascular smooth muscle and endothelial cells, is considered a crucial event in Alzheimer's disease (AD) pathogenesis. AD patients exhibit increased membrane-bound SSAO levels and enzyme activities in their brains are co-localized with β-amyloid protein (Ferrer et al., 2002).

Increased plasma levels of BzAO in AD patients may indicate insulin resistance, and membrane-bound SSAO inhibitors may trigger β-amyloid peptide polymerization (Munch et al., 1997) due to increased production of formaldehyde, methylglyoxal, and malonyldialdehyde from endogenous membrane-bound SSAO substrate degradation. Membrane-bound SSAO activity may act as an initiating factor for protein fibrillation, a typical manifestation of this disease. Therefore, inhibition of membrane-bound SSAO may prevent local increases of aldehydes, thus reducing the pro-inflammatory potential of such compounds. If that hypothesis is correct, AD therapy including inhibitors of membrane-bound SSAO may become a primary prevention strategy.

If the aldehydes play a role in generating protein cross linkage, diabetic patients may be at risk for AD. The clinical relationship between Type 2 diabetes mellitus and AD has been debated for over a decade. Several studies have failed to show a clear clinical correlation; others have demonstrated that Type 2 diabetes is an independent risk factor for inflammatory-based neurodegenerative diseases including AD (Beeri et al., 2005; Watson and Craft, 2006).

Increasing evidence suggests that insulin contributes to normal brain functioning and that peripheral insulin abnormalities increase risks of memory loss and neurodegenerative disorders such as AD. Potential mechanisms of these effects include the role of insulin in cerebral glucose metabolism, peptide regulation, modulation of neurotransmitter levels and other aspects of the inflammatory network. AD patients are not routinely evaluated for Type 2 diabetes or hyperinsulinemia.

Current AD treatments produce modest benefits, and several drugs that target metabolic and inflammatory pathways are being evaluated, most notably the statins that reduce low density lipoproteins and inflammation but may not influence amyloid deposition—an important precursor of AD. Although some evidence supports a potentially important role for peroxisome-proliferative activated receptor agonists such as glitazones (Craft, 2007), no current reports cover randomized clinical trials in AD patients of drugs that target insulin or insulin resistance (Williamson et al., 2007). Such patients may benefit from treatment with statins or antidiabetic drugs that target the insulin cascade involved in glucose homeostasis. No data currently indicate that statins or other therapies including amine oxidase inhibitors show potential in ameliorating the severity of AD. On the other hand, no experimental or clinical data demonstrate that inhibition of tissue or plasma SSAO is included in the pharmacological profiles of cholinomimetic or anti-glutamatergic drugs used to treat AD. This point must be considered in assigning a precise role for SSAO in AD pathogenesis and care.

15.1.4 Antiinflammatory Activity of Molecule Inhibiting VAP-1/SSAO Activity

BzAO activity is essential in VAP-1 to mediate leukocyte adhesion to endothelial cells (Yegutkin et al., 2004). Inhibition of SSAO activity reduces the production of reactive and toxic compounds in the endothelium (Yu and Zuo, 1996). In theory, specific and selective inhibitors of VAP-1/SSAO have the potential to reduce tissue damage by reducing accumulation of reactive oxygen species including hydrogen peroxide and toxic aldehydes. In addition, VAP-1/SSAO inhibitors may prevent tissue leukocyte infiltration, an initiating factor for immuno-mediated diseases including Type 1 diabetes.

Whether SSAO plays a role in the preferential entry of leukocytes into an organ or tissue remains unknown. However, based on their possible double role, VAP-1/SSAO inhibitors have been designed, synthesized (Wang et al., 2006), proposed as anti-inflammatory drugs (Salter-Cid et al., 2005), and included in the LJP series. LJP 1586 is a potent amine-based inhibitor for VAP-1/SSAO with good oral bioavailability and therapeutic window. Interestingly, after LJP 1586 supplementation in animals, the enzyme activity recovery time was approximately 72 hr. The pharmacokinetic activity of this compound thus predicts low turnover of enzyme at the cell surface and/or slow reversibility of the interaction with the enzyme (O'Rourke et al., 2008).

From the first study, LJP 1586 has shown benefits as an anti-inflammatory treatment in acute and chronic pulmonary diseases—above and beyond its effects on leukocyte migration and possibly including a favorable impact on fibrosis secondary to local formaldehyde production (O'Rourke et al., 2008). Other SSAO inhibitors of the LJP series have been tested. LJP 1207 has shown potential in a mouse model of relapsing–remitting experimental autoimmune encephalomyelitis, a model that shares many characteristics with human multiple sclerosis (O'Rourke et al., 2007). Animal treatment with LJP 1017 led to a dramatic reduction in adhesion and transmigration across pial vessels of leukocytes, predominantly neutrophils, and also significantly improved neurological outcomes in diabetic ovariectomized female rats given chronic estrogen replacement therapy—a model associated with increased postischemic inflammation (Xu et al., 2005). In a mouse model of ulcerative colitis, administration of LJP 1207 significantly reduced mortality, body weight loss, and colonic cytokine levels and revealed highly significant suppressions of inflammation, injury, and ulceration scores (Salter-Cid et al., 2005). Inhibition of VAP-1/SSAO may also represent a novel strategy for reducing ocular inflammation of uveitis, macular degeneration, and diabetes retinopathy (Noda et al., 2008).

15.1.5 Assessment of Physiopharmacological Role of Membrane-Bound SSAO Endogenous Substrates: Inhibition of Enzyme as Strategy to Increase Substrate Levels

Endogenous substrates for membrane-bound SSAO include methylamine (MET), aminoacetone (AA), and β-phenylethylamine (β-PEA). The latter is considered a trace amine precursor of neurotransmitters involved in neuronal plasticity and drug abuse (Gass and Olive, 2008), but the significance of MET and AA is poorly understood. Substrate degradation by membrane-bound SSAO produces ammonia as a secondary product. Although many reports cover the effects of aldehydes and hydrogen peroxide, the consequences of membrane-bound SSAO-dependent production of ammonia are largely unknown.

Recent evidence suggests that MET may belong to a series of small endogenous molecules endowed with signaling features. MET supplementation in rodents modifies feeding behavior, producing species-specific effects without eliciting amphetamine-like effects. Interestingly, ammonia supplementation produced similar effects, suggesting a common mechanism of action and inclusion of ammonia effects in those of MET.

Although, MET and ammonia produce hypophagia in mice (Pirisino et al., 2004), MET directly delivered to the central nervous system by intracerebroventricular injection (i.c.v.) induced different effects in rats, depending on dose (Raimondi et al., 2007). Moreover, MET reduced feeding when administered intraperitoneally (i.p.), suggesting that it can freely cross the blood–brain barrier even in the absence of inflammation.

MET can therefore be included in the plethora of endogenous compounds controlling the hypothalamus—a site virtually devoid of SSAO activity. Peripheral membrane-bound SSAO actively controls MET levels that reach the hypothalamus. In conditions of SSAO inhibition by aminoguanadine, a potentiation of MET hypophagia was produced in both healthy and diabetic mice (Pirisino et al., 2001; Cioni et al., 2006). MET would have exerted its own effects independently of membrane-bound SSAO. In fact, MET hypophagia is potentiated by inhibiting the oxidative deamination by SSAO that represents main MET metabolic pathway. These results suggest that MET may be of use in resolving diabetes hyperphagia.

Alimentary disorders are usually included in the clinical aspects of patients suffering neurodegenerative pathologies including AD and senile dementia (Mamhidir et al., 2007; Tamura et al., 2007). Thus, inhibition of membrane-bound SSAO, which is increased in the brain vessels of such patients, would reduce oxidative and carbonyl stress in the central nervous system and help correct alimentary disturbances regulating MET levels. Unfortunately, no information to date indicates the effects of endogenous levels of MET in physiological and pathological conditions.

MET effects in the hypothalamus are linked to the expression of a particular type of voltage-dependent potassium channel of the Shaker-like family known as Kv1.6. This channel is not involved in the hypophagic effects of ammonia. The abilities of ammonia and derivatives to interact with potassium channels are well known (Choi et al., 1993). The identification of Kv1.6 as a possible target for MET effects presents new perspectives. Interacting in these channels in the hypothalamus, MET modulates the releases of NO and dopamine, two key mediators of animal feeding. These and other observations (Carpéné et al., 2007; Prévot et al., 2007) suggest that membrane-bound SSAO inhibitors may be effective as anti-obesity drugs because of their effects on the hypothalamus and also resulting from SSAO localization.

Preferential expression at adipocytes (Raimondi et al., 1991) guarantees a link between enzyme expression and adipose differentiation (Raimondi et al., 1990). Oxidative deamination of SSAO substrates producing hydrogen peroxide in adipocytes induces and sustains adipose differentiation in rodents (Carpéné et al., 2001) and humans (Bour et al., 2007a, b). Again, the double face of this enzyme suggests a crucial role for its ability to regulate factors involved in energy intake and storage.

15.1.6 Drugs Interacting with Plasma BzAO Activity

BzAO, membrane-bound SSAO, and the mitochondrial MAOs may also represent nonmicrosomial phase I enzymes involved in drug metabolism. Although several drugs have shown capacities to inhibit BzAO plasma activity, whether this feature is clinically relevant is still unknown. BzAO inhibition may, however, explain certain drug-related side effects. Two classes can be identified: (1) drugs bearing aminoguanidine-like moieties, and (2) drugs bearing BzAO or SSAO substrate-like moieties.

Aminoguanadine belongs to the first class. Clearly, BzAO and membrane-bound SSAO inhibitions are integral parts of aminoguanidine therapeutic activity. In addition to aminoguanadine, other drugs such as benserazide, a decarboxylase inhibitor used in anti-Parkinson therapy, inhibits plasma BzAO activity. Patients suffering from Parkinson's disease and treated with benserazide exhibited lower plasma BzAO activity than controls. It is possible that side effects following long-term therapy may include modification of circulating levels of the amine substrate for BzAO (Coelho et al., 1985). Whether benserazide therapy reduces oxidative stress of patients or ameliorates their alimentary disorders in consequence of plasma BzAO inhibition is not known.

Phenelzine, a nonselective and irreversible mitochondrial inhibitor of MAO-A and -B, has been used for many years as an antidepressant to treat panic disorders and social anxiety. Its efficacy is the result of MAO inhibition leading to increased sympathetic amine levels at the synaptic cleft and γ-aminobutyric (GABA) acid transaminase inhibition that markedly increases GABA brain levels. Phenelzine also exerts inhibitory activity on BzAO and tissue-bound SSAO. This feature has a role in its neuronal protective effects.

Hydralazine is a guanidine-like drug used in antihypertensive therapy, although its use has been discouraged based on the availability of more selective drugs. Hydralazine is also a potent relaxant of smooth muscle cells; the mechanism of action remained unknown. Hydralazine is a potent and irreversible inhibitor of BzAO and tissue-bound SSAO activities. Recent studies in aortic rings of rats indicate that BzAO and SSAO substrates such as benzylamine, phenylethylamine, and methylamine, by producing H_2O_2, magnify the vasodilation activity of hydralazine. It is speculated that this mechanism may be novel for hydralazine-dependent vasodilation (Vidrio and Medina, 2007).

Isoniazid is a hydrazine derivative used in antitubercular therapy. Because of its structure, isoniazid is an inhibitor of copper-containing amine oxidases (CAOs), including diamine oxidases (DAOs) and membrane-bound semicarbazide-sensitive amine oxidases (SSAOs). No current evidence indicates that inhibition of these enzyme activities is involved in isoniazid's antimicrobial activity. Instead, histamine intoxication after ingestion of histamine-rich foods has been described in isoniazid-treated patients (Uragoda and Lodha, 1979; Uragoda and Kottegoda, 1977). These adverse drug effects result from histamine accumulation, a condition that reduces histamine catabolism.

15.1.7 Diamine Oxidase

15.1.7.1 Role of Diamine Oxidase in Anaphylaxis

Histamine plays a fundamental role in anaphylaxis and is involved in allergic and pseudoallergic reactions. At variance with other metabolic pathways, histaminase activity is not directly upregulated by endogenously released histamine. Plasma histaminase activity increases in anaphylactic shock, but not during histamine injection. In some cases, plasma levels of histaminase may be intrinsically low and its activity further decreased by exogenous histamine, thereby predisposing to anaphylactic reactions.

Enhanced histamine levels in humans may be related to various endogenous and/or exogenous factors. Food-induced histaminosis has been described as the result of high histamine content or histamine releasers in food (Sattler et al., 1989). The first symptom of excess histamine intake and/or release is an increase in gastric secretion followed by tachycardia, headache, and hypotension (Slorach, 1991). The largest amounts of histamine and tyramine (that have similar vasoactive properties) are found in fermented foods such as cheeses, red wines, tinned fish including tuna, fish sauces, sauerkraut, cured pork, and sausages. The histamine content of French cheeses can reach values >800 μg/g and can cause toxic symptoms (Taylor, 1986). High levels of histamine have also been detected in Oriental food, accounting for the so-called Chinese restaurant syndrome (Chin et al., 1989).

De novo chemical formation of amines may also occur during normal cooking and food storage (Taylor 1986). The concept of pseudoallergic reactions, or false food allergies, was initially hypothesized and recently confirmed due to abnormal intake of biogenic amines like histamine (Reese et al., 2008). The abnormal intake of biogenic amines like histamine and tyramine is one of the major mechanisms involved. The main clinical symptoms include a drop in blood pressure, angioedema, vasomotor headache, alteration of intestinal functions, and cutaneous and respiratory symptoms. High levels of histamine may also derive from bacterial flora of the colon, especially in subjects with colonic dysmicrobism due to excessive intake of foods rich in cellulose and starch.

Other histamine-dependent mechanisms of pseudoallergic reactions to food may include changes in the activities of the enzymes involved in the metabolism of histamine. Decreases in histaminase and MAO-B were found in patients with atopical dermatitis (Juhlin, 1981). Histaminase may also be inhibited by food toxins, sodium nitrite, antibiotics (i.e., clavulanic acid), and viral hepatitis (Sattler et al., 1985). Reduced DAO activity and histamine catabolism may be involved in bronchoconstriction episodes occurring after intake of red wine or histamine-rich food.

Intestinal DAO is thought to be required for clearance of diamine and putrescine (Gang et al., 1976b). Elevated mucosal histamine content and secretion were observed in the guts of patients with allergic enteropathy and in women with Crohn's disease and ulcerative colitis. Evidence indicates that mast cell-derived inflammatory mediators including histamine play major pathogenic roles in these diseases (Amon et al., 1999). In a model of ulcerative colitis in rats, the treatment with DAO from pig kidney immobilized on concanavalin A–Sepharose produced a significant reduction in the inflammatory process (Fogel et al., 2006). Histopathologic changes in the colons of DAO-pretreated rats were markedly blunted, as were plasma ceruloplasmin activity and tissue myeloperoxidase. These studies highlight the pivotal role of histamine in inflammatory bowel diseases and the role of histaminase in controlling histamine levels in tissues and blood.

All these findings support the concept of a possible use of plant-derived histaminase, characterized by high specific activity and ability to degrade various biogenic amines including histamine and polyamines, to treat allergic and pseudoallergic diseases, as discussed in detail in Chapter 16.

15.1.7.2 Plant-Derived DAOs

Plant DAO is involved in oxidative deamination of various biogenic amines to the corresponding aminoaldehydes. DAO is the most abundant soluble protein detected in the extracellular fluids from the Fabaceae, in particular, pea (*Pisum sativum*), lentil (*Lens culinaris*), and chickpea (*Cicer arietinum*) seedlings (Rea et al., 1998). Plant DAOs (histaminase) differ from mammalian and prokaryotic enzymes in a number ways including high turnover rate of catalysis, binding affinity for histamine, and chemical stability. This enzyme can be isolated to a high degree of purity in two simple and inexpensive chromatographic steps.

An international patent (PCT/EP01/13770) has been obtained for a drug based on plant-derived histaminase for the treatment of histamine-mediated diseases such as cardiac anaphylaxis, allergic asthma, allergic and septic shock, urticaria, rhinitis, and conjunctivitis. The main sources of histaminase are etiolated seedlings of leguminous plants such as *Pisum sativum*, *Lens culinaris*, *Cicer arietinum*, and *Latirus sativus* in which the enzyme is present in high concentrations, up to 4% of total protein content. Despite the promising pharmacological potential of plant DAOs, their therapeutic use is limited by immunogenicity and short half-lives in blood. To overcome these problems, purified wild pea DAO has been modified by binding to polyethylene glycol (PEG), a polymer that masks the protein surface, giving rise to a semi-synthetic derivative with decreased immunogenicity and catabolism, resulting in a prolonged half-life.

Animal studies have shown that PEGylation with a 20-kDa branched polymer totally eliminated DAO immunogenicity as judged by determination of IgM and IgG raised against the native enzyme. It also caused a marked prolongation of detectable DAO levels in blood, thus markedly improving its pharmacokinetic profile. PEGylation does not interfere with DAO biological activity (Wisniewski et al., 2000). These findings offer a good argument for the use of DAO/histaminase as a novel therapy for histamine-related disorders. The advantages of administering vegetable histaminase and corresponding pharmacological preparations to humans include: (1) a direct (unmediated) action on the allergic process since histaminase acts directly on the histamine, eliminating it by oxidation, unlike currently available drugs that act only by reducing the allergic response after it starts; (2) absence of toxicity and substantial side effects; and (3) presumable maintenance of therapeutic efficacy in long-term treatment, because no tolerance has been demonstrated. To validate these assumptions, it is useful to describe studies performed with purified pea seedling histaminase in animal models of anaphylactic and inflammatory diseases.

15.1.7.3 Plant DAO/Histaminase Modulates Allergic and Anaphylactic Responses

The massive release of histamine from tissue mast cells elicited by the cross linking of antigen with IgE bound to FCε receptors at cell surfaces is considered the paramount event in type I allergic reactions. The most common experimental animal models of type I allergic reactions involve cardiac anaphylaxis *ex vivo* and asthma-like reactions in sensitized guinea pigs *in vivo*. Pig kidney DAO has been demonstrated to exert antihistaminic activity *in vivo* and a protective role in guinea pig anaphylactic shock (Mondovì et al., 1975).

The protective effect of purified pea seedling histaminase, both free and immobilized on CNBr–Sepharose, in active cardiac anaphylaxis was demonstrated by Masini et al. (2002). Guinea pigs were actively sensitized; 2 to 3 weeks later, the hearts were removed and perfused in a Langendorff apparatus that allows an accurate determination of heart rate, contraction strength, and coronary flow. Challenge with the antigen to the isolated hearts induced typical histamine-related changes in function: (1) transient increase followed by long-lasting reduction of myocardial contractility; (2) increased heart rate and severe arrhythmias; and (c) transient marked reduction in coronary flow followed by a less pronounced long-lasting coronary constriction. After antigen challenge in the presence of free or immobilized histaminase, the positive inotropic and chronotropic responses including a dramatic decrease in ventricular tachyarrhythmias and levels of histamine in the perfusates were fully blocked.

As expected, histaminase preparations did not reduce mast cell degranulation in response to the allergen presenting a chromoglycate-like effect on anaphylactic degranulation of mast cells, clearly demonstrating that the cardioprotective effect was chiefly dependent on the inactivation of endogenously released histamine by free or immobilized histaminase (Masini et al., 2002).

Antigen challenge caused a slight decrease in cardiac cGMP and increase in tissue Ca^{2+} levels, both prevented by histaminase treatment. The mechanisms underlying these effects remain a matter of speculation, but we can correctly hypothesize that this effect may result from interaction with the nitric oxide (NO) generation system since NO can increase cardiac cGMP levels and decrease tissue Ca^{2+} concentrations (Masini et al., 2002). Pea seedling histaminase was able to significantly increase cardiac NO synthase activity (Masini et al., 2002).

Allergic asthma is a major respiratory disease on the increase worldwide. Despite decades of research, its complex pathogenetic mechanisms are not completely understood. Nevertheless, scientists agree that histamine is a crucial mediator of inflammation and bronchospasm, two key features of allergic asthma. Based on this activity, a study to evaluate the effects of free and immobilized histaminases on asthma-like reactions induced in antigen-sensitized guinea pigs by aerosol exposure to the allergen was recently published. This animal model is known to reproduce respiratory abnormalities, airway hyperresponsiveness, and leukocyte lung infiltration resembling the functional and histopathological hallmarks of human allergic asthma (Masini et al., 2004; Masini et al., 2005b).

Histaminase, free or immobilized on BrCN–Sepharose, was injected i.p. The pharmacokinetic profiles showed higher amounts of histaminase in lungs from animals treated with the immobilized enzyme than in those treated with the free enzyme that exhibited lower blood levels. The unexpected higher levels of immobilized histaminase in lungs could be ascribed to the facilitation of the homing of the enzyme conjugate in those organs by the galactose units in the Sepharose vehicle. Challenge of ovalbumin-sensitized guinea pigs with aerosolized antigen produced severe abnormalities of respiratory patterns consisting of reduced latency time for the appearance of cough, greater cough severity, and increased occurrence of dyspnea and gasping, indicating the onset of respiratory failure. Aerosol or i.p. pretreatment of animals with histaminases resulted in marked reductions in breathing abnormalities and prevention of respiratory failure (Figure 15.1).

Histaminase treatment also prevented the histopathological lung changes induced by antigen-induced bronchospasm and reduced the inflammatory processes characterized by leukocyte infiltration and oxygen free radical generation. Histaminase markedly reduced myeloperoxidase activity, an indication of tissue leukocyte infiltration, and malonyldialdehyde production, an indication of peroxidation of cell membrane lipids by reactive oxygen species (ROS).

15.1.7.4 Protective Effect of Plant DAO/Histaminase against Ischemia–Reperfusion Injury

Besides its pivotal involvement in anaphylaxis and allergic reaction, histamine plays a role in exacerbation of inflammatory tissue damage; many proinflammatory mediators such as prostanoids and cytokines are potent mast cell activators. In particular,

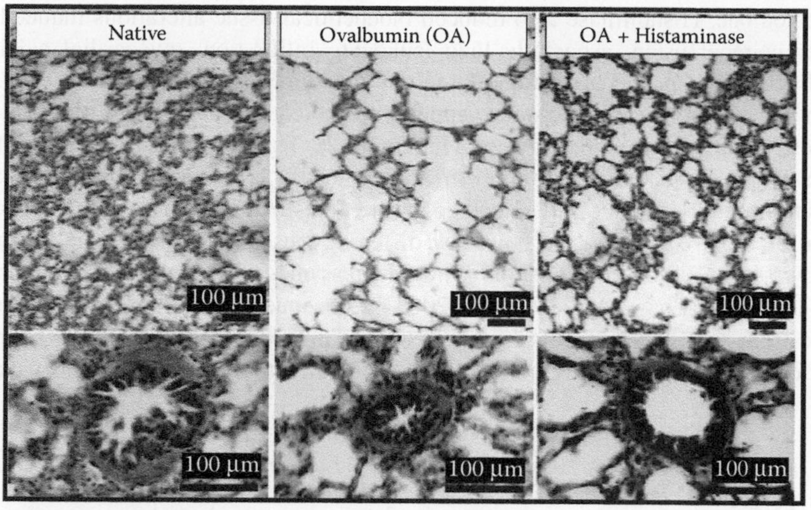

FIGURE 15.1 **(See color insert following page 202.)** Representative micrographs of lungs from naive guinea pig, ovalbumin-sensitized guinea pig challenged with the antigen, and sensitized guinea pig pretreated with aerosolized plant histaminase given 30 min before antigen challenge.

histamine has been found to contribute to endothelial dysfunction, hampering blood–tissue exchange and functional tissue impairment in organs undergoing ischemia and reperfusion. Accordingly, pea seedling histaminase may effectively blunt the adverse effects of cardiac and intestinal ischemia–reperfusion.

The heart is particularly susceptible to tissue damage by several key agents involved in the pathophysiological mechanisms of ischemia–reperfusion such as ROS, histamine, and NO (Masini et al., 1999; Valen et al., 1996). A close relationship between histamine, acting as a pro-oxidant, and ROS was observed in patients with coronary heart disease (Valen et al., 1996). In turn, histamine release by mast cells is amplified by excess superoxide generation and concurrent decrease in local NO amounts because superoxide and NO react promptly, giving rise to harmful peroxynitrite (Masini et al., 1999). Histamine may induce spasm of coronary vessels, thereby causing or worsening acute coronary insufficiency and myocardial ischemia; histamine levels in the coronary sinus are paralleled by the occurrence and severity of ventricular arrhythmias. Pea seedling histaminase demonstrated cardioprotective effects against postischemic reperfusion damage, in an *in vivo* model of ischemia–reperfusion in rats. Anesthetized artificially ventilated rats were subjected to 30-min ischemia by temporary occlusion of the left anterior descending coronary artery, followed by 60-min reperfusion. Intravenous histaminase at a dose of 80 IU/kg body weight given 10 min before reperfusion or during reperfusion reduced the size of myocardial infarction caused by ischemia–reperfusion. No cardiac protection was afforded by semicarbazide-inactivated histaminase (Masini et al., 2003).

Analysis of electrocardiogram recordings showed that histaminase reduced the occurrence of ventricular arrhythmias. Overall, the number of animals that survived the 60-min reperfusion was higher in the histaminase-treated groups than in the untreated one. Histaminase also reduced biochemical tissue alterations induced by ischemia–reperfusion, especially lung leukocyte infiltration, myocardial calcium overload, and caspase-3 activation, the initiator enzyme of the apoptotic cascade.

In conclusion, plant histaminase appears to protect the heart from the deleterious effects of ischemia–reperfusion (Figure 15.2). The mechanisms of this effect mainly rely on histamine catabolism and removal of the proinflammatory and arrhythmogenic actions of this biogenic amine. Intestinal ischemia may result from impaired blood supply to the bowel by different causes including cardiac insufficiency, sepsis, vasodepressant and cardiodepressant drugs, and complications of long-lasting surgery (Mallick et al., 2004). Consequences of intestinal ischemia range from persistent bleeding and symptomatic intestinal strictures to bowel perforation and peritonitis. Surgical resection of the affected bowel segment is usually

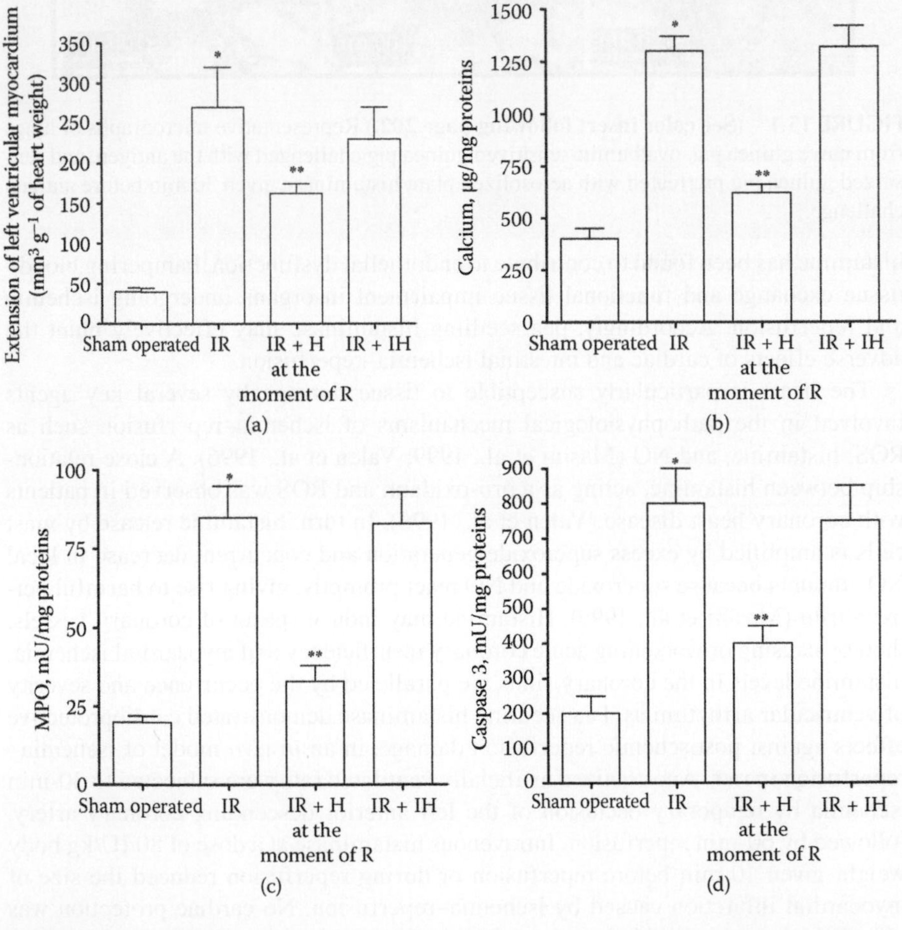

FIGURE 15.2 Effect of plant histaminase (H, 80 U/Kg body weight) and inactivated histaminase (IH) on (a) extension of injured left ventricular myocardium; (b) calcium overload; (c) myeloperoxidase (MPO) activity; and (d) caspase-3 activity induced by 30-min ischemia followed by 60-min reperfusion (R).

required to minimize adverse outcomes. The pathophysiology of intestinal ischemia has been widely investigated in laboratory animals undergoing surgical occlusion of the splanchnic circulation followed by reperfusion. This led to intestinal injury and circulatory shock, characterized by severe hypotension and hemoconcentration associated with a high mortality rate (Lefer and Lefer, 1993).

Endothelial dysfunction plays a key role in intestinal ischemia and reperfusion; it predisposes to vasospasm, platelet activation, and increased neutrophil adherence that exacerbate local bowel injury and general cardiocirculatory failure (Cuzzocrea et al., 1997). Endothelial–leukocyte interaction is known to involve specific surface glycoproteins known as endothelial cell adhesion molecules. These include early-phase molecules such as P-selectin that play a role in leukocyte tethering and rolling. P-selectin is rapidly translocated from the Weibel-Palade bodies to the endothelial cell surface upon stimulation by histamine, hypoxia, ROS, and late-phase molecules such as E-selectin, vascular cell adhesion molecule (VCAM)-1, and intercellular adhesion molecule (ICAM)-1 involved in leukocyte adhesion and extravasation into tissue, whose expression is induced by most inflammatory mediators including histamine and cytokines (Masini et al., 2007).

At reperfusion, neutrophil accumulation and activation in the ischemic bowel results in the local generation and release of free radicals, lysosomal hydrolases, and chemotactic factors such as leukotrienes (Deitch et al., 1990) that make major contributions to tissue injury by causing further endothelial damage and leukocyte recruitment, plasma membrane peroxidation, and DNA damage (Dix et al., 1987). Paradoxically, activation of DNA repair enzyme poly-(ADP-ribose) synthetase can exacerbate intestinal mucosal injury and dysfunction (Cuzzocrea et al., 1997). Free radical-mediated bowel tissue injury is largely attributable to peroxynitrite, the reaction product of superoxide anion and nitric oxide (NO), a potent cytotoxic and proinflammatory molecule (Dix et al., 1987 and 1996). A synergism between ROS and histamine acting as a pro-oxidant was also observed in ischemic diseases. The release of histamine from mast cells in the intestinal mucosa was increased by excess superoxide anion and concurrent decreases of NO (Mannaioni et al., 1991).

Histamine secretion from storage and producing cells, mainly intestinal mast cells, in the extracellular compartment and blood contributes to lethal circulatory shock occurring upon intestinal ischemia–reperfusion. Histamine is released by the intestinal mucosa, mostly during reperfusion (Masini et al., 2007) and sparks a vicious cycle that leads to further endothelial activation and leukocyte adhesion and extravasation. Along with the release of histamine, the activity of diamine oxidase, a key enzyme involved in histamine catabolism, is significantly reduced after 60-min ischemia (Befani et al., 1999). In keeping with these findings, mast cell-deficient (W/Wv) mice proved to be refractory to ischemia–reperfusion-induced mucosal damage. In the search for new interventions to interrupt the cycle underlying the pathophysiology of mesenteric shock, researchers have focused on antioxidants and free radical scavengers, many of which show protective effects in experimental models of ischemia–reperfusion injury, even though none of them has achieved clinical use. In particular, copper oxidases such as ceruloplasmin, the plasma blue-copper protein and ROS scavenger, and bovine serum amine oxidase have been found to afford fair cardioprotection against myocardial injury by electrolysis-generated oxygen radicals and/or by ischemia–reperfusion.

Pea seedling histaminase exhibited potent beneficial effects when given to sensitized guinea pigs subjected to type I allergic reactions such as the asthma-like reaction induced by allergen challenge and cardiac anaphylaxis (Masini et al., 2002 and 2003). Since histamine is heavily involved in the pathophysiology of reperfusion-induced tissue injury, histaminase is also able to exert cardioprotective effects in myocardial ischemia–reperfusion in a rat model in vivo (Masini et al., 2003).

Recent results indicate that pea seedling histaminase may also have protective effects on intestinal ischemia in a rodent model of intestinal ischemia–reperfusion, paying attention to its effects on mean arterial blood pressure, endothelial cell adhesion protein expression, neutrophil infiltration, free radical-mediated bowel tissue injury, and cell apoptosis. Histaminase, 80 IU/kg body weight/0.5 ml phosphate buffer, administered 15 min before reperfusion, significantly reduced the drop in blood pressure and high mortality rate caused by intestinal shock. Histaminase also reduced histopathological changes, leukocyte infiltration, and expression of endothelial cell adhesion molecules in the ileum (Masini et al., 2007). Besides reducing local tissue inflammation through acceleration of histamine catabolism, histaminase also seems to counteract free radical-mediated tissue injury, based on the significant decrease in tissue levels of peroxidation and nitration products and DNA damage and consumption of tissue antioxidant enzymes such as superoxide dismutase. Histaminase caused a reduction in ileal cell apoptosis, as assessed by analysis of caspase 3-activity and of the number of terminal deoxynucleotidyl transferase biotin-dUTP nick end labelling (TUNEL) cells.

In agreement with the findings from cardiac ischemia–reperfusion, these recent results offer evidence that histaminase may afford protection against ischemia–reperfusion-induced splanchnic injury, possible due to oxidative catabolism of proinflammatory histamine and antioxidant effects, resulting in hindrance of free radical-mediated tissue injury endothelial dysfunction and leukocyte recruitment.

15.2 POSSIBLE FUTURE PERSPECTIVES

The control of CAO activity represents a field of growing interest in pharmacology research. A clear understanding of the time of appearance of VAP-1/SSAO activity in mesenchymal cells committed to differentiate into specific phenotypes represents a crucial point to exploit the physiological significance of this enzyme. In this respect, the discovery of drug modulating VAP-1/SSAO activity in differentiated cells could open new clinical perspectives in the cell therapy of damaged organs.

The therapeutic potential of plant-derived histaminase in allergic diseases and anaphylaxis and its ability to inhibit inflammatory and ROS-mediated tissue injury arising from ischemia–reperfusion has been studied. Recent progress with polymeric conjugates of histaminase should enhance the pharmacological profile of this enzyme and possibly allow alternative routes of administration (e.g., aerosol) to be as effective as parenteral ones.

16 Copper Amine Oxidases as Antioxidant and Cardioprotective Agents

Mircea Alexandru Mateescu and Réginald Nadeau

CONTENTS

16.1 ANTIOXIDANT PROPERTIES OF SAO

To evaluate the antioxidant properties of copper amine oxidases (CAOs), the *in vitro* scavenging effects of serum amine oxidases (SAOs) were compared with those of ceruloplasmin (CP) and superoxide dismutase (SOD), both known as reactive oxygen species (ROS) scavengers (Mateescu et al., 1997), and of BSA on the free radicals and oxidant by-products (OCl^-, HOCl, and Cl_2) generated by electrolysis (10 mA DC current. 1 min) of Krebs-Henseleit (KH) buffer (Jackson et al., 1986) and quantified by the N,N-diethyl-p-phenylenediamine (DPD) colorimetric method (Chahine et al., 1991; Mateescu et al., 1995; Dumoulin et al., 1996). The electrolysis-induced ROS react with DPD (colorless when reduced, red when oxidized) allowing the quantification of oxidant species at $\lambda = 515$ nm.

Purified BSAO and CP (1 μM) were the best antioxidants *in vitro* (~90% scavenging capacity). At equal concentration, SOD presented only 12.8% scavenging (Table 16.1). Unexpectedly, a relatively high scavenging activity (72.1%) was observed *in vitro* for serum albumin (initially used as a control) and, due to its particular effect, later replaced with α-amylase (showing practically no scavenging: 0.4 % antioxidant capacity). The high antioxidant value for albumin seems nonspecific since BSA affords no cardioprotection *ex vivo* (Mateescu et al., 1995 and 1997).

TABLE 16.1
Scavenging Capacities of Various Antioxidants Revealed by *in vitro* Electrolysis-Induced ROS Antioxidant

	%*
Serum amineoxidase [1*μM*]	91.9 ± 0.2
Ceruloplasmin [1*μM*]	89.2 ± 0.2
Ceruloplasmin denatured [1*μM*]	87.2 ± 0.2
Catalase [1*mM*]	71.5 ± 0.3
Albumin [1*μM*]	72.1 ± 0.3
SOD [1*μM*]	12.8 ± 0.2
Alpha-Amylase [1*μM*]	0.4 ± 0.2
Deferoxamin mesylate [1*μM*]	29.5 ± 0.3
Mannitol [25 m*M*]	44.7 ± 0.2

Notes: * n = 3; values expressed as percentage of scavenged species quantified by DPD method.

16.2 CARDIOPROTECTION AGAINST ELECTROLYSIS-INDUCED OXIDATIVE DAMAGE ON ISOLATED RAT HEART

The cardioprotective and scavenging effects of SAO, CP, SOD, and BSA against free radicals and by-products (OCl^-, HOCl, and Cl_2) generated by electrolysis (10 mA DC current, 1 min) of KH buffer were evaluated *ex vivo* by quantifying cardiodynamic variables of isolated perfused rat heart. Jackson et al. (1986) proposed electrolysis-induced oxidative damage as a model mimicking the oxidative stress in isolated heart at postischemia reperfusion.

Isolated hearts (Langendorff model) perfused with KH buffer (15 mL/min) were submitted to electrolysis only (control). For the treated groups, the perfusing (15 mL/min) KH buffer was supplemented by injection (0.5 mL/min) in the in-flow cannula above the heart with SAO, CP, SOD, or BSA (1 *μM*) for 5 min before, during, and 5 min after electrolysis.

Concentration-dependent cardioprotective effects were found with SAO and CP (Mateescu et al., 1997). The best cardioprotection with high recovery (80 to 95%) for all cardiodynamic variables (left ventricular pressure, heart rate, and coronary flow) was recorded with SAO (0.5 *μM*) and with CP (1 *μM*), significantly higher than that afforded by the same optimal concentration of 1 *μM* SOD (Figure 16.1). In fact, SOD, well known as antioxidant, exhibited moderate scavenging activity (38%) *in vitro* and a relatively low cardioprotective effect (20 to 45%) *ex vivo*.

Although BSA exhibits good antioxidant capacities *in vitro* (71%), in *ex vivo* electrolysis-induced oxidative damage to isolated heart (Figure 16.1), BSA showed no cardiac protection (Dumoulin et al., 1996; Mateescu et al., 1997). Similar to BSA, heat-denatured SAO (Figure 16.1) showed a relatively high antioxidant capacity (80%) *in vitro* and no protection *ex vivo* (same cardiodynamic alterations with low values of cardiodynamic variables, as for unprotected control hearts).

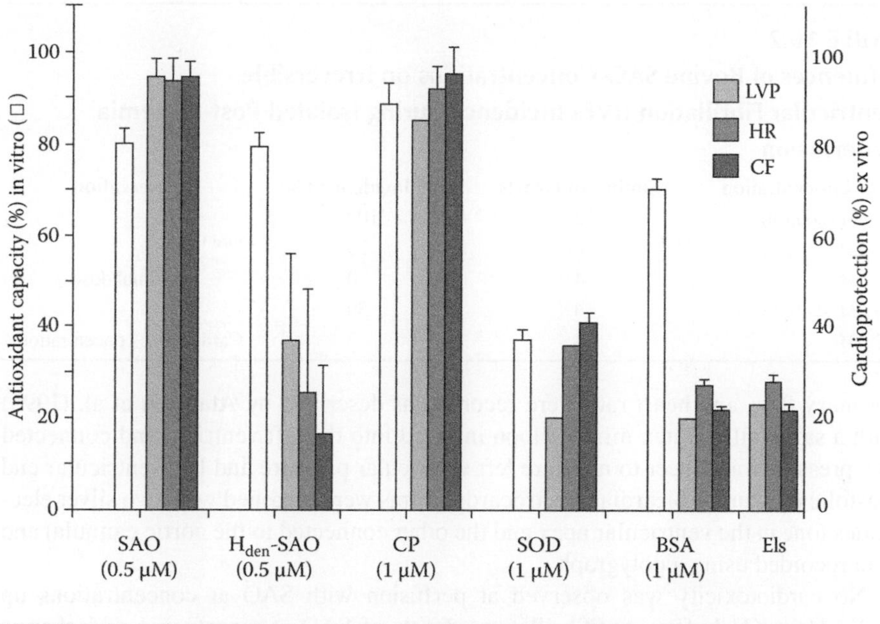

FIGURE 16.1 Comparative *in vitro* antioxidant and *ex vivo* cardioprotective effects of BSAO, heat-denatured (Hden) BSAO, CP (ceruloplasmin), SOD (superoxide dismutase), and bovine serum albumin (BSA) against damages generated by electrolysis-induced (10 mA DC, 1 min) reactive oxygen species (ROS) and by-products, measured 15 min after electrolysis. Cardiodynamic variables (Els) measured under electrolysis: left ventricular pressure (LVP), heart rate (HR), and coronary flow (CF) without treatment. (Adapted from Mateescu et al., 1997. With permission from the publisher.)

Particularities in scavenging specificities and mechanisms seem to explain the important differences in *in vitro* and *ex vivo* behaviors of these proteins (Mateescu et al., 1997) and probably a major role of the integrity of the copper center (lost in heat-denatured SAO and CP).

16.3 ANTIARRHYTHMIC EFFECT OF SAO ON ISOLATED RAT HEART IN ISCHEMIA–REPERFUSION

As a continuation of the ischemia–reperfusion studies showing an antifibrillatory effect of CP (Atanasiu et al., 1995) and the cardioprotection study with SAO (Mateescu et al., 1997) in an electrolysis model (Jackson et al., 1986) to produce oxidative damage, it was of interest to evaluate and better reveal the effect of SAO on heart functions under *ex vivo* conditions of the ischemia–reperfusion model that are closer to genuine physiopathological conditions. Isolated rat hearts were damaged by regional ischemia (15 min) through left anterior descending (LAD) artery ligature and treated for 10 min (last 5 min ischemia and first 5 min reperfusion) with purified SAO at different concentrations (Mondovì et al., 1997).

A modified Langendorff heart perfusion system allowed the continuous measurement of major cardiodynamic variables. Electrocardiogram, left ventricular pressure,

TABLE 16.2
Influences of Bovine SAO Concentrations on Irreversible Ventricular Fibrillation (IVF) Incidence during Isolated Post-Ischemia Reperfusion

SAO Concentration	Number of Hearts	IVF Incidence (%)	Observation
0.0 *μM* (control)	12	100	
0.2 *μM*	4	25	
0.3 *μM*	4	0	Optimal dose
0.5 *μM*	4	50	
1.0 *μM*	4		Cardiotoxic concentration

coronary flow, and heart rate were recorded as described by Atanasiu et al. (1995) with a saline-filled latex mini balloon inserted into the left ventricle and connected to a pressure transducer to measure left ventricular pressure and left ventricular end diastolic pressure. Epicardial electrocardiograms were obtained with two silver electrodes (one in the ventricular apex and the other connected to the aortic cannula) and data recorded using a polygraph.

No cardiotoxicity was observed at perfusion with SAO at concentrations up to 1 *μM* in KH buffer. Antifibrillatory effects of SAO at reperfusion of ischemic hearts showed a bell-shaped dependency (Table 16.2) on the enzyme concentration in the range 0 to 1.5 *μM*. A high degree of cardioprotection, with almost no irreversible ventricular fibrillation was observed for 0.3 *μM* SAO (Table 16.2). At concentrations of 1.5 *μM* and higher, native SAO presented *ex vivo* cardiotoxicity (without oxidative damage) with an arrest of cardiac functions. Similar cardiotoxicity was found with CP at 4 *μM* (Atanasiu et al., 1995). Without SAO treatment, all the ischemic hearts fibrillated immediately at reperfusion (100% irreversible ventricular fibrillation) and arrhythmia continued during reperfusion (lasting 15 min and more, until the end of the experiment). Typical tracings of electrocardiogram and left ventricular pressure covering the ischemia and reperfusion periods are presented in Figure 16.2. In the absence of protective SAO, ventricular fibrillation and flat left ventricular pressure (heart dying) were observed (Figure 16.2a). Under SAO cardioprotective treatment, the normalized left ventricular pressure and electrocardiogram (with no fibrillation) show full cardiac recovery (Figure 16.2b) upon reperfusion.

The time of normal sinus rhythm over the first 5 min of reperfusion was 25 sec for untreated hearts (controls) and 223 sec for the group treated with 0.5 *μM* SAO. Time of normal sinus rhythm was considered as the time (sec) of normal cardiac function. Hearts were considered recovered when sinus rhythm lasted more than 60 sec. When SAO was administered at optimal point (seventh minute after starting ischemia), the ventricular fibrillation was eliminated or reversed within 2 min of reperfusion.

To better elucidate the cardioprotective mechanisms, the effect of native bovine SAO was compared with bovine SAO inactivated by various treatments (Table 16.3):

1. Heat-denatured (enzymatically inactive) SAO
2. Phenylhydrazine (PHy)-treated SAO (blocking carbonyl prosthetic group)
3. Diethydithiocarbamate (DDC)-treated SAO (copper center chelating agent)

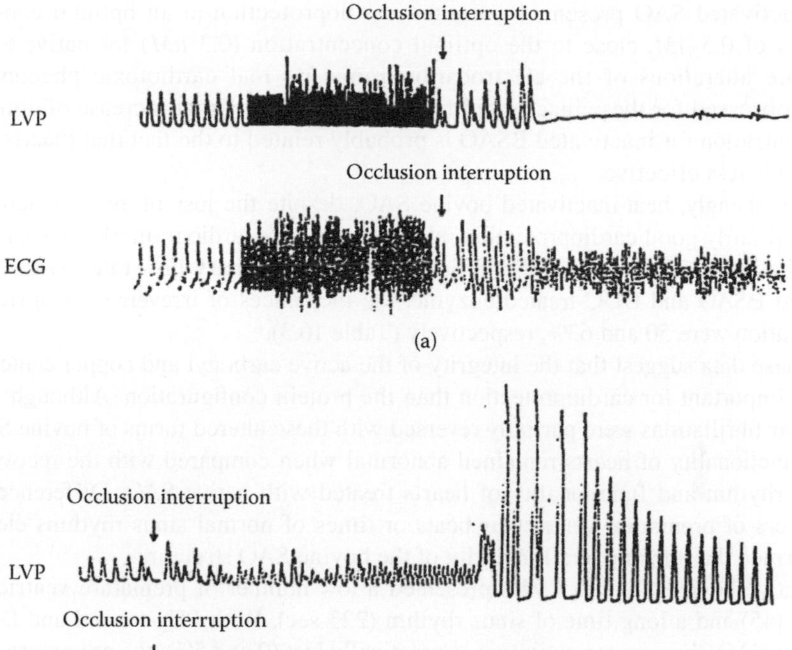

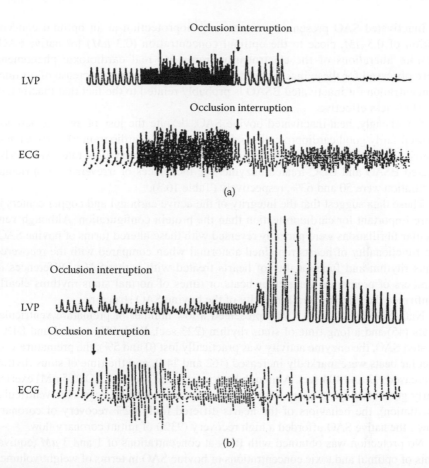

FIGURE 16.2 Tracings of epicardial electrocardiogram (ECG) and left ventricular pressure (LVP) of isolated heart from ischemia to reperfusion in the absence of SAO protection: (a) heart fibrillated irreversibly (control); (b) under treatment. Ventricular fibrillation reversed within 2 min. (Adapted from Mondovì et al., 1997. With permission from publisher.)

TABLE 16.3
Antifibrillatory Effects of Native and Altered SAO Forms

SAO Form and Concentration	Number of Hearts	IVF Incidence (%)	Observation
0.5 μM (native)	4	0	
0.5 μM (heat denatured)	4	0	No recovery for flow rate
0.5 μM (PHy-treated)	8	50	Abnormal ECG (tachycardia)
0.5 μM (DDC-treated)	8	63	Abnormal ECG (tachycardia)
1.5 μM (heat denatured)	4	–	Cardiotoxicity

Inactivated SAO presented a limited cardioprotection at an optimal concentration of 0.5 *μM*, close to the optimal concentration (0.3 *μM*) for native SAO. Despite alterations of the electrocardiograms, no real cardiotoxic phenomena were observed for these inactivated forms of SAO. The slight increase of optimal concentration for inactivated BSAO is probably related to the fact that inactivated BSAO is less effective.

Surprisingly, heat-inactivated bovine SAO, despite the loss of enzyme activity, showed fairly good cardioprotection in terms of electrocardiogram (0% irreversible ventricular fibrillation, but only partial recovery of coronary flow rate). With PHy-treated BSAO and DDC-treated enzyme, the incidences of irreversible ventricular fibrillation were 50 and 63%, respectively (Table 16.3).

These data suggest that the integrity of the active carbonyl and copper centers is more important for cardioprotection than the protein configuration. Although ventricular fibrillations were partially reversed with these altered forms of bovine SAO, the functionality of hearts remained abnormal when compared with the recovered sinus rhythm and functionality of hearts treated with native SAO. Differences in numbers of premature ventricular beats or times of normal sinus rhythms clearly confirmed the role of overall integrity of the bovine SAO structure.

Native bovine SAO (0.3 *μM*) presented a low number of premature ventricular beats (45) and a long time of sinus rhythm (223 sec). With PHy-treated and DDC-treated SAO, the enzyme activity was practically lost (0 and 5%), the premature ventricular beats were markedly increased (162 and 144), and the time of sinus rhythm decreased (92 and 104 sec). Although heat-inactivated bovine SAO (0.5 μM) exerted fairly good cardioprotection (Table 16.3) at reperfusion (0% irreversible ventricular fibrillation), the behaviors of the hearts differed (only 42% recovery of coronary flow); the native SAO afforded a high recovery (93%) of initial coronary flow.

No protection was obtained with BSA at concentrations of 1 and 3 *μM* (equivalents of optimal and toxic concentrations of bovine SAO in terms of weight/volume). Under similar conditions, the antifibrillatory action of CP (1 to 2 *μM* in the perfusing medium) was maximal (0% irreversible ventricular fibrillation) and without toxicity (Atanasiu et al., 1995). Since the best antifibrillatory effects of bovine SAO were found at concentrations of 0.3 to 0.5 *μM* (Table 16.2), it appears that 0.5 *μM* SAO (90 mg/L perfusing buffer) affords an antiarrhythmic effect similar that afforded by 1 *μM* CP (132 mg/L).

It is noteworthy that the bovine SAO concentration (0.3 *μM*) found optimal for cardioprotection *ex vivo* (Table 16.2) is near the normal concentration of circulatory SAO (0.2 *μM*) *in vivo* (Mondovì et al., 1997). Higher concentrations are less cardioprotective. Similar decreases of cardioprotective effects at increased concentration were also found with other proteins (CP, SOD) involved in cardioprotection. SOD showed an optimal cardioprotective effect at a 20 mg/mL concentration; at higher concentrations a loss of cardioprotection associated with pro-oxidant effects and cardiotoxicity was found (Omar et al., 1990). For SAO, cardiotoxicity was found at 1.5 *μM* (seven times higher than the normal value in bovine plasma).

The protection observed with heat-denatured, enzymatically inactive SAO (0.5 *μM*) represents an interesting aspect, suggesting particular structural characteristics involved in cardioprotection rather than an overall conformational assembly of the protein. This aspect appears characteristic for denaturated SAO and differs from denaturated CP (enzymatically inactivated and conformationally modified) which is unable to protect the ischemic heart at reperfusion.

Plant DAO totally loses cardioprotective effects when it is heat-denatured (Masini et al., 2003). These differences may relate to the fact that the SAO and DAO enzymes come from animal and vegetal sources. Another difference to consider is that histamine is a good substrate for the plant enzyme but not for bovine SAO. This issue is important (Masini et al., 2003) based on the role of histamine in ischemia–reperfusion injury. Also different are the model approaches: *ex vivo* on isolated rat heart model (this study) and *in vivo* in guinea pigs (Masini et al., 2003).

The toxicity of bovine SAO may arise from its enzymatic oxidase activity that releases cytotoxic products as aldehydes and hydrogen peroxide (Averill-Bates et al., 1993). However, this cannot explain the cardiotoxicity of 1.5 μM BSAO because heat-denatured BSAO (1.5 μM) is cardiotoxic and enzymatically inactive (Table 16.3). The partial recovery of hearts (Table 16.3) observed with DDC–SAO was lower (63% irreversible ventricular fibrillations) than that obtained with PHy–SAO (50% irreversible ventricular fibrillation). These differences suggest that the integrity of the copper center is even more important than that of the carbonyl site and seems in line with our hypothesis of copper–protein structures involved in cardioprotection.

Masini et al. (2003) reported the first *in vivo* study showing cardioprotection with purified plant diamine oxidase that prevented arrhythmia associated with ischemia–reperfusion. This *in vivo* report is important because it completes the previous *in vitro* and *ex vivo* studies with BSAO (Mateescu et al., 1997 and Mondovì et al., 1997), suggesting that several copper proteins of various origins may exert effects in cardiac protection.

All the data presented here were obtained *in vitro* and *ex vivo* in the absence of additional biogenic amines substrates for amine oxidases. No biogenic amines were added to the KF buffer for *in vitro* or *ex vivo* perfusing models. Thus, most of the observed antioxidant and cardioprotective effects were probably related to the direct action of the protein rather than to its enzyme activity. CP was found to present its own electrophysiological action on isolated heart in the absence of oxidative damage, inducing a prolongation of effective refractory period and action potential, acting as a Class III antiarrhythmic agent (Atanasiu et al., 1996). It was therefore of interest to evaluate the electrophysiological effects of copper SAO.

16.4 ELECTROPHYSIOLOGICAL PROPERTIES OF BSAO: PROLONGATION OF EFFECTIVE REFRACTORY PERIOD

In case of CAOs (with no ferroxidase activity), the influences on membrane electrophysiological properties may be relevant for explaining the mechanisms involved in the antiarrhythmic effects. Therefore, it was of interest to evaluate whether BSAO can exert a prolongation of ventricular effective refractory periods and compare its action with the effects of CP, SOD, catalase, deferoxamine, and BSA.

First evaluations were obtained *ex vivo* on isolated rat hearts cannulated *via* the aorta and retrogradely perfused at a coronary flow of 14 mL/min in the absence of oxidative damage, pro-oxidant agents, and biogenic amine substrates. The bovine SAO was infused in the cannula with an injector as in case of heart treatment to achieve desired concentrations (up to 2 μM) in the perfusing KH buffer. Simultaneous epicardial stimulation and recordings were accomplished with two bipolar electrodes (1.5 mm in diameter), one positioned at the apex and the other at the base of the left ventricle. Data acquisition occurred under conditions similar to those previously described (Atanasiu et al., 1996).

Purified SAO (0.25 to 0.50 μM) induced a prolongation of effective refractory periods and conduction time. The bovine SAO-induced increase of effective refractory period was 15 to 33%, seemingly regardless of SAO concentration in the mentioned range of 0.25 to 0.50 μM. The SAO-induced prolongation of effective refractory period appeared at 3 min of infusion, remained constant during the 10-min infusion period, and disappeared 3 min after the SAO infusion was stopped.

CP increased the effective refractory period (Atanasiu et al., 1996) in a concentration-dependent manner from 33% (at 0.25 μM) to 94% (at 2 μM). Other antioxidants and antifibrillatory agents such as SOD, catalase, deferoxamine, and BSA had no significant effect on effective refractory period or conduction time at any dose within the mentioned range (Atanasiu et al., 1996). The lack of impact of antioxidants and BSA on effective refractory periods suggests that the effects of SAO and CP may be very specific for these two copper proteins. The prolongation of effective refractory periods by BSAO may explain its antiarrhythmic action by a mechanism related, to a certain extent, to that of Class III type antifibrillatory drugs that act on the prolongation of cardiac repolarization.

In conclusion, these studies showing antioxidant, cardiomodulatory, and cardioprotective actions of BSAO appear as complementary to the reports of Boomsma et al. (2005b) and Göktürk et al. (2007) describing the cardiovascular involvement of tissual amine oxidase from smooth muscle cells in various pathological or risk conditions and contribute to our knowledge of the cardioprotective functions of this copper enzyme, in addition to its main role of controlling blood biogenic amine metabolism.

ACKNOWLEDGMENTS

We gratefully acknowledge the financial support provided by the National Science and Engineering Research Council (NSERC) and by the Canadian Institutes for Health Research (CIHR). We also thank Drs. R. Atanasiu and R. Chahine for their valuable contributions to this research.

17 Biotechnological Aspects of Copper Amine Oxidases

Roberto Stevanato

CONTENTS

17.1 INTRODUCTION

Several oxygen accepting oxidoreductases have been diffusely used for some time in the food, pharmaceutical, and chemical industries (Cheetham, 1985; Peppler and Reed, 1987; Soda and Yonaha, 1987; Chibata et al., 1987; Godfrey and West, 1996; Faber, 2000; Schafer et al., 2007) and also for clinical, process, and environmental analyses (Bergmeyer, 1974; Karube, 1987). In the case of copper amine oxidases (CAOs), few biotechnological applications have been reported. More than likely, progress has been limited by the scant availability and questionable purity of commercial preparations. For these reasons, most biotechnological applications reported in literature were carried out using CAOs purified by the authors. The main concerns related to biotechnological applications of CAOs are (1) analytical determination of polyamines, (2) biocatalysis, and (3) therapeutic uses.

17.2 IMMOBILIZED CAOS

Most applications use immobilized amine oxidases, in particular to (1) set up amperometric biosensors, disposable test strips, small flow reactors, and enzyme arrays for biogenic amine determinations; (2) use them as practical heterogeneous catalysts in biotransformations; and (3) increase AO plasmatic half-lives of therapeutic preparations.

The immobilization of enzymes to soluble or insoluble supports can increase their stability to heat, organic solubility, and pH without losing too much catalytic activity. Furthermore, the change of physical form as a consequence of the enzyme linkage to an insoluble support makes an enzyme suitable in certain applications such as biosensors and analytical or scale-up bioreactors. Finally, immobilized enzymes can be used repeatedly as long as they remain active, thereby minimizing cost and analysis time and making them suitable to operate in a continuous mode (Gacesa et al., 1987).

It has been reported that the products of the AO-catalyzed oxidative deamination of polyamines, spermine in particular, may exert curative effects in cancer therapy. This aspect is discussed in greater detail in the Chapters 5 and 10. We must point out that only immobilized BSAO was found to show antitumor activity *in vivo* (Mondovì et al., 1982). Based on that report, several biocompatible polymers coupling to BSAO have been proposed in order to assay the potential use of the enzyme in antitumor applications. Immobilization produces better thermal stability (Befani et al., 1989; Mondovì et al., 1992; Šebela et al., 2005), prolonged life, and slower activity. The reduced activity levels seem to allow a slow and prolonged release of cytotoxic products compared to the more rapid generation of higher levels of cytotoxic products generated by the native enzyme (Averill-Bates et al., 2005).

High molecular weight polyvinylalcohol (100 kDa) has been proposed for coupling via hydrogen bonds to PKAO which, when so linked, shows higher thermal stability than the free enzyme (Befani et al., 1989; Mondovì et al., 1992). Similarly, a poly(ethylene glycol) (PEG) biocompatible matrix activated by 4-nitrophenyl chloroformate and mixed with serum albumin was utilized to immobilize the BSAO (Demers et al., 2001; Averill-Bates et al., 2005), showing a better operational stability.

Šebela et al. (2005) found that β-cyclodextrin covalently linked to garden pea and grass pea AO improved the thermostability and biological lifetimes of these enzymes.

Also when linked to insoluble supports, AOs reveal improved lifetimes and physicochemical characteristics, although a decrease of activity is generally observed. This is due to partitioning effects or modification of the active site configuration, as found by linkage of BSAO on CH–Sepharose by carbodiimide and by physical absorption of PKAO on concanavalin (Con A)–Sepharose (Stevanato et al., 1989).

Agostinelli et al. (2007) studied a polyketone polymer obtained by copolymerization of ethene and carbon monoxide as an insoluble support of BSAO and LSAO. The immobilization occurs via hydrogen bonds only and the resulting linkage is very strong. LSAO completely retained its activity and physicochemical characteristics, but cannot be removed from the support, even with acidic or high ionic strength solutions. Pea seedling histaminase immobilized on CNBr–Sepharose was proposed to treat histaminosis by eliminating the excess histamine in foods (Federico et al., 2000).

Enzyme immobilization is used for AO purification by affinity chromatography because of the abilities of these enzymes to link to aminoalkyl matrices or Con A. The affinity of AO toward aminoalkyl residues has been extensively investigated by Houen et al. (1993) who considered different matrices, generally characterized by the diaminohexyl active residue linked to the Agarose or Sepharose insoluble support. Purification factors of 100 to 1000, depending on the enzyme and experimental conditions, have been obtained by affinity chromatographic (Toraya et al., 1976; Svenson and Hynning, 1981; Pettersson, 1985; Vianello et al., 1992). One consequence of the immobilization of substrate-like structures as ligands for AO affinity purification is the oxidative deamination of matrix-bound amino groups to aldehydes. These new functional groups can react with the amino groups of proteins via a reversible Schiff base, causing immobilization of the enzyme. The immobilized amine oxidase obtained in this way retains 80% activity (Befani et al., 1989).

Some CAOs are glycoproteins (Pettersson, 1985; Rinaldi et al., 1985) containing ~10% carbohydrate such as mannose, galactose, sialic acid, and glucosamine that can be linked by Con A—a lectin that binds molecules containing α-D-glucopyranosyl and α-D-mannopyranosyl and sterically related residues (Goldstein et al., 1965; Remy et al., 1998). The Con A–BSAO complex is characterized by a Con A-to-enzyme ratio of 2 and it completely retains its oxidative deamination activity, indicating that the carbohydrate is not involved in the catalytic mechanism and the link location is far from the active site. The enzyme can be released by eluating with α-methyl-D-mannoside (Ishizaki and Yasunobu, 1980; Turini et al., 1982).

Other immobilization procedures generally utilized for protein linking (Kennedy and White, 1985) and are based on covalent bonds created by glutaraldehyde, carbodiimide (Stevanato et al., 1990), difluoride polyvinylidene (Draisci et al., 1998), N-hydroxysuccinimide active ester (Frébort et al., 2000c), and epoxy-activated porous cellulose containing formil groups (Yoshida et al., 1997). Adsorption on graphite (Niculescu et al., 2000a) and entrapment on polyacrylamide (Hasebe et al., 1997) have been adopted for physical immobilization as well. More particular linkages are those based on the adsorption and affinity on carbon paste supported BAT–Silasorb: (2-[4,6-*bis* (aminoethylamine)-1,3,5-triazine]–Silasorb; Wimmerová and Macholán, 1999), on the interaction with cysteine-derivatized Nafion (Hong et al., 2006), on cross linking with PEG 400 diglycidyl ether and poly(vinylimidazole) derivative (Castillo et al., 2003) and with Os-bipyridine modified redox polymer (Niculescu et al., 2000c).

17.3 ENZYMATIC ANALYSIS OF AMINES BY CAOS

Biogenic amines in food and beverages can affect human health because of their toxicological manifestations (Draisci et al., 1998):

- Histamine poisoning arising from ingestion of foods with high levels of histamine induces a variety of gastrointestinal, cutaneous, hemodynamic, and neurological symptoms.
- Tyramine displays vasoactive effects, such as the "cheese reaction" caused by high levels contained in cheese.
- Ingested biogenic amines may provoke hypertensive crises in patients treated with MAO inhibitors; the amines can react with nitrites to produce mutagenic N-nitrosamines.

Furthermore, polyamine biosynthesis and accumulation are closely associated with cellular growth and proliferation and their levels in urine are frequently higher than usual in patients with cancer (Yoshida et al., 1997; Frébort et al., 2000c). For all these reasons, most biotechnological applications of CAOs concern the rapid determination of polyamines in foods and biological fluids and tissues. Other analytical techniques such as HPLC are expensive, time consuming, and not suitable for routine use. Analytical determinations of polyamines by CAOs are carried out usually by batch analysis, by flow injection analysis (FIA) systems including small bioreactors containing immobilized AO, and by AO-based amperometric biosensors or solid-phase assays (disposable test strips).

17.3.1 Batch and FIA Analysis

Most of batch and FIA assays of polyamines are correlated to the determination of H_2O_2 produced by AO-catalyzed oxidative deamination of polyamines. Fluorimetric measurements by homovanillic acid (Matsumoto et al., 1981b) and chemiluminescence-based measurements by Luminol (Fagerström et al., 1984; Bachrach and Plesser, 1986) have been proposed, but the spectrophotometric method based on the Trinder reaction (Barham and Trinder, 1972) appears more widely applied. In the presence of peroxidase (generally horseradish peroxidase, HRP), the hydrogen peroxide reacts with a reduced leuko-dye such as *o*-dianisidine (Bernt and Bergmeyer, 1974), guaiacol (Frébort et al., 2000c), or 4-aminoantipyrine and an aromatic amine or substituted phenol (Isobe et al., 1980; Matsumoto et al., 1981a; Suzuki et al., 1982; Stevanato et al., 1990; Bouvrette et al., 1997) to yield a quinone–imine colored product whose absorbance is linearly dependent on the H_2O_2 and consequently on polyamine concentration.

This colorimetric method has been also utilized for studying CAO activity. In fact, it achieves good sensitivity due to the high molar absorbance of the quinone–imine adduct, for example, 46000 $M^{-1}cm^{-1}$ using 3,5-dichloro-2-hydroxybenbensulfonic acid as the phenol (Agostinelli et al., 2007). Peroxidase and the reagents utilized do not affect the AO activity. The rate of this reaction is higher than the rate for CAO-catalyzed oxidative deamination and the rate of formation of the colored band is directly correlated to the rate of AO-catalyzed oxidative deamination (Corazza et al., 1992; Vianello et al., 1992 and 1993; Stevanato et al., 1994; Di Paolo et al. 1995a, b; Vianello et al., 1999).

Obviously, in all methods discussed, catalase activity and other hydrogen peroxide-consuming catalysts must be negligible in the solution. Otherwise, not all the hydrogen peroxide produced can be detected by the assay, leading to underestimation of the polyamine concentration and/or AO activity.

The insertion of small enzymatic reactors containing immobilized CAOs into FIA systems has been proposed for the determination of polyamines in food and biological tissues with sensitivity similar to that obtained with electrochemical biosensors. Our laboratory (Stevanato et al., 1990) utilized BSAO and HRP co-immobilized on CH–Sepharose and a spectrophotometric detector. Tombelli and Mascini (1998) and Compagnone et al. (2001) used PKAO or LSAO linked to aminopropyl-controlled pore glass and amperometric detectors. Ohashi et al. (1994) determined in-batch polyamine concentration by measuring oxygen consumption. They used a Clark electrode and based their method on the oxidative deamination of polyamine-containing food samples catalyzed by soluble *Aspergillus niger* AO.

17.3.2 CAO-Based Biosensors

Major advantages of using biosensors for the analytical determinations of metabolites include simplicity of use and rapidity of measurement. Furthermore, biosensors may be used as long as the immobilized biocatalyst remains active. This becomes particularly important for enzymes that are expensive and require laborious purification. Due to the increased life conferred by immobilization, an enzymatic biosensor can be inserted in a flow system for FIA or continuous monitoring of analytes. CAO biosensors are almost exclusively enzymatic electrodes based on amperometric measurements. Table 17.1 reports the most important works on CAO amperometric biosensors for the analytical determinations of biogenic amines. The main purposes of such studies include:

1. Increasing CAO biosensor sensitivity and usable life
2. Reduction of interference, particularly in experiments with complex matrices
3. Quantifying substrate analogs and alkaloids through inhibitory effects
4. Determination, through combinations with other enzymes, of precursors of the polyamines

Csoregi et al. (2001) analyzed the various existing biosensor designs utilizing electron transfer pathways; they also specifically studied CAO-based biosensors. The first enzymatic biosensors were based on amperometric determination of co-substrate decreases (Toul and Macholan 1975; Macholán and Slanina, 1991; Hasebe et al., 1997;) or formation of co-products that acted as "electron shuttles" (Draisci et al., 1998; Hasebe et al., 1997; Bouvrette et al., 1997; Tombelli and Mascini, 1998; Compagnone et al., 2001; Gasparini et al., 1991; Botrè et al., 1993; Male et al., 1996; Esti et al., 1998; Lange and Wittmann, 2002; Carelli et al., 2007), i.e. O_2 or H_2O_2 in the case of AO-catalyzed oxidative deamination (Scheme 17.1a).

One disadvantage of this method is the electrochemical interference originating from the high overpotential necessary to produce an electrochemical reaction when complex matrices must be analyzed. Artificial electron-transfer mediators are added to allow operation within an optimal potential window (about −0.10 to +0.05 V versus a saturated calomel electrode) at which electrochemical interference is minimal. The mediators directly interact with the redox form of the enzyme (Scheme 17.1b).

In an improved design of this biosensor, $Os^{2+/3+}$ as a redox mediator is complexed on a polyvinylimidazole nonconducting polymeric backbone cross linked by PEG 400 diglycidyl ether (PEGDGE) to grass pea AO (Niculescu et al., 2000c). Sample contamination is often a problem for these biosensors; thus, their application to complex matrices such as food or biological samples may result only in adding extra protection membranes.

A similar but more complex design is based on the ability of peroxidase, generally horseradish peroxidase (HRP), to directly exchange electrons with an electrode via electron tunneling. These sensors are used to determine oxidase co product H_2O_2 and are known as bi-enzyme electrodes based on their use of coupled oxidase–peroxidase enzymes (Scheme 17.1c). A simple CAO biosensor involved adsorbing grass pea AO and HRP on a graphite electrode (Niculescu et al., 2000a). The small current response and slowness of this sensor design suggested the use of an additional mediator (Scheme 17.1d), such as ferrocene (Gasparini et al., 1994), ferrocene carboxylic acid (Wimmerovà and Macholán, 1999) or osmium ion complexed with 4,4′-dimethylbipyridine (Niculescu et al., 2000a; Castillo et al., 2003; Niculescu et al., 2001).

TABLE 17.1
CAO Amperometric Biosensors

CAO Source	Biosensor Design[a]	Transducer/ Mediator	E (vs. Ag/AgCl) [mV]	Immobilization/ Technique/Support/ Ligand	LR[μM]	DL[μM]	Analytes	pH/ Buffer/ °C	Note	Reference
PSAO	1a	Oxygen cell	–	Covalent bond, Polyamide network, BSA, Glutaraldehyde	0 to 400	17	Put, Cad, Hex, His, Spd	7.0 P 30		Toul and Macholan, 1975
Soybean seedling AO	1b	Pt	+500 (vs. SCE)	Covalent bond, AH–Sepharose, BSA, Glutaraldehyde	0.5 to 10	0.2	Cad, Put, Spm, Spd	7.0 P 22		Gasparini et al., 1991
PSAO	1a	Clark-type oxygen electrode	–	Covalent bond, Polyamine mesh, Cross-linking agent	1 to 1000 on inhibitor	0.005	Amines, alkaloids, drugs, PSAO inhibitors	7.0 P 30	b	Macholán et al., 1991
PSAO	1b	Pt	–	Covalent bond	0.5 to 320	0.25	Put	6.6 P 37	c	Botrè et al., 1993
LSAO				Immunodyne	0.5 to 200		Cad			
PKAO	1b	Pt	+400	Covalent bond, Immunodyne, BSA, Glutaraldehyde	0 to 600	25	His, Cad, Put	7.2 = =		Male et al., 1996

BSAO	1a	Clark-type oxygen electrode (in presence of ascorbate)	-700	Entrapment, Polyacrylamide gel	0.5 to 50 5 to 1000	0.3 3	Hisd Hism	8.5 Tris =	d	Hasebe et al., 1997
PKAO	1b	Pt	+700	Covalent bond, Immunodyne, Glutaraldehyde	0 to 6000	25	Hism, Cad, Put	7.4 P	e	Bouvrette et al., 1997
PKAO	1b	Pt	+700	Covalent bond, Cellulose acetate, Glutaraldehyde	1 to 100	0.6	His, Put, Cad, Benz, Spd	7.4 P =	f	Tombelli and Mascini, 1998
Cicer seedlings AO	1b	Pt	+650	Covalent bond, Nylon, Glutaraldehyde	2 to 50	1	Put, Spd, Spm	8 P room temp	g	Esti et al., 1998
Cicer seedlings AO	1b	Pt	+650	Covalent bond, Immobilon affinity Nylon-net derivatized Nylon net, Glutaraldehyde	1 to 50 to 1 to 2000	0.5	Put, Cad, Hism, Spm, Tyr, Spd, Tyr	8.0 P room temp	h	Draisci et al., 1998
LSAO	1b	Pt	+650	Nylon-net, Glutaraldehyde	10 to 100 to 5 to 2000	1 to 10	Put, Cad, PhEtA, Try, Hism, Try	8.0 P =	i	Compagnone et al., 2001

(*Continued*)

TABLE 17.1 (Continued)

CAO Source	Biosensor Design[a]	Transducer/ Mediator	E (vs. Ag/AgCl) [mV]	Immobilization/ Technique/Support/ Ligand	LR[μM]	DL[μM]	Analytes	pH/ Buffer/ °C	Note	Reference
BSAO, PSAO, (tyramine oxidase)	1b	Pt; three electrodes working in parallel	+700	Silanization, cross linking, Glutaraldehyde, BSA	0 to 100 to 0 to 200 mg/Kg	5 to 10 mg/Kg	Hism, Tyr Put	6.5 P room temp	j	Lange and Wittmann, 2002
PKAO	1b	Pt; RDE	+700	Covalent bond, Glutaraldehyde, Poly-pyrrole, Poly-β-naphthol	100 200 500	6 to 12	His Put Cad	7.0 P =	k	Carelli et al., 2007
Grass pea AO	3	Graphite HRP		Absorption	1 to 100 1 to 100	0.16 0.06	His Put	7.2 P room temp	l	Niculescu et al., 2000a
	4	Graphite HRP/ PVI_{13}-dmeOs	−50	Cross linking, PVI_{13}-dmeOs + PEGDGE	1 to 150 1 to 400	0.33 0.17				
BSAO, soybean seedlings AO	4	Carbon paste, HRP, ferrocene	+200	Covalent bond, AH–Sepharose, Carbodiimide	0 to 17 0 to 6.7	1 to 5	Spm, Cad	7.0 HEPES 22		Gasparini et al., 1994
PSAO	4	Carbon paste HRP, FeCOOH	0.00	Affinity, BAT–Silasorb	0.01 to 100 (Put)	0.01 (Put)	Put, 54 other amines	7.0 or 8.5 P 30		Wimmerová and Macholán, 1999

Grass pea AO	4	Graphite PVI_7-dmeOs HRP or SPP	−50	Cross linking, PVI_7-dmeOs + PEGDGE	–	0.5 (HRP) 0.3 (SPP)	Put	7.0 P (with or without CH_3CN) room temp	m	Castillo et al., 2003
Grass pea AO	2	Graphite PVI_{13}-dmeOs	+200	Cross linking, PVI_{13}-dmeOs + PEGDGE	10 to 200	2.2	Hism	7.0 P room temp	n	Niculescu et al., 2000c
	5	Graphite		Absorption	10 to 150	3.7				
Grass pea AO	4	Graphite HRP	−50	Cross linking, PVI_{13}-dmeOs	–	–	His, Tyr, Spd, Put	7.2 P	o	Niculescu et al., 2001
	5	Graphite	+200	Absorption	–	–	Cad, et al.,	room temp		
Euphorbia latex AO	5	Gold	35	Absorption Nafion–cysteine	–	–	–	7.0 P 25	p	Hong et al., 2006

Notes: BAT–Silasorb = (2-[4,6-bis (aminoethylamine)-1,3,5-triazine]–Silasorb; BSA = bovine serum albumin; Cad = cadaverine; DL = detection limit; FeCOOH = ferro cene carboxylic acid; Hex = hexamethylenediamine; Hisd = histidine; Hism = Histamine; HRP = horseradish peroxidase; LR = linear range; n.a. = not available; P = phosphate buffer; PEGDGE = PEG 400 diglycidyl ether; PhEtA = phenylethylamine; Put = putrescine; PVI_n-dmeOs = poly(1-vinylimidazole)–[osmium (4,4′- dimethylbipyridine)$_2$Cl]$^{+/2+}$; RDE = rotating disk electrode; S = sensitivity; Spd = spermidine; Spm = spermine; SCE = saturated calomel electrode; SPP = sweet potato peroxidase; Try = tryptamine; Tyr = tyramine.

[a] Numbers correspond to those reported for biosensor designs: (1) electron transfer pathway by "electron shuttle" (a = oxygen; b = H_2O_2); (2) electron transfer pathway with electron transfer mediator; (3) electron transfer pathway in coupled enzyme electrodes; final electron transfer step via direct electron tunneling; (4) electron transfer pathway in coupled enzyme electrodes; final electron transfer step via electron transfer mediator; (5) direct electron transfer between enzyme active site and polarized electrode.

[b] Biosensor assembled to quantify on basis of inhibition effect, substrate analog of AO, certain alkaloids and drugs.

[c] LSAO biosensor used in combination with lysine decarboxylase membrane for realization of hybrid enzyme electrode for determination of lysine. Only the lentil seedling AO biosensor was used to determine ornithine, adding adequate ornithine decarboxylase directly into measuring cell.

(*Continued*)

TABLE 17.1 (Continued)

[d] Upon addition of histidine and histamine, CAO catalyzes the oxidation by dissolved oxygen of ascorbic acid, producing H_2O_2 that can be recleaved by the enzyme-activated oxygen electrode; in Tris buffer and in presence of histidine and histamine, BSAO generates ascorbate oxidase activity.

[e] Artificial mediators, e.g., ferrocene derivatives and others, were not observed to facilitate electron transfer from reduced enzyme to electrode.

[f] Analytical characteristics of AO biosensor are compared with those of AO reactor inserted in FIA arrangement. Enzyme reactor performances are investigated based on (1) AO alone, and (2) AO co-immobilized with HRP using a FeCOOH mediator dissolved in the carrier stream.

[g] Two biosensors are proposed to discriminate the contributions of SPM and SPD to total amine contents of packaged fruits. The first is assembled with immobilized maize AO, the second with cicer seedling AO.

[h] Comparison of performances of three biosensors assembled using three different activated membranes; glutaraldehyde-based procedure is selected for simplicity and lower enzyme requirement.

[i] Performances of enzyme electrode are compared with those shown by enzyme reactor containing same AO immobilized.

[j] BSAO, PSAO, and *Arthrobacter* spp. tyramine oxidase immobilized individually on a separate screen-printed thick-film electrode comprising an enzyme sensor array for simultaneous determination of Hism, Tyr, and Put by pattern recognition via artificial neural network.

[k] Several electroproduced anti-interference mono- and bi-layer films were tested to minimize fouling and the interference caused by the direct electrochemical oxidation of both the analytes and the common interferents usually present in food.

[l] Characteristics of bienzyme electrode containing grass pea and HRP cross linked to Os-based redox polymer are compared with those obtained by analogous biosensor in which enzymes are simply adsorbed onto graphite electrode. The first electrode shows better biosensor performance.

[m] SPP-based electrodes display higher sensitivity and better detection limit for Put than those using HRP and retained activity in organic phase much better than HRP-based electrodes.

[n] Two different biosensor designs are considered: one based on adsorbed AO on graphite electrode with detection based on DET mechanism; second based on an Os bipyridine-modified redox polymer using a mediated electron transfer (MET) pathway. PVI_{13}-dme Os-based electrodes showed superior stability, selectivity, and linear range.

[o] Comparison of monoenzymatic AO biosensor based on direct electron transfer (DET) and bienzymatic biosensor co-immobilizing AO and HRP and based on MET. The bienzymatic biosensor showed superior electrode characteristics; the monoenzymatic one can work anaerobically.

[p] Promotion of DET rate of the Nafion–cysteine functional membrane verified by cyclic voltammetry using CAO and three other redox proteins immobilized on a gold electrode.

To improve selectivity, more selective and stable peroxidases were sought, such as a peroxidase purified from sweet potato (Castillo et al., 2003). Niculescu et al. (2000c and 2001) hypothesize a direct electron transfer mechanism with grass pea AO-based biosensor method. The observed unmediated catalytic current was assumed to be actually caused by (1) a direct electron transfer process (Scheme 17.1e); (2) the electro-oxidation of the formed product (Scheme 17.1f); or (3) a combination of both processes.

Hong et al. (2006) proposed a similar electron transfer mechanism for ELAO. A functional membrane composed of Nafion, a perfluorinated anionic polyelectrolyte with good electrical conductivity, and cysteine was developed to promote electron transfer from the protein to the gold electrode. According to the direct electron transfer (DET) hypothesis, a mechanism of AO-catalyzed oxidative deamination (Figure 17.1) revealed a possible "short circuit" caused by the DET between the enzyme active site and the oxidative electrode (Niculescu et al., 2000b).

It is important to stress that most measurements were carried out in phosphate buffer (Table 17.1). However, the phosphate complexes the polyamines—spermine and spermidine in particular—that in this form are not recognizable as substrates by CAOs. A dissociation constant of 17 *mM* for the spermine–phosphate and spermidine–phosphate complex has been found (Corazza et al., 1992). As a consequence, the analytical determination of spermine and spermidine using phosphate as buffer may underestimate the amounts of these polyamines. Similar considerations must be included when measuring AO activity. The use of a HEPES buffer is recommended.

An interesting solid-phase assay (test strip/dipstick) for determining putrescine and cadaverine in tuna has been proposed (Hall et al., 1999). The system involves coupling PSAO or soybean seedling AO to a peroxidase–dye system. Test strips revealed a minimum detectable concentration of 0.5 μM and were stable at 4°C for at least 12 mo.

17.4 CAO BIOCATALYSIS

The research produced to date is sufficient to indicate interesting uses of CAOs in biocatalysis. The oxidative deamination of amines to aldehydes may be particularly attractive in cases in which the presence of other oxidable functional groups in a structurally complex molecule do not allow chemical modifications producing good yields without undesirable side reactions. For example, Yoshida et al. (1997) proposed the production of vanillin from vanillylamine by *Aspergillus niger* AO, either free or immobilized on an epoxy-activated porous cellulose containing formyl groups.

The AO-catalyzed oxidative deamination of diamine residues of appropriate length can produce a chemically interesting cyclization of structure, such as occurs with putrescine and cadaverine from which Δ^1-pyrroline and Δ^1-piperideine, respectively, are obtained (Perin et al., 1985). This possibility was extensively investigated by Cragg et al. (1990) using PSAO as a catalyst and α-ω diamines of different chain lengths substituted into the linear hydrocarbon structure (Figure 17.2). Similarly, Cheng et al. (1996) studied the PKAO-catalyzed oxidation of nazlinin, a serotonergic

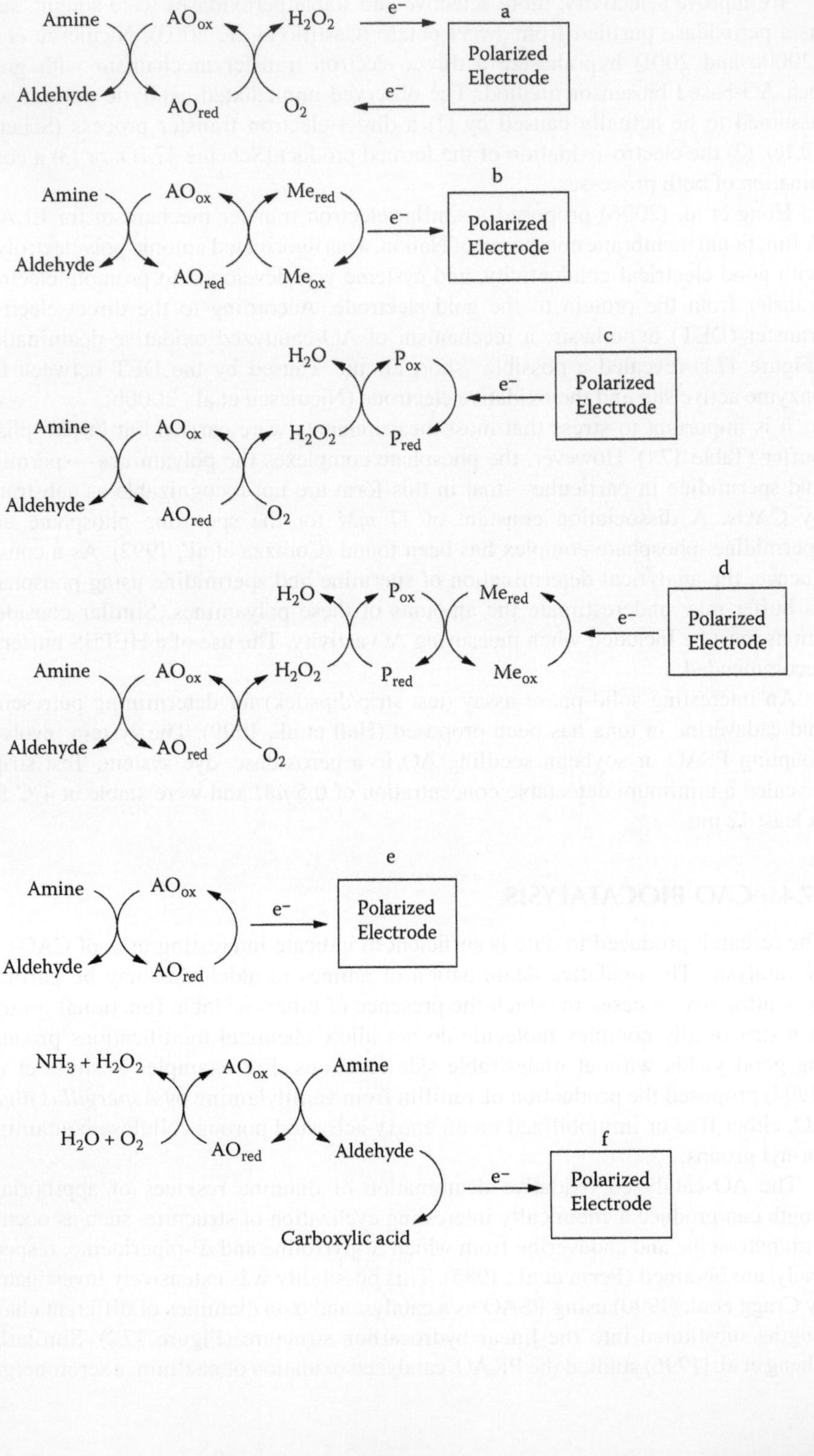
a
Amine
Aldehyde
AO_ox
AO_red
H_2O_2
O_2
e−
Polarized Electrode
b
Me_red
Me_ox
c
H_2O
P_ox
P_red
d
e
f
NH_3 + H_2O_2
H_2O + O_2
Carboxylic acid

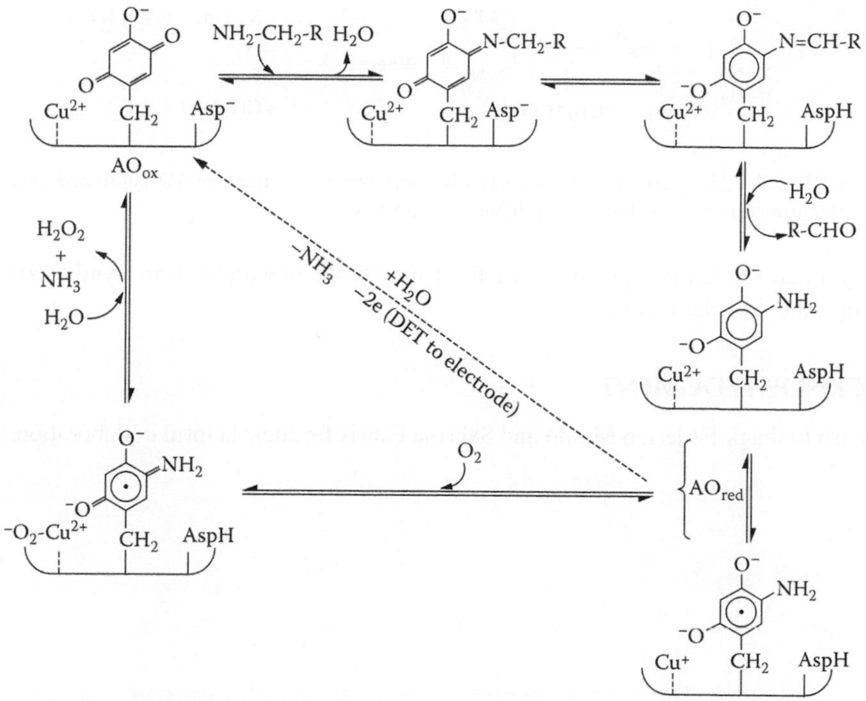

FIGURE 17.1 Mechanism of AO catalyzed reaction. Dashed line indicates possible "short circuit" caused by a DET between the enzyme active site and an oxidative electrode. (Adapted from Niculescu et al., 2000b.)

alkaloid, and its derivatives to the oxidized form of indoloquinolizidine. They demonstrated that, under appropriate conditions, the cyclization into structurally complex molecules can take place.

Chaplin et al. (2001) found that BASO and PKAO in nonaqueous media catalyze the complete oxidative deamination of benzylamine used as a general substrate, when the reaction is carried out in a biphasic system. Best results were obtained using highly hydrophobic solvents such as benzene, toluene, and cyclohexane, probably based on the more efficient extraction of the hydrophobic inhibitory products. Enzymes immobilized on celite have been used for the biocatalysis to overcome the rate-limiting step due to the mass transfer of substrate and/or products between phases. With a 5% water content, BSAO produced a high rate of conversion of benzylamine to benzaldehyde in all tested solvents. These results

---

SCHEME 17.1 (Opposite page) Possible electron transfer pathways in CAOs biosensors. (a) By "electron shuttle" oxygen or hydrogen peroxide. (b) With an electron transfer mediator (ME). (c) In coupled enzyme electrodes; final electron transfer step is via direct electron tunneling by peroxidase (P). (d) In coupled enzyme electrodes; final electron transfer step is via electron transfer mediator. (e) Between enzyme active site and polarized electrode. (f) By electro-oxidation of biogenerated aldehyde. (Adapted from Csoregi et al., 2001.)

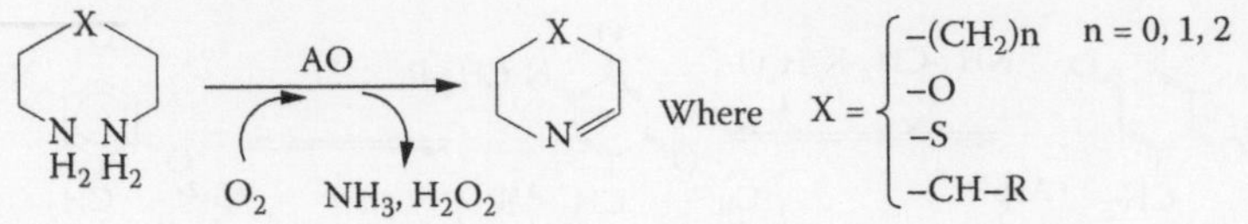

FIGURE 17.2 Δ^1-pyrroline and Δ^1-piperideine derivatives obtained by AO-catalyzed oxidative deamination with cyclization of different diamines.

may open interesting perspectives for using CAO biocatalysis in synthesizing drugs and fine chemicals.

ACKNOWLEDGMENT

I wish to thank Federico Momo and Sabrina Fabris for their helpful collaboration.

18 Concluding Remarks

David M. Dooley

It is fascinating to compare our current understanding of the structures, mechanisms, properties, and biological roles of amine oxidases, as summarized in this volume, to information available in the mid-1980s. Two influential sources—a review by Knowles and Yadav (1984) and a volume edited by Mondovì (1985)—thoroughly assessed the state of knowledge at that time. The following quotations from these volumes are revealing.

> From the biochemical point of view, one of the most interesting aspects of the AOs is the nature of the cofactors present and the mechanism of their action. Most AOs have flavin or metal cofactors. However, in general, the role of these cofactors in the mechanism of electron transfer is not clear and, in some of the AOs, the nature of the cofactors present is unknown or uncertain. (Mondovì, 1985)

> In summary, the articles in this volume, plus those in the many recent symposia, clearly indicate the many developments in this area in recent years. However, they also indicate that even though these enzymes have been studied for over 50 years, much more basic biochemical work is needed before we can understand the mechanism of action, physiological significance, and interrelationship of these functionally related series of enzymes. (Mondovì, 1985)

> The future of amine oxidase research seems very promising. On the molecular side, it is important the structure of at least one amine oxidase be determined; at present we know far more about the mechanism of action than we do about the structure. Evolutionary aspects of the amine oxidases have been neglected and such studies could improve our knowledge of how hormone and transmitter mechanisms are developed. Biological studies on how copper-containing amine oxidases are involved in the control of hormone or transmitter metabolism and in cross-linking reactions during the formation of connective tissue have not received sufficient attention. It is clear that linking the molecular and biological studies will provide a logic for more efficient drug design and an understanding of pathological conditions. (Knowles and Yadav, 1984)

As cogently documented in this volume, research on amine oxidases during the ensuing years has proven more fascinating and more productive than any investigator envisioned in 1985. Amine oxidases are now recognized as the first case of the formation of a redox cofactor (TPQ) by a self-processing reaction (Matsuzaki et al., 1994; Ruggiero et al., 1997). Although many different enzymes are now recognized to contain cofactors composed of modified amino acids, the six-electron oxidation of a tyrosine residue to produce TPQ, requiring only the protein, copper, and dioxygen, is one of the most remarkable chemical transformations in biology (Dooley, 1999).

Mechanistic (Cai and Klinman, 1994; Matsuzaki et al., 1994; Ruggiero et al., 1997; Ruggiero and Dooley, 1999), spectroscopic (Hirota et al., 2001; Nakamura et al., 1996), and structural (Kim et al., 2002; Wilce et al., 1997) studies of TPQ biogenesis have revealed many details of the reaction mechanism, but the reaction has not yet been reproduced in model studies, suggesting that the three-dimensional structure of the active site in the unprocessed protein controls the chemistry in important ways.

As described herein, multiple amine oxidase structures have been elucidated over the past several years, including bacterial, plant, fungal, and mammalian enzymes. Among many insights into amine oxidase structure and function that these crystal structures have provided is the realization that the properties of the various active site "channels" or "funnels" are important in determining substrate specificity (Duff et al., 2003; Langley et al., 2006; Parsons et al., 1995; Wilmot et al., 1997). Amine oxidases may also provide a protein binding site for O_2 as well (Duff et al., 2004; Johnson et al., 2007). It is intriguing to note that the characteristics of the amine recognition channel may be exploited to design amine oxidase inhibitors with remarkable selectivity (Contakes et al., 2005; Langley et al., 2008a; Shepard et al., 2002).

Substantial progress has been achieved in elucidating the mechanisms of amine oxidation and O_2 reduction (Mure, 2004). Metal substitution and site-directed mutagenesis experiments have been particularly informative (Hirota et al., 2001; Kishishita et al., 2003; Mills et al., 2002; Mills and Klinman, 2000; Murray et al., 1999; Plastino et al., 1999) as noted in this volume. Although the available evidence indicates that the overall mechanism for amine oxidation is very similar among amine oxidases (although stererospecificities may vary) (Alton et al., 1995; Uchida et al., 2003), it currently appears that both outer-sphere and inner-sphere mechanisms may operate for O_2 reduction, depending on the specific amine oxidase (Johnson et al., 2007; Juda et al., 2006; Mukherjee et al., 2008; Shepard et al., 2008; Su and Klinman, 1998).

In the outer-sphere mechanism, electron transfer occurs directly between reduced TPQ (aminoquinol form) and O_2, whereas in the inner-sphere mechanism O_2 reduction is facilitated by binding to Cu(I). Recent experiments demonstrate that the inner-sphere mechanism is operative for amine oxidases from pea seedlings (Mukherjee et al., 2008) and *Arthrobacter globiformis* (Shepard et al., 2008), but strong evidence in favor of the outer-sphere mechanism has been presented for amine oxidases from bovine plasma and the *Hansenula polymorpha* yeast (Mure et al., 2002; Su and Klinman, 1998). Thermodynamic analysis indicates that the inner-sphere pathway should be the lower energy reaction (Mukherjee et al., 2008), but it is conceivable that the active site may modulate the energetics. This possibility must be investigated on a case-by-case basis.

The discovery that vascular adhesion protein-1 (VAP-1) is an amine oxidase (Smith et al., 1998), and is probably the source of soluble SSAO activity (Jaakkola et al., 2000a, b), has stimulated a great deal of new research on the physiological roles of amine oxidases. Although a VAP-1 knockout mouse displayed no overt phenotypes in a pathogen-free environment (Stolen et al., 2005), indicating that VAP-1 is not required for normal development, it is possible that soluble SSAO may participate in multiple physiological processes, with the understanding that other enzymes may substitute for its critical roles (if any, beyond cellular immunity). Moreover, the discoveries of the potential diverse expression and broad substrate specificity of the human diamine oxidase and multiple genes for lysyl-oxidase-like proteins, suggests that the physiological role of copper-containing amine oxidases may be more diverse than previously considered.

Much of the knowledge that Bruno Mondovì suggested was important to obtain is now in hand. But it is still fair to say, as Knowles and Yadav did, that the future of amine oxidase research appears very promising. The modern tools of biochemistry, structural and molecular biology, proteomics, and informatics should enable investigators in the field to elucidate the diverse roles of the amine oxidase and lysyl oxidase families in the growth, development, and metabolism of multiple organisms in which these enzymes are clearly important.

References

Abboud, H.E. 1983. Catabolism of histamine in the isolated glomeruli and tubules of the rat kidney. *Kidney Int* 24: 534.

Abdel-Rahman, E. and W.K. Bolton. 2002. Pimagedine: a novel therapy for diabetic nephropathy. *Expert Opin Invest Drugs* 11: 565.

Abeles, R.H. and A.L. Maycock. 1976. Suicide enzyme inactivators. *Acc Chem Res* 9: 313.

Abella, A.L. et al. 2003. Semicarbazide-sensitive amine oxidase/vascular adhesion protein-1 activity exerts an antidiabetic action in Goto-Kakizaki rats. *Diabetes* 52: 1004.

Abella, A. et al. 2004. Adipocytes release a soluble form of VAP-1/SSAO by a metalloprotease-dependent process and in a regulated manner. *Diabetologia* 47: 429.

Adámková, Š. et al. 2001. Probing of active site of pea seedlings amine oxidase with optical antipodes of sedamine alkaloids. *J Enzym Inhib* 16: 367.

Adams, J.L., T.M. Chen, and B.W. Metcalf. 1985. 4-Amino-4,5-dihydrothiophene-2-carboxylic acid. *J Org Chem* 50: 2730.

Adams, J.D. Jr. and L.K. Klaidman. 1993. Acrolein-induced oxygen radical formation. *Free Rad Biol Med* 15: 187.

Agostinelli, E. et al. 1994a. Properties of cobalt-substituted bovine serum amine oxidase. *Eur J Biochem* 222: 727.

Agostinelli, E. et al. 1994b. Heat enhancement of cytotoxicity induced by oxidation products of spermine in Chinese hamster ovary cells. *Biochem Pharmacol* 48: 1181.

Agostinelli, E. et al. 1997. Reactions of the oxidized organic cofactor in copper-depleted bovine serum amine oxidase. *Biochem J* 324: 497.

Agostinelli, E. et al. 1998. Reconstitution of Cu^{2+}-depleted bovine serum amine oxidase with Co^{2+}. *Biochem J* 330: 383.

Agostinelli, E. et al. 2007. Polyketone polymer: a new support for direct enzyme immobilization. *J Biotechnol* 127: 670.

Airas, L. et al. 2006. Elevated serum soluble vascular adhesion protein 1 (VAP-1) in patients with active relapsing remitting multiple sclerosis. *J Neuroimmunol* 177: 132.

Airas, L. et al. 2007. Vascular adhesion protein-1 in human ischaemic stroke. *Neuropathol Appl Neurobiol* http: //www.blackwell-synergy.com/doi/pdf/10.1111/j.1365-2990.2007.00911.x.

Airenne, T.T. et al. 2005. Crystal structure of the human vascular adhesion protein-1: unique structural features with functional implications. *Protein Sci* 14: 1964.

Alarcon, R.A. 1970. Acrolein. IV. Evidence for formation of the cytotoxic aldehyde acrolein from enzymatically oxidized spermine or spermidine. *Arch Biochem Biophys* 137: 365.

Alarcon, R.A. 1972. Acrolein, a component of a universal cell-growth regulatory system: a theory. *J Theor Biol* 37: 159.

Alhonen, L. et al. 2000. Activation of polyamine catabolism in transgenic rats induces acute pancreatitis. *Proc Natl Acad Sci USA* 97: 8290.

Alton, G. et al. 1995. Stereochemistry of benzylamine oxidation by copper amine oxidases. *Arch Biochem Biophys* 316: 353.

Altschul, S.F et al. 1997. Gapped BLAST and PSI-BLAST: a new generation of protein database search programs. *Nucleic Acids Res* 25: 3389.

Ameyama, M. et al. 1984. Microbial production of pyrroloquinoline quinone. *Agri Biol Chem* 48: 561.

Amon, U. et al. 1999. Enteral histaminosis: clinical implications. *Inflamm Res* 48: 291.

An, Z. et al. 2008. Hydrogen peroxide generated by copper amine oxidase is involved in abscisic acid-induced stomatal closure in *Vicia faba*. *J Exp Bot* 59: 815.

Andree, T.H. and D.E. Clarke. 1981. The isolated perfused rat brain preparation in the study of monoamine oxidase and benzylamine oxidase: lack of selective brain perfusion. *Biochem Pharmacol* 30: 959.

Andree, T.H. and D.E. Clarke. 1982a. Characteristics and specificity of phenelzine and benserazide as inhibitors of benzylamine oxidase and monoamine oxidase. *Biochem Pharmacol* 31: 825.

Andree, T.H. and D.E. Clarke. 1982b. Characteristics of rat skull benzylamine oxidase. *Proc Soc Exp Biol Med* 171: 298.

Andres, N. et al. 2001. Tissue activity and cellular localization of human semicarbazide-sensitive amine oxidase. *J Histochem Cytochem* 49: 209.

Ansar, M.M. and M. Ansari. 2006. Nitric oxide involvement in pancreatic beta cell apoptosis by glibenclamide. *Nitric Oxide* 14: 39.

Ardanaz, N. and P.J. Pagano. 2006. Hydrogen peroxide as a paracrine vascular mediator: regulation and signaling leading to dysfunction. *Exp Biol Med (Maywood)* 231: 237.

Artico, M. et al. 1992. Aromatic hydrazides as specific inhibitors of bovine serum amine oxidase. *Eur J Med Chem* 27: 219.

Arvilommi, A.M. et al. 1996. Lymphocyte binding to vascular endothelium in inflamed skin revisited: a central role for vascular adhesion protein-1 (VAP-1). *Eur J Immunol* 26: 825.

Arvilommi, A.M., M. Salmi, and S. Jalkanen. 1997. Organ-selective regulation of vascular adhesion protein-1 expression in man. *Eur J Immunol* 27: 1794.

Asthir, B. et al. 2002. Diamine oxidase is involved in H_2O_2 production in the chalazal cells during barley grain filling. *J Exp Bot* 53: 677.

Atanasiu, R. et al. 1995. The antiarrhythmic effects of ceruloplasmin during reperfusion in the ischemic isolated rat heart. *Can J Physiol Pharmacol* 73: 1253.

Atanasiu, R. et al. 1996. Class III-type antiarrhythmic effects of ceruloplasmin on rat heart. *Can J Physiol Pharmacol* 74: 652.

Averill-Bates, D. et al. 1993. Cytotoxicity and kinetic analysis of purified bovine serum amine oxidase in the presence of spermine in Chinese hamster ovary cells. *Arch Biochem Biophys* 300: 75.

Averill-Bates, D.A. et al. 1994. Aldehyde dehydrogenase and cytotoxicity of purified bovine serum amine oxidase and spermine in Chinese hamster ovary cells. *Biochem Cell Biol* 72: 36.

Averill-Bates, D.A. et al. 2005. Anti-tumoral effect of native and immobilized bovine serum amine oxidase in a mouse melanoma model. *Biochem Pharmacol* 69: 1693.

Averill-Bates, D.A. et al. 2008. Mechanism of cell death induced by spermine and amine oxidase in mouse melanoma cells. *Int J Oncol* 32: 79.

Aydin, S. et al. 2008. Influence of microvascular endothelial cells on transcriptional regulation of proximal tubular epithelial cells. *Am J Physiol Cell Physiol* 294: C543.

Azakami, H. et al. 1994. Nucleotide sequence of the gene for monoamine oxidase (Maoa) from *Escherichia coli*. *J Ferment Bioeng* 77: 315.

Baba, S. et al. 1984. High-performance liquid chromatographic determination of serum aliphatic amines in chronic renal failure. *Clin Chim Acta* 136: 49.

Bachrach, U. 1985. Copper amine oxidases and amines as regulators of cellular processes. In *Structure and Functions of Amine Oxidases*, Mondovì, B., Ed. Boca Raton: CRC Press, p. 5.

Bachrach, U. and Y.M. Plesser. 1986. A sensitive, rapid, chemilumunescence-based method for the determination of diamines and polyamines. *Anal Biochem* 152: 423.

Bachrach, U., I. Ash, and E. Rahamim. 1987. Effect of microinjected amine and diamine oxidases on the ultrastructure of eukaryotic cultured cells. *Tissue Cell* 19: 39.

Balaña-Fouce, R. et al. 1986. Inhibition of diamine oxidase and S-adenosylmethionine decarboxylase by diminacene aceturate (berenil). *Biochem Pharmacol* 35: 1597.

Ball, W.J. Jr. and M. Balis. 1976. Changes in ornithine decarboxylase activity in rat intestines during aging. *Cancer Res* 36: 3312.

Banchelli, G. and F. Buffoni. 1981. Histochemistry of porcine tissues using antibodies to pig plasma amine oxidase. *Br J Pharmacol* 72: 146.

Banchelli, G. et al. 1986. Guanabenz as inhibitor of copper-containing amine oxidases. *Agents Actions* 18: 46.

Banchelli, G. et al. 1990. A study of the biochemical pharmacology of 3,5-ethoxy-4-amino-methylpyridine (B24), a novel amine oxidase inhibitor with selectivity for tissue bound semicarbazide-sensitive amine oxidase enzymes. *Neurochem Int* 17: 215.

Bang, L. et al. 1998. Hydralazine-induced vasodilation involves opening of high conductance Ca^{2+}-activated K^{+} channels. *Eur J Pharmacol* 361: 43.

Barbouti, A. et al. 2002. DNA damage and apoptosis in hydrogen peroxide-exposed Jurkat cells: bolus addition versus continuous generation of H_2O_2. *Free Radic Biol Med* 33: 691.

Barbry, P. et al. 1990. Human kidney amiloride-binding protein: cDNA structure and functional expression. *Proc Natl Acad Sci USA* 87: 7347.

Bardsley, W.G., C.M. Hill, and R.W. Lobley. 1970. A reinvestigation of the substrate specificity of pig kidney diamine oxidase. *Biochem J* 117: 169.

Bardsley, W.G., J.S. Ashford, and C.M. Hill. 1971. Synthesis and oxidation of aminoalkyl ammonium compounds by pig kidney diamine oxidase. *Biochem J* 122: 557.

Bardsley, W.G. and J.S. Ashford. 1972. Inhibition of pig kidney diamine oxidase by substrate analogues. *Biochem J* 128: 253.

Bardsley, W.G., M.J.C. Crabbe, and I.V. Scott. 1974a. The amine oxidases of human placenta and pregnancy plasma. *Biochem J* 139: 169.

Bardsley, W.G., R.E. Childs, and M.J.C. Crabbe. 1974b. Inhibitions of enzymes by metal-ion chelating reagents: the action of copper-chelating reagents on diamine oxidase. *Biochem J* 137: 61.

Bardsley, W.G. et al. 1980. Deviations from Michaelis-Menten kinetics: the possibility of complicated curves for simple kinetic schemes and the computer fitting, etc. *Biochem J* 187: 739.

Bardsley, W.G. 1985. Inhibitors of copper amine oxidases. In *Structure and Functions of Amine Oxidases*, Mondovì, B., Ed. Boca Raton: CRC Press, p. 135.

Barham, D. and P. Trinder. 1972. An improved color reagent for the determination of blood glucose by the oxidase system. *Analyst* 97: 142.

Barrand, M.A. and A. Callingham. 1982. Monoamine oxidase activities in brown adipose tissue of the rat: some properties and subcellular distribution. *Biochem Pharmacol* 31: 2177.

Barrand, M.A., B.A. Callingham, and S.A. Fox. 1984. Amine oxidase activities in brown adipose tissue of the rat: identification of semicarbazide-sensitive (clorgyline-resistant) activity at the fat cell membrane. *J Pharm Pharmacol* 36: 652.

Barri, Y.M. 2006. Hypertension and kidney disease: a deadly connection. *Curr Cardiol Rep* 8: 411.

Battaglia, V., M. Salvi, and A. Toninello. 2005. Oxidative stress is responsible for mitochondrial permeability transition induction by salicylate in liver mitochondria. *J Biol Chem* 280: 33864.

Battaglia, V. et al. 2007. Different behavior of agmatine in liver mitochondria: inducer of oxidative stress or scavenger of reactive oxygen species? *Biochim Biophys Acta* 1768: 1147.

Baylin, S.B., S.A. Stevens, and K.M. Shakir. 1978. Association of diamine oxidase and ornithine decarboxylase with maturing cells in rapidly proliferating epithelium. *Biochim Biophys Acta* 541: 415.

Baylin, S.B. and G.D. Luk. 1985. Diamine oxidase activity in human tumors: clinical and biological significance. In *Structure and Functions of Amine Oxidases,* Mondovì, B., Ed. Boca Raton: CRC Press, p. 187.

Bechara, E.J. et al. 2007. The dual face of endogenous alpha-aminoketones: pro-oxidizing metabolic weapons. *Comp Biochem Physiol C Toxicol Pharmacol* 146: 88.

Beeri, M.S. et al. 2005. Type 2 diabetes is negatively associated with Alzheimer's disease neuropathology. *J Gerontol A Biol Sci Med Sci* 60: 471.

Befani, O. et al. 1989. Peculiar effects of temperature and polyvinylalcohol on the activity of bovine serum amine oxidase. *Biochem Biophys Res Commun* 163: 1168.

Befani, O. et al. 1995. Inhibition of diamine oxidase by metronidazole. *Biochem Biophys Res Commun* 212: 589.

Befani O. et al. 1998. Serum amine oxidase can specifically recognize and oxidize aminohexyl chains on AH-Sepharose support: single step affinity immobilization. *Biotechnol Appl Biochem* 28: 99.

Befani, O., F. Missiroli, and B. Mondovì. 1999. Histaminase (diamine oxidase) activity in human cataract. *Inflamm Res* 48: S77.

Behl, C. et al. 1994. Hydrogen peroxide mediates amyloid beta protein toxicity. *Cell* 77: 817.

Bellelli, A. et al. 1985. Transient kinetics of copper-containing lentil (*Lens culinaris*) seedling amine oxidase. *Biochem J* 232: 923.

Bellelli, A. et al. 1991. On the mechanism and rate of substrate oxidation by amine oxidase from lentil seedlings. *J Biol Chem* 266: 20654.

Bellelli, A. et al. 2000. The oxidation and reduction reactions of bovine serum amine oxidase: a kinetic study. *Eur J Biochem* 267: 3264.

Bergeret, B., H. Blaschko, and R. Hawes. 1957. Occurrence of an amine oxidase in horse serum. *Nature* 180: 1127.

Bergmeyer, H.U. 1974. Determination of D-glucose with glucose oxidase and peroxidase. In *Methods of Enzymatic Analysis*. Vol. 3, Bergmeyer, H.U., Ed. Weinheim: Verlag Chemie, p. 1205.

Berman, H., K. Henrick, and H. Nakamura. 2003. Announcing the worldwide Protein Data Bank. *Nat Struct Biol* 10: 980.

Bernt, E. and H.U. Bergmeyer. 1974. Inorganic peroxides. In *Methods of Enzymatic Analysis*. Vol. 4, Bergmeyer, H.U., Ed. Weinheim: Verlag Chemie, p. 2246.

Berry, M.D. 2004. Mammalian central nervous system trace amines : pharmacologic amphetamines, physiologic neuromodulators. *J Neurochem* 90: 257.

Bertini, V. et al. 2005. Alkylamino derivatives of 4-aminomethylpyridine as inhibitors of copper-containing amine oxidases. *J Med Chem* 48: 664.

Best, C.H. 1929. Disappearance of histamine from autolysing lung tissue. *J Physiol* 67: 256.

Bhatti A.R. et al. 1988. Synthesis and nuclear magnetic resonance spectroscopic, mass spectroscopic, and plasma amine oxidase inhibitory properties of analogues of 1-methyl-4-phenyl-1,2,3,6-tetrahydropyridine. *J Neurochem* 59: 1097.

Bieganski, T., M.Z. Blasinska, and J. Kusche. 1977. Determination of histaminase (diamine oxidase) activity by o-dianisidine test: interference of ceruloplasmin. *Agents Actions* 7: 85.

Bieganski, T. et al. 1980. Human intestinal diamine oxidase: substrate specificity and comparative inhibitor study. *Agents Actions* 10: 108.

Bieganski, T., Z. Osinska, and C. Maslinski. 1982a. Inhibition of plant and mammalian diamine oxidases by hydrazine and guanidine compounds. *Int J Biochem* 14: 949.

Biegański, T., Z. Osińska, and C. Masliński. 1982b. Inhibition of plant and mammalian diamine oxidase by substrate analogues. *Agents Actions* 12: 41.

Bieganski, T., J. Kusche, and W. Lorenz. 1983. Distribution and properties of human intestinal diamine oxidase and its relevance for the histamine catabolism. *Biochim Biophys Acta* 756: 196.

Binda, C. et al. 2008. Structural and mechanistic studies of arylalkylhydrazine inhibition of human monoamine oxidases A and B. *Biochemistry* 47: 5616.

Bird, D.W., J.E. Savage, and B.L. O'Dell. 1966. Effect of copper deficiency and inhibitors on the amine oxidase activity. *Proc Soc Exp Biol Med* 23: 250.

Bisby, R.H., S.A. Johnson, and A.W. Parker. 2000. Radicals from one-electron oxidation of 4-aminoresorcinol: models for the active site radical intermediate in copper amine oxidases. *J Phys Chem B* 104: 5832.

Blaschko, H. et al. 1959. The amine oxidase of mammalian plasma. *J Physiol* 145: 384.
Blaschko, H. 1962. The amine oxidases of mammalian blood plasma. *Comp Physiol Biochem* 1: 67.
Blaschko, H. and R. Bonney. 1962. Spermine oxidase and benzylamine oxidase: distribution, development and substrate specificity. *Proc Roy Soc B* 156: 268.
Blaschko, H. and F. Buffoni. 1965. Pyridoxal phosphate as a constituent of the histaminase (benzylamine oxidase) of pig plasma. *Proc R Soc London B* 163: 45.
Blaschko, H. 1974. The natural history of amine oxidases. *Rev Physiol Biochem* 70: 83.
Bodmer, S., C. Imark, and M. Kneubühl. 1999. Biogenic amines in foods: histamine and food processing. *Inflamm Res* 48: 296.
Boér, K. et al. 2004. Histamine metabolism and CD8(+) T cell infiltration in colon adenomas. *Inflamm Res* 53: S83.
Boér, K. et al. 2008. Decreased expression of histamine H1 and H4 receptors suggests disturbance of local regulation in human colorectal tumours by histamine. *Eur J Cell Biol* 87: 227.
Bolt, H.M. 1987. Experimental toxicology of formaldehyde. *J Cancer Res Clin Oncol* 113: 305.
Bonder, C. et al. 2005. Rules of recruitment of trafficking Th1 and Th2 cells in inflamed liver. *Immunity* 23: 153.
Bonneau, M.J. and R. Poulin. 2000. Spermine oxidation leads to necrosis with plasma membrane phosphatidylserine redistribution in mouse leukemia cells. *Exp Cell Res* 259: 23.
Bono, P. et al. 1998a. Isolation, structural characterization, and chromosomal mapping of the mouse vascular adhesion protein-1 gene and promoter. *J Immunol* 161: 2953.
Bono, P. et al. 1998b. Cloning and characterization of mouse vascular adhesion protein-1 reveals a novel molecule with enzymatic activity. *J Immunol* 160: 5563.
Bono, P., S. Jalkanen, and M. Salmi. 1999. Mouse vascular adhesion protein-1 (VAP-1) is a sialoglycoprotein with enzymatic activity and is induced in diabetic insulitis. *Am J Pathol* 155: 1613.
Boomsma, F. et al. 1995. Plasma semicarbazide-sensitive amine oxidase activity is elevated in diabetes mellitus and correlates with glycosylated haemoglobin. *Clin Sci (Lond)* 88: 675.
Boomsma, F. et al. 1997. Plasma semicarbazide-sensitive amine oxidase is elevated in patients with congestive heart failure. *Cardiovasc Res* 33: 387.
Boomsma, F. et al. 1999. Circulating semicarbazide-sensitive amine oxidase is raised both in Type I (insulin-dependent), in Type II (non-insulin-dependent) diabetes mellitus and even in childhood Type I diabetes at first clinical diagnosis. *Diabetologia* 42: 233.
Boomsma, F. et al. 2000a. Variation in semicarbazide-sensitive amine oxidase activity in plasma and tissues of mammals. *Comp Biochem Physiol [C]* 126: 69.
Boomsma, F. et al. 2000b. Plasma semicarbazide-sensitive amine oxidase (SSAO) is an independent prognostic marker for mortality in chronic heart failure. *Eur Heart J* 21: 1859.
Boomsma, F., K. Ekberg, and G.J. Bruining. 2001. C Peptide and insulin do not influence plasma semicarbazide-sensitive amine oxidase activity. *Diabetologia* 44: 388.
Boomsma, F. et al. 2003. Plasma semicarbazide-sensitive amine oxidase in human (patho) physiology. *Biochim Biophys Acta Prot Struct Mol Enzymol* 1647: 48.
Boomsma, F. et al. 2005a. Semicarbazide-sensitive amine oxidase (SSAO): from cell to circulation. *Med Sci Monit* 11: RA122.
Boomsma, F. et al. 2005b. Association between plasma activities of semicarbazide-sensitive amine oxidase and angiotensin-converting enzyme in patients with type 1 diabetes mellitus. *Diabetologia* 48: 1002.
Boor, P.J. and T.J. Nelson. 1980. Allylamine cardiotoxicity III. Protection by semicarbazide and *in vivo* derangements of monoamine oxidase. *Toxicology* 18: 87.
Boor, P.J. and V.J. Ferrans. 1985. Ultrastructural alterations in allylamine cardiovascular toxicity: late myocardial and vascular lesions. *Am J Pathol* 121: 39.
Boor, P.J. and R.M. Hysmith. 1987. Allylamine cardiovascular toxicity. *Toxicology* 44: 129.

Boor, P.J., R. Hysmith, and R. Sanduja. 1990. A role for a new vascular enzyme in the metabolism of xenobiotic amines. *Circ Res* 66: 249.

Boor, P. et al. 2007. Treatment targets in renal fibrosis. *Nephrol Dial Transplant* 22: 3391.

Bossa, M., G.O. Morpurgo, and L. Morpurgo. 1994. Models and molecular orbital semi-empirical calculations in the study of the spectroscopic properties of bovine serum amine oxidase quinone cofactor. *Biochemistry* 33: 4425.

Botrè, F. et al. 1993. Plant tissue for the determination of biogenic diamines and of their amino acid precursors: effect of carbonic anhydrase. *Sensor Actuat B* 15: 135.

Bouchereau, A. et al. 1999. Polyamines and environmental challenges: recent development. *Plant Sci* 140: 103.

Bour, S. et al. 2006. The imidazoline I2-site ligands BU 224 and 2-BFI inhibit MAO-A and MAO-B activities, hydrogen peroxide production, and lipolysis in rodent and human adipocytes. *Eur J Pharmacol* 552: 20.

Bour, S. et al. 2007a. Adipogenesis-related increase of semicarbazide-sensitive amine oxidase and monoamine oxidase in human adipocytes. *Biochimie* 89: 916.

Bour, S. et al. 2007b. Semicarbazide-sensitive amine oxidase substrates fail to induce insulin-like effects in fat cells from AOC3 knockout mice. *J Neural Transm* 114: 829.

Bouvrette, P. et al. 1997. Amperometric biosensor for diamine using diamine oxidase purified from porcine kidney. *Enzyme Microb Technol* 20: 32.

Boyce, S. and K.F. Tipton. 2000. History of the enzyme nomenclature system. *Bioinformatics* 16: 34.

Bruinenberg, P.G. et al. 1989. Cloning and sequencing of the peroxisomal amine oxidase gene from *Hansenula polymorpha*. *Biochim Biophys Acta* 1008: 157.

Buckman, T.D., M.S. Sutphin, and B. Mitrovic. 1993. Oxidative stress in a clonal cell line of neuronal origin: effects of antioxidant enzyme modulation. *J Neurochem* 60: 2046.

Buffoni, F. and H. Blaschko. 1964. Benzylamine oxidase and histaminase: purification and crystallization of an enzyme from pig plasma. *Proc Res Soc (London) B* 161: 153.

Buffoni, F. and H. Blaschko. 1965. The amine oxidase of pig plasma in copper deficiency. *Biochem J* 96: 4c.

Buffoni, F. 1966. Histaminase and related amine oxidases. *Pharm Rev* 18: 1163.

Buffoni, F., L.D. Corte, and P.F. Knowles. 1968. The nature of copper in pig plasma benzylamine oxidase. *Biochem J* 106: 575.

Buffoni, F. and L. Della Corte. 1972. Pig plasma benzylamine oxidase. *Adv Biochem Psychopharmacol* 5: 133.

Buffoni, F. and G. Ignesti. 1975. Active-site titration of pig plasma benzylamine oxidase with phenylhydrazine. *Biochem J* 145: 369.

Buffoni, F., L. Della Corte, and D.B. Hope. 1977a. Immunofluorescence histochemistry of porcine tissues using antibodies to pig plasma amine oxidase. *Proc Roy Soc B* 195: 417.

Buffoni, F. et al. 1977b. Platelets monoamine oxidases and liver mixed function oxidases in cirrhotic patients. *Med Biol* 55: 109.

Buffoni, F. et al. 1989. Effect of pyridoxamine on semicarbazide-sensitive amine oxidase activity of rabbit lung and heart. *J Pharm Pharmacol* 41: 469.

Buffoni, F. et al. 1990. *In vivo* effect of 3,5-ethoxy-4-aminomethylpyridine (B24), a novel amine oxidase inhibitor with selectivity for tissue bound semicarbazide-sensitive amine oxidase enzymes. *Pharmacol (Life Sci Adv)* 9: 209.

Buffoni, F. et al. 1994. The role of semicarbazide-sensitive amine oxidase with a high affinity for benzylamine (Bz. SSAO) in the catabolism of histamine in the mesenteric arterial bed of the rat. *Agents Actions* 42: 1.

Buffoni, F. et al. 2000. Semicarbazide-sensitive amine oxidases in heart and bovine serum. *Neurobiology* 8: 17.

Burke, W.J. et al. 2004. Neurotoxicity of MAO metabolites of catecholamine neurotransmitters: role in neurodegenerative diseases. *Neurotoxicology* 25: 101.

Butcher, E.C. and L.J. Picker. 1996. Lymphocyte homing and homeostasis. *Science* 272: 60.

Cai, D. and J.P. Klinman. 1994. Copper amine oxidase: heterologous expression, purification, and characterization of an active enzyme in *Saccharomyces cerevisiae*. *Biochemistry* 33: 7647.

Cai, D. and J.P. Klinman. 1994. Evidence for a self-catalytic mechanism of 2,4,5- trihydroxyphenylalanine quinone biogenesis in yeast copper amine oxidase. *J Biol Chem* 269: 32039.

Cai, D. et al. 1997a. Mechanism-based inactivation of a yeast methylamine oxidase mutant: implications for the functional role of the consensus sequence surrounding topaquinone. *Biochemistry* 36: 11472.

Cai, D., N.K. Williams, and J.P. Klinman. 1997b. Effect of metal on 2,4,5-trihydroxyphenylalanine (topa) quinone biogenesis in the *Hansenula polymorpha* copper amine oxidase. *J Biol Chem* 272: 19277.

Calcabrini, A. et al. 2002. Enzymatic oxidation products of spermine induce greater cytotoxic effects on human multidrug-resistant colon carcinoma cells (LoVo) than on their wild-type counterparts. *Int J Cancer* 99: 43.

Calderone, V. et al. 2003. Crystallization and preliminary x-ray data of amine oxidase from bovine serum. *Acta Cryst* 59: 727.

Callingham, B.A., J. Elliott, and R.B. Williams. 1988. Amine oxidase interactions in the cardiovascular system. *Neurol Neurobiol* 42A: 109.

Callingham, B.A., A.E. Crosbie, and B.A. Rous. 1995. Some aspects of the pathophysiology of semicarbazide-sensitive amine oxidase enzymes. *Progr Brain Res* 106: 305.

Cameron, N.E. and M.A. Cotter. 1993. Potential therapeutic approaches to the treatment or prevention of diabetic neuropathy: evidence from experimental studies. *Diabet Med* 10: 593.

Carelli, D. et al. 2007. An interference free amperometric biosensor for the detection of biogenic amines in food products. *Biosens Bioelectron* 15: 640.

Carnes, W.H. 1971. Role of copper in connective tissue metabolism. *Fed Proc* 30: 995.

Carpéné, C. et al. 1995. Inhibition of amine oxidase activity by derivatives that recognize imidazoline I2 sites. *J Pharmacol Exp Ther* 272: 681.

Carpéné, C. et al. 2001. Substrates of semicarbazide-sensitive amine oxidase mimic diverse insulin effects in adipocytes. *Inflamm Res* 50: S142.

Carpéné, C. et al. 2003. Characterization of semicarbazide-sensitive amine oxidase in human subcutaneous adipocytes and search for novel functions. *Inflammopharmacology* 11: 119.

Carpéné, C. et al. 2006. Short- and long-term insulin-like effects of monoamine oxidase and semicarbazide-sensitive amine oxidase substrates in cultured adipocytes. *Metabolism* 55: 1397.

Carpéné, C. et al. 2007. Reduction of fat deposition by combined inhibition of monoamine oxidases and semicarbazide-sensitive amine oxidases in obese Zucker rats. *Pharmacol Res* 56: 522.

Carter, S.R. et al. 1994. Purification and active-site characterization of equine plasma amine oxidase. *J Inorg Biochem* 56: 127.

Casero, R.A. Jr. and L.J. Marton. 2007. Targeting polyamine metabolism and function in cancer and other hyperproliferative diseases. *Nat Rev Drug Discov* 6: 373.

Castillo, J. et al. 2003. Bienzyme biosensors for glucose, ethanol and putrescine built on oxidase and sweet potato peroxidase. *Biosens Bioelectron* 18: 705.

Castillo, V. et al. 1998. Semicarbazide-sensitive amine oxidase (SSAO) from human and bovine cerebrovascular tissues: biochemical and immunohistological characterization. *Neurochem Int* 33: 415.

Castillo, V., J.M. Lizcano, and M. Unzeta. 1999. Presence of SSAO in human and bovine meninges and microvessels. *Neurobiology (Bp)* 7: 263.

Catalano, S.M. et al. 2006. The role of amyloid-beta derived diffusible ligands (ADDLs) in Alzheimer's disease. *Curr Top Med Chem* 6: 597.

Chahine, R. et al. 1991. Protective effect of ceruloplasmin against electrolysis-induced oxygen free radicals in rat heart. *Can J Physiol Pharmacol* 69: 1459.

Chakravarti, R.N. 1955. Chemical identity of moringine. *Bull Calcutta School Trop Med* 3: 162.

Chaplin, J.A., C.L. Budde, and Y.L. Khmelnitsky. 2001. Catalysis by amine oxidases in nonaqueous media. *J Mol Cat B* 13: 69.

Chapman, J.R., P.J. O'Connell, and B.J. Nankivell. 2005. Chronic renal allograft dysfunction. *J Am Soc Nephrol* 16: 3015.

Chassande, O. et al. 1994. The human gene for diamine oxidase: an amiloride binding protein. *J Biol Chem* 269: 14484.

Cheetham, P.S.J. 1985. The applications of enzyme in industry. In *Handbook of Enzyme Biotechnology*, Wiseman, A., Ed. Chichester: Ellis Horwood, p. 274.

Chen, K., J. Maley, and P.H. Yu. 2006. Potential inplications of endogenous aldehydes in beta-amyloid misfolding, oligomerization and fibrillogenesis. *J Neurochem* 99: 1413.

Chen, K., M. Kazachkov, and P.H. Yu. 2007. Effect of aldehydes derived from oxidative deamination and oxidative stress on beta-amyloid aggregation: pathological implications to Alzheimer's disease. *J Neural Transm* 114: 835.

Chen, Y. et al. 2001. Apoptotic signaling in polyamine analogue-treated SK-MEL-28 human melanoma cells. *Cancer Res* 61: 6437.

Chen, Z. et al. 2000. Crystal structure at 2.5 Å resolution of zinc-substituted copper amine oxidase of *Hansenula polymorpha* expressed in *Escherichia coli*. *Biochemistry* 39: 9709.

Cheng, E. et al. 1995. Inhibition of pig kidney diamine oxidase by nazlinin and nazlinin derivatives. *Biochim Biophys Acta* 1253: 189.

Cheng, E. et al. 1996. Enzyme catalysed oxidation of nazlinin derivatives: characterisation of the reaction products. *Tetrahedron* 52: 6725.

Chibata, I., T. Tosa, and T. Sato. 1987. Application of immobilized biocatalysts in pharmaceutical and chemical industries. *Enzyme Technol* 7: 653.

Chin, K.W., M.M. Garriga, and D.D. Metcalfe. 1989. The histamine content of oriental foods. *Food Chem Toxicol* 27: 283.

Chiu, Y.C. et al. 2006. Kinetic and structural studies on the catalytic role of the aspartic acid residue conserved in copper amine oxidase. *Biochemistry* 45: 4105.

Choi, K.L. et al. 1993. The internal quaternary ammonium receptor site of Shaker potassium channels. *Neuron* 10: 533.

Choi, Y.H. et al. 1995. Copper/topa quinone-containing histamine oxidase from *Arthrobacter globiformis*. Molecular cloning and sequencing, overproduction of precursor enzyme, and generation of topa quinone cofactor. *J Biol Chem* 270: 4712.

Chow, W.S., J.E. Savage, and B.L. O'Dell. 1968. Role of copper in biosynthesis of intramolecular cross-links in chick tendon collagen. *J Biol Chem* 244: 5785.

Cioni, L. et al. 2006. Activity and expression of semicarbazide-sensitive benzylamine oxidase in a rodent model of diabetes: interactive effects with methylamine and alpha-aminoguanidine. *Eur J Pharmacol* 529: 179.

Clarke, D.E., G.A. Lyles, and B.A. Callingham. 1982. A comparison of cardiac and vascular clorgyline-resistant amine oxidase and monoamine oxidase. Inhibition by amphetamine, mexiletine and other drugs. *Biochem Pharmacol* 31: 27.

Coelho, H., M. Azevedo, and C. Manso. 1985. Inhibitory effect of drugs used in the treatment of Parkinson's disease on plasma monoamine oxidase activity. *J Neural Transm* 61: 271.

Cohen, G. and R.E. Heikkila. 1974. The generation of hydrogen peroxide, superoxide radical, and hydroxyl radical by 6-hydroxydopamine, dialuric acid, and related cytotoxic agents. *J Biol Chem* 249: 2447.

Cohen, S.M. et al. 1992. Acrolein initiates rat urinary bladder carcinogenesis. *Cancer Res* 52: 3577.

Collison, D. et al. 1989. Studies on the active site of pig plasma amine oxidase. *Biochem J* 264: 663.

Compagnone, D. et al. 2001. Amperometric detection of biogenic amines in cheese using immobilised diamine oxidase. *Anal Lett* 34: 841.

Cona, A. et al. 2006. Functions of amine oxidases in plant development and defence. *Trends Plant Sci* 11: 80.

Conesa, A. et al. 2001. The secretion pathway in filamentous fungi: a biotechnological view. *Fungal Genet Biol* 33: 155.

Conforti, L. et al. 1995. Semicarbazide-sensitive amine oxidase activity in white adipose tissue of the insulin-deficient rat. *J Pharm Pharmacol* 47: 420.

Conklin, D.J. et al. 2001. Amine metabolism: a novel path to coronary artery vasospasm. *Toxicol Appl Pharmacol* 175: 149.

Conklin, D.J. et al. 2006. Acrolein generation stimulates hypercontraction in isolated human blood vessels. *Toxicol Appl Pharmacol* 217: 277.

Conner, J.W., Z.M. Yuan, and P.S. Callery. 1992. Active-site directed irreversible inhibition of diamine oxidase by a homologous series of aziridinylalkylamines. *Biochem Pharmacol* 44: 1229.

Contakes, S.M. et al. 2005. Reversible inhibition of copper amine oxidase activity by channel-blocking ruthenium(II) and rhenium(I) molecular wires. *Proc Natl Acad Sci USA* 102: 13451.

Coquil, J.F. et al. 1973. Monoamine oxidase in rat arteries: evidence for different forms and selective localization. *Br J Pharmacol* 48: 590.

Corazza, A. et al. 1992. Effect of phosphate ion on the activity of bovine plasma amine oxidase. *Biochem Biophys Res Commun* 189: 722.

Corazza, G.R. et al. 1988. Decreased plasma postheparin diamine oxidase levels in celiac disease. *Dig Dis Sci* 33: 956.

Coronas, V. et al. 1997. *In vitro* induction of apoptosis or differentiation by dopamine in an immortalized olfactory neuronal cell line. *J Neurochem* 69: 1870.

Couser, W.G. 1998. Pathogenesis of glomerular damage in glomerulonephritis. *Nephrol Dial Transpl* 13, Suppl. 1: 10.

Crabbe, M.J. and W.G. Bardsley. 1974a. The inhibition of human placental diamine oxidase by substrate analogues. *Biochem J* 139: 183.

Crabbe, M.J. and W.G. Bardsley. 1974b. Monoamine oxidase inhibitors and other drugs as inhibitors of diamine oxidase from human placenta and pig kidney. *Biochem Pharmacol* 23: 2983.

Crabbe, M.J.C., R.E. Childs, and W.G. Bardsley. 1975. Time-dependent inhibition of diamine oxidase by carbonyl-group reagents and urea. *Eur J Biochem* 60: 325.

Craft, S. 2007. Insulin resistance and Alzheimer's disease pathogenesis: potential mechanisms and implications for treatment. *Curr Alzheimer Res* 4: 147.

Cragg, J.E., R.B. Herbert, and M. Kgaphola. 1990. Pea-seedling diamine oxidase: applications in synthesis and evidence relating to its mechanism of action. *Tetrahedron Lett* 31: 6907.

Csoregi, E. et al. 2001. Amperometric enzyme-based biosensors for application in food and beverage industry. *Phys Chem Basis Biotechnol* 7: 105.

Cubría, J.C. et al. 1991. Inhibition of diamine oxidase from porcine kidney by pentamidine and other aminoguanidine compounds. *Comp Biochem Physiol B* 100: 543.

Cubría, J.C. et al. 1993. Aromatic diamidines are reversible inhibitors of porcine kidney diamine oxidase. *Biochem Pharmacol* 45: 1355.

Cuzzocrea, S. et al. 1997. Role of peroxynitrite and activation of poly (ADP-ribose) synthase in the vascular failure induced by zymosan-activated plasma. *Br J Pharmacol* 122: 493.

Czech, M.P., J.C.J. Lawrence, and W.S. Lynn. 1974. Hexose transport in isolated brown fat cells. *J Biol Chem* 249: 5421.

D'Agostino, L. et al. 1987a. Postheparin plasma diamine oxidase in subjects with small bowel mucosal atrophy. *Dig Dis Sci* 32: 313.

D'Agostino, L. et al. 1987b. Postheparin plasma diamine oxidase increases in patients with coeliac disease during gluten free diet. *Gut* 28, Suppl. 1: 131.

D'Agostino, L. et al. 1988a. Postheparin plasma diamine oxidase in patients with small bowel Crohn's disease. *Gastroenterology* 95: 1503.

D'Agostino, L. et al. 1988b. Postheparin plasma diamine oxidase in subjects with small bowel disease: diagnostic efficiency of a simplified test. *Digestion* 41: 46.

Dar, M.S., P.L. Morselli, and E.R. Bowman. 1985. The enzymatic systems involved in the mammalian metabolism of methylamine. *Gen Pharmacol* 16: 557.

Davidson, V.L. 1993. Methylamine dehydrogenase. In *Principles and Application of Quinoprotein*, Davidson, V.L., Ed. New York: Marcel Dekker, p. 73.

Davis, J. and R.S. de Ropp. 1961. Metabolic origin of urinary methylamine in the rat. *Nature* 190: 636.

Dawkes, H.C. and S.E. Phillips. 2001. Copper amine oxidase: cunning cofactor and controversial copper. *Curr Opin Struc Biol* 11: 666.

Day, C. and C. Savage. 2001. Primary systemic vasculitis. *Minerva Med* 92: 349.

De Biase, D. et al. 1996. Half-of-the-sites reactivity of bovine serum amine oxidase: reactivity and chemical identity of the second site. *Eur J Biochem* 237: 93.

De Matteis, G. et al. 1999. The metal function in the reactions of bovine serum amine oxidase with substrates and hydrazine inhibitors. *J Biol Inorg Chem* 4: 348.

Deane, R. et al. 2004. LRP/amyloid beta-peptide interaction mediates differential brain efflux of A beta isoforms. *Neuron* 43: 333.

Deitch, E.A. et al. 1990. Hemorrhagic shock-induced bacterial translocation: role of neutrophils and hydroxyl radicals. *J Trauma* 30: 942.

del Mar-Hernandez, M. et al. 2005. Human plasma semicarbazide sensitive amine oxidase (SSAO), beta-amyloid protein and aging. *Neurosci Lett* 384: 183.

Delis, C. et al. 2006. A root- and hypocotyl-specific gene coding for copper-containing amine oxidase is related to cell expansion in soybean seedlings. *J Exp Bot* 57: 101.

Demers, N. et al. 2001. Immobilization of native and poly(ethylene glycol)-treated (PEGylated) bovine serum amine oxidase into a biocompatible hydrogel. *Biotechnol Appl Biochem* 33: 201.

Deng, Y., F. Boomsma, and P.H. Yu. 1998. Deamination of methylamine and aminoacetone increases aldehydes and oxidative stress in rats. *Life Sci* 63: 2049.

Deng, Y. and P.H. Yu. 1999. Assessment of the deamination of aminoacetone, an endogenous substrate for semicarbazide-sensitive amine oxidase. *Anal Biochem* 270: 97.

Desiderio, M.A., A. Sessa, and A. Perin. 1982. Induction of diamine oxidase activity in rat kidney during compensatory hypertrophy. *Biochim Biophys Acta* 714: 243.

Desiderio, M.A., A. Sessa, and A. Perin. 1985. Regulation of diamine oxidase expression by beta 2-adrenoceptors in normal and hypertrophic rat kidney. *Biochim Biophys Acta* 845: 463.

Desiderio, M.A. et al. 1995. Involvement of ornithine decarboxylase and polyamines in glucocorticoid-induced apoptosis of rat thymocytes. *Cell Growth Differ* 6: 505.

Devoto, G. et al. 1986. Inhibitory activity of bivalent transition-metal complexes with diamines toward a diamine oxidase. *Polyhedron* 5: 1023.

Dewar, K.M. et al. 1988. Involvement of brain trace amines in the behavioural effects of phenelzine. *Neurochem Res* 13: 113.

Dial, E.J. and D.E. Clarke. 1978. Phenylethylamine deamination by multiple types of monoamine oxidase. *Biochem Pharmacol* 27: 2374.

Dickson, B.C., C.J. Streutker, and R. Chetty. 2006. Celiac disease: update for pathologists. *J Clin Pathol* 59: 1008.

Dimaculangan, D.D. et al. 1994. Retinoic acid prevents downregulation of ras recision gene/lysyl oxidase early in adipocyte differentiation. *Differentiation* 58: 47.

Di Donato, A. et al. 1997. Lysyl oxidase expression and collagen cross-linking during chronic adriamycin nephropathy. *Nephron* 76: 192.

Di Paolo, M.L. et al. 1995a. Effect of polyphosphates on the activity of amine oxidases. *Biochim Biophys Acta* 1247: 246.

Di Paolo, M.L. et al. 1995b. Kinetic characterization of soybean seedling amine oxidase. *Arch Biochem Biophys* 323: 329.

Di Paolo, M.L. et al. 2002. Binding of cations of group IA and IIA to bovine serum amine oxidase: effect on the activity. *Biophys J* 83: 2231.

Di Paolo, M.L. et al. 2003. Electrostatic compared with hydrophobic interactions between bovine serum amine oxidase and its substrates. *Biochem J* 371: 549.

Di Paolo, M.L. et al. 2004. Phosphonium compounds as new and specific inhibitors of bovine serum amine oxidase. *Biochem J* 384: 551.

Di Paolo, M.L. et al. 2007. N-alkanamines as substrates to probe the hydrophobic region of bovine serum amine oxidase active site: a kinetic and spectroscopic study. *Arch Biochem Biophys* 465: 50.

Di Silvestro, R.A. et al. 1997. Plasma diamine oxidase activities in renal dialysis patients, a human with spontaneous copper deficiency and marginally copper-deficient rats. *Clin Biochem* 30: 559.

Dix, T.A., D.M. Kuhn, and S.J. Benkovic. 1987. Mechanism of oxygen activation by tyrosine hydroxylase. *Biochemistry* 26: 3354.

Dix, T.A. et al. 1996. Mechanism of site-selective DNA nicking by the hydrodioxyl (perhydroxyl) radical. *Biochemistry* 35: 4578.

Dooley, D.M. et al. 1991. A Cu(I)-semiquinone state in substrate-reduced amine oxidases. *Nature* 349: 262.

Dooley, D.M. et al. 1993. Structure and reactivity of copper-containing amine oxidases. In *Bioinorganic Chemistry of Copper*, Karlin, K.D. and Tyeklar, Z., Eds. New York: Chapman & Hall, p. 459.

Dooley, D.M. 1999. Structure and biogenesis of topaquinone and related cofactors. *J Biol Inorg Chem* 4: 1.

Dostert, P.L., B.M. Strolin, and K.F. Tipton. 1989. Interactions of monoamine oxidase with substrates and inhibitors. *Med Res Rev* 9: 45.

Dove, J.E. et al. 2000. Investigation of spectroscopic intermediates during copper-binding and TPQ formation in wild-type and active-site mutants of a copper-containing amine oxidase from yeast. *Biochemistry* 39: 3690.

Draisci, R. et al. 1998. Determination of biogenic amines with an electrochemical biosensor and its application to salted anchovies. *Food Chem* 62: 225.

DuBois, J.L. and J.P. Klinman. 2005. The nature of O_2 reactivity leading to topa quinone in the copper amine oxidase from *Hansenula polymorpha* and its relationship to catalytic turnover. *Biochemistry* 44: 11381.

Duff, A.P. et al. 2003. The crystal structure of *Pichia pastoris* lysyl oxidase. *Biochemistry* 42: 15148.

Duff, A.P. et al. 2004. Using xenon as a probe for dioxygen-binding sites in copper amine oxidases. *J Mol Biol* 344: 599.

Duff, A.P. et al. 2006a. The 1.23 Å structure of *Pichia pastoris* lysyl oxidase reveals a lysine-lysine cross-link. *Acta Cryst D* 62: 1073.

Duff, A.P. et al. 2006b. A C-terminal disulfide bond in the copper-containing amine oxidase from pea-seedlings violates the two-fold symmetry of the molecular dimer. *Acta Cryst F* 62: 1168.

Dullaart, R.P., S.C. Riemens, and F. Boomsma. 2006. Plasma semicarbazide-sensitive amine oxidase is moderately decreased by pronounced exogenous hyperinsulinemia but is not associated with insulin sensitivity and body fat. *Scand J Clin Lab Invest* 66: 559.

Dumoulin, M.J. et al. 1996. Comparative cardioprotective and antioxidant properties of ceruloplasmin, superoxide dismutase and bovine serum albumin. *Arzneim Forsch Drug Res* 46: 855.

Dvorak, A.M., A.B. Connell, and G.R. Dickersin. 1979. Crohn's disease: a scanning electron microscopic study. *Human Pathol* 10: 165.

Dyrløv-Bendtsen, J. et al. 2004. Feature based prediction of non-classical and leaderless protein secretion. *Protein Eng Des Sel* 17: 349.

Eguchi, Y., S. Shimizu, and Y. Tsujimoto. 1997. Intracellular ATP levels determine cell death fate by apoptosis or necrosis. *Cancer Res* 57: 1835.

Ekblom, J. et al. 1999. Elevated activity of semicarbazide-sensitive amine oxidase in blood from patients with skeletal metastases of prostate cancer. *Clin Sci* 97: 111.

El Hadri, K. et al. 2002. Semicarbazide-sensitive amine oxidase in vascular smooth muscle cells: differentiation-dependent expression and role in glucose uptake. *Arterioscler Thromb Vasc Biol* 22: 89.

Eliakim, R. et al. 1992. Effect of drugs on colonic eicosanoid accumulation in active ulcerative colitis. *Scand J Gastroenterol* 27: 968.

Elliott, W.H. 1960. Aminoacetone formation by *Staphylococcus aureus*. *Biochem J* 74: 478.

Elliott, J., B.A. Callingham, and D. Sharman. 1989. Metabolism of amines in the isolated perfused mesenteric arterial bed of the rat. *Br J Pharmacol* 98: 507.

Elliott, J. et al. 1991. Physiological and pathological influences on sheep blood plasma amine oxidase: effect of pregnancy and experimental alloxan-induced diabetes mellitus. *Res Vet Sci* 50: 334.

Elmore, B.O., J.A. Bollinger, and D.M. Dooley. 2002. Human kidney diamine oxidase: heterologous expression, purification, and characterization. *J Biol Inorg Chem* 7: 565.

Emanuelsson, O. et al. 2007. Locating proteins in the cell using TargetP, SignalP, and related tools. *Nature Protocols* 2: 953.

Enrique-Tarancón, G. et al. 1998. Role of semicarbazide-sensitive amine oxidase on glucose transport and GLUT4 recruitment to the cell surface in adipose cells. *J Biol Chem* 273: 8025.

Enrique-Tarancón, G. et al. 2000. Substrates of semicarbazide-sensitive amine oxidase co-operate with vanadate to stimulate tyrosine phosphorylation of insulin-receptor-substrate proteins, phosphoinositide 3-kinase activity and GLUT4 translocation in adipose cells. *Biochem J* 350: 171.

Enzyme Nomenclature (1992) *Recommendations of the Nomenclature Committee of the International Union of Biochemistry and Molecular Biology on the Nomenclature and Classification of Enzymes*. New York: Academic Press.

Erdman, S.H. et al. 1989. Suppression of diamine oxidase activity enhances postresection ileal proliferation in the rat. *Gastroenterology* 96: 1533.

Eremina, V., H.J. Baelde, and S.E. Quaggin. 2007. Role of the VEGF-a signaling pathway in the glomerulus: evidence for crosstalk between components of the glomerular filtration barrier. *Nephron Physiol* 106: 32.

Eriksson, M. and C.J. Fowler. 1984. Inhibition of monoamine oxidase and semicarbazide-sensitive amine oxidase by mexiletine and related compounds. *Naun Schmied Arch Pharmacol* 327: 273.

Erler, J.T. and A.J. Giaccia. 2006. Lysyl oxidase mediates hypoxic control of metastasis. *Cancer Res* 66: 10238.

Esti, M. et al. 1998. Determination of amines in fresh and modified atmosphere packaged fruits using electrochemical biosensors. *J Agric Food Chem* 46: 4233.

Faber, K. 2000. In *Biotransformations in Organic Chemistry*, Faber, K. Ed. Berlin: Springer, p. 298.

Fagerström, R., P. Seppänen, and J. Jänne. 1984. A rapid chemiluminescence-based method for the determination of total polyamines in biological samples. *Clin Chim Acta* 143: 45.

Falk, M.C., A.J. Staton, and T.J. Williams. 1983. Heterogeneity of pig plasma amine oxidase: molecular and catalytic properties of chromatographically isolated forms. *Biochemistry* 22: 3746.

Farnum, M., M.M. Palcic, and J.P. Klinman. 1986. pH Dependence of deuterium isotope effects and tritium exchange in the bovine plasma amine oxidase reaction: a role for single-base catalysis in amine oxidation and imine exchange. *Biochemistry* 25: 1898.

Federico, R. et al. 1997. Competitive inhibition of swine kidney copper amine oxidase by drugs. *Biochem Biophys Res Comm* 240: 150.

Federico, R. et al. 2000. Immobilization of plant histaminase for medical applications. *Inflamm Res* 49, Suppl.1: S60.

Ferrer, I. et al. 2002. Overexpression of semicarbazide sensitive amine oxidase in the cerebral blood vessels in patients with Alzheimer's disease and cerebral autosomal dominant arteriopathy with subcortical infarcts and leukoencephalopathy. *Neurosci Lett* 321: 21.

Feurle, J. and H.G. Schwelberger. 2007. Porcine plasma amine oxidase has a broad substrate specificity and efficiently converts histamine. *Inflamm Res* 56: 55.

Finazzi-Agrò, A. et al. 1977. On the nature of chromophore in pig kidney diamine oxidase. *Eur J Biochem* 74: 435.

Fitzgerald, D.H. and K.F. Tipton. 2002. Inhibition of monoamine oxidase modulates the behaviour of semicarbazide-sensitive amine oxidase (SSAO). *J Neural Transm* 109: 251.

Floris, G., R. Medda, A. Padiglia et al. 2000. The physiopathological significance of ceruloplasmin: a possible therapeutic approach. *Biochem Pharmacol* 60: 1735.

Fogel, W.A. 1986. GABA and polyamine metabolism in peripheral tissues. In *GABAergic Mechanisms in the Mammalian Periphery*, Erdo, S.L. and Bowery, N.G., Eds. New York: Raven, p. 35.

Fogel, W.A., A. Lewinski, and J. Jochem. 2005a. Histamine in idiopathic inflammatory bowel diseases: not a standby player. *Folia Med Cracov* 66: 107.

Fogel, W.A. et al. 2005b. The role of histamine in experimental ulcerative colitis in rats. *Inflamm Res* 54: S68.

Fogel, W.A. and A. Lewinski. 2006. The effect of diamine oxidase administration on experimental ulcerative colitis in rats. *Inflamm Res* 55: S63.

Fogel, W.A., A. Lewinski, and J. Jochem. 2007. Histamine in food: is there anything to worry about? *Biochem Soc Trans* 35: 349.

Foligne, B. et al. 2007. Prevention and treatment of colitis with *Lactococcus lactis* secreting the immunomodulatory Yersinia LcrV protein. *Gastroenterology* 133: 862.

Fontana, E. et al. 2001. Amine oxidase substrates mimic several of the insulin effects on adipocyte differentiation in 3T3 F442A cells. *Biochem J* 356: 769.

Forget, P. et al. 1985. Serum diamine oxidase activity in acute gastroenteritis in children. *Pediatric Res* 19: 26.

Forget, P. et al. 1986. Diamine oxidase in serum and small intestinal biopsy tissue in childhood celiac disease. *J Pediatr Gastroenterol Nutr* 5: 379.

Fowler, C.J. et al. 1984. Stereoselective inhibition of monoamine oxidase and semicarbazide-sensitive amine oxidase by 4-dimethylamino-2,alpha-dimethylphenethylamine (FLA 336). *Naun Schmied Arch Pharmacol* 327: 279.

Frébort, I. et al. 1994. Active-site covalent modifications of quinoprotein amine oxidases from *Aspergillus niger*: evidence for binding of the mechanism based inhibitor, 1,4-diamino-2-butyne, to residue Lys 356 involved in the catalytic cycle. *Eur J Biochem* 225: 959.

Frébort, I. and O. Adachi. 1995. Copper/containing amine oxidases, an exciting class of ubiquitous enzymes. *J Ferment Bioeng* 80: 625.

Frébort, I. et al. 1996. Two distinct quinoprotein amine oxidases are induced by *n*-butylamine in the mycelia of *Aspergillus niger* AKU 3302. Purification, characterisation, cDNA cloning and sequencing. *Eur J Biochem* 237: 255.

Frébort, I., K. Matsushita, and O. Adachi. 1997. Involvement of multiple copper/topa quinone-containing and flavin-containing amine oxidases, and $NAD(P)^+$ aldehyde dehydrogenases in amine degradation by filamentous fungi. *J Ferment Bioeng* 84: 200.

Frébort, I. et al. 1999. Purification and characterization of methylamine oxidase induced in *Aspergillus niger* AKU 3302. *Biosci Biotechnol Biochem* 63: 125.

Frébort, I. et al. 2000a. Molecular mode of interaction of plant amine oxidase with the mechanism-based inhibitor 2-butyne-1,4-diamine. *Eur J Biochem* 267: 1423.

Frébort, I. et al. 2000b. Cellular localization and metabolic function of *n*-butylamine induced amine oxidases in the fungus *Aspergillus niger* AKU 3302. *Arch Microbiol* 173: 358.

Frébort, I., L. Skoupa, and P. Peč. 2000c. Amine oxidase-based flow biosensor for the assessment of fish freshness. *Food Control* 11: 13.

Frébort, I. et al. 2003. Gene organisation and molecular modeling of copper amine oxidase from *Aspergillus niger*: re-evaluation of the cofactor structure. *Biol Chem* 384: 1451.

Friedman, E.A. 1995. Potential of aminoguanidine in diabetic CAPD patients. *Per Dial Int* 15: 110.

Fu, Y.H. and G.A. Marzluf. 1990. Nit-2, the major positive-acting nitrogen regulatory gene of *Neurospora crassa*, encodes a sequence-specific DNA-binding protein. *Proc Natl Acad Sci USA* 87: 5331.

Fu, M. and R.B. Silverman. 1999. Isolation and characterization of the product of inactivation of gamma-aminobutyric acid aminotransferase by gabaculine. *Bioorg Med Chem* 7: 1581.

Fubara, B. et al. 1986. Purification and properties of aminoacetone synthetase from beef liver mitochondria. *J Biol Chem* 261: 12189.

Gacesa, P. and J. Hubble. 1987. Effects of immobilization on enzyme stability and use. In *Enzyme Technology*, Gacesa, P. and Hubble, J., Eds. New York: Taylor & Francis, p. 77.

Gacheru, S.N. et al. 1989. Vicinal diamines as pyrroloquinoline quinone-directed irreversible inhibitors of lysyl oxidase. *J Biol Chem* 264: 12963.

Gahl, W.A. and H.C. Pitot. 1982. Polyamine degradation in foetal and adult bovine serum. *Biochem J* 202: 603.

Galuszka, P. et al. 1998. Cytokinins as inhibitors of copper amine oxidase. *J Enzym Inhib* 13: 457.

Gang, V. et al. 1976a. Diamine oxidase (histaminase) in chronic renal disease and its inhibition *in vitro* by methylguanidine. *Clin Nephrol* 3: 171.

Gang, V., M. Baldus, and M. Kadereit. 1976b. Serum level changes of endogenous and postheparin diamine oxidase (histaminase) in clinical and experimental hepatitis. *Acta Hepatogastroenterol (Stuttg)* 23: 104.

García-Vicente, S. et al. 2005. The release of soluble VAP-1/SSAO by 3T3-L1 adipocytes is stimulated by isoproterenol and low concentrations of TNF-alpha. *J Physiol Biochem* 61: 395.

García-Vicente, S. et al. 2007. Oral insulin-mimetic compounds that act independently of insulin. *Diabetes* 56: 486.

Gardini, G. et al. 2001. Agmatine induces apoptosis in rat hepatocyte cultures. *J Hepatol* 35: 482.

Garpenstrand, H. et al. 1999a. Elevated plasma semicarbazide-sensitive amine oxidase (SSAO) activity in Type 2 diabetes mellitus complicated by retinopathy. *Diabet Med* 16: 514.

Garpenstrand, H. et al. 1999b. Plasma semicarbazide-sensitive amine oxidase in stroke. *Eur Neurol* 41: 20.

Gasparini, R. et al. 1991. Amino oxidase amperometric biosensor for polyamines. *Bioelectroch Bioenerg* 25: 307.

Gasparini, R. et al. 1994. Renewable miniature enzyme-based sensing devices. *Anal Chim Acta* 294: 299.

Gass, J.T. and M.F. Olive. 2008. Glutamatergic substrates of drug addiction and alcoholism. *Biochem Pharmacol* 75: 218.

Gemperlová, L., J. Eder, and M. Cvikrová. 2005. Polyamine metabolism during the growth cycle of tobacco BY-2 cells. *Plant Physiol Biochem* 43: 375.

Giarnieri, D. et al. 1985. Diamine oxidase content in urine of patients with renal failure. *Agents Actions* 16: 249.

Giartosio, A., E. Agostinelli, and B. Mondovì. 1988. Domains in bovine serum amine oxidase. *Biochem Biophys Res Commun* 154: 66.

Gill, D.S. et al. 1991. Plasma histamine in patients with chronic renal failure and nephrotic syndrome. *J Clin Pathol* 44: 243.

Godfrey, T. and S. West. 1996. Listing of Industrial enzymes and sources. In *Industrial Enzymology,* Godfrey, T. and West, S., Eds. London: Macmillan, p. 583.

Göktürk, C. et al. 2003a. Overexpression of semicarbazide-sensitive amine oxidase in smooth muscle cells leads to an abnormal structure of the aortic elastic laminas. *Am J Pathol* 163: 1921.

Göktürk, C. et al. 2003b. Studies on semicarbazide-sensitive amine oxidase in patients with diabetes mellitus and in transgenic mice. *Biochim Biophys Acta Protein Struct Mol Enzymol* 1647: 881.

Göktürk, C. et al. 2004. Semicarbazide-sensitive amine oxidase in transgenic mice with diabetes. *Biochem Biophys Res Commun* 325: 1013.

Göktürk, C. et al. 2007. Macrovascular changes in mice overexpressing human semicarbazide-sensitive amine oxidase in smooth muscle cells. *Am J Hypertens* 20: 743.

Goldstein, I.J., C.E. Hollerman, and J.M. Merrick. 1965. Protein-carbohydrate interaction. I. The interaction of polysaccarides with concanavalin A. *Biochim Biophys Acta* 97: 68.

Goldstein, I.M. et al. 1982. Ceruloplasmin: an acute phase reactant that scavenges oxygen derived free radicals. *Ann NY Acad Sci* 389: 368.

Goto, Y. and J.P. Klinman. 2001. Binding of dioxygen to non-metal sites in proteins: exploration of the importance of binding site size versus hydrophobicity in the copper amine oxidase from *Hansenula polymorpha. Biochemistry* 41: 13637.

Goto, Y. et al. 2005. Transforming growth factor-beta1 mediated up-regulation of lysyl oxidase in the kidneys of hereditary nephrotic mouse with chronic renal fibrosis. *Virchows Arch* 447: 859.

Grassilli, E. et al. 1995. Is polyamine decrease a common feature of apoptosis? Evidence from gamma ray- and heat shock-induced cell death. *Biochem Biophys Res Commun* 216: 708.

Green, E.L. et al. 2002. Rate of oxygen and hydrogen exchange as indicators of TPQ cofactor orientation in amine oxidases. *Biochemistry* 41: 687.

Gronvall-Nordquist, J.L. et al. 2001. Follow-up of plasma semicarbazide-sensitive amine oxidase activity and retinopathy in Type 2 diabetes mellitus. *J Diabetes Compl* 15: 250.

Gubisne-Haberle, D. et al. 2004. Protein cross-linkage induced by formaldehyde derived from semicarbazide-sensitive amine oxidase-mediated deamination of methylamine. *J Pharmacol Exp Ther* 310: 1125.

Guffroy, C., T. Boucher, and M. Strolin Benedetti. 1985. Further investigations of the metabolism of two trace amines, beta-phenylethylamine and p-tyramine by rat aorta semicarbazide-sensitive amine oxidase. In *Neuropharmacology of the Trace Amines*, Boulton, A.A. et al., Eds. Clifton, NJ: Humana, p. 39.

Guida, B. et al. 2000. Histamine plasma levels and elimination diet in chronic idiopathic urticaria. *Eur J Clin Nutr* 54: 155.

Ha, H.C., P.M. Woster, and R.A. Casero, Jr. 1998. Unsymmetrically substituted polyamine analogue induces caspase-independent programmed cell death in Bcl-2-overexpressing cells. *Cancer Res* 58: 2711.

Hall, M. et al. 1999. A test strip for diamines in tuna. *J AOAC Int* 82: 1102.

Hall, T.A. 1999. BioEdit: a user-friendly biological sequence alignment editor and analysis program for Windows 95/98/NT. *Nucl Acids Symp Ser* 41: 95.

Han, J. et al. 1997. Activation of the transcription factor MEF2C by the MAP kinase p38 in inflammation. *Nature* 386: 296.

Hansson, R. et al. 1966. Heparin-induced diamine oxidase increase in human blood plasma. *Acta Med Scand* 180: 533.

Hartmann, C. and J.P. Klinman. 1991. Structure–function studies of substrate oxidation by bovine serum amine oxidase: relationship to cofactor structure and mechanism. *Biochemistry* 30: 4605.

Hartmann, C., P. Brzovic, and J.P. Klinman. 1993. Spectroscopic detection of chemical intermediates in the reaction of para-substituted benzylamines with bovine serum amine oxidase. *Biochemistry* 32: 2234.

Hasebe, Y. et al. 1997. Specific and amplified current responses to histidine and histamine using immobilized copper-monoamine oxidase membrane electrode, based on novel ascorbate oxidase activity induced by exogenous ligands. *Anal Biochem* 251: 32.

Hayashi, K. et al. 2004. Comparative immunocytochemical localization of lysyl oxidase (LOX) and the lysyl oxidase-like (LOXL) proteins: changes in the expression of LOXL during development and growth of mouse tissues. *J Mol Histol* 35: 845.

Hayes, G.R. and D.H. Lockwood. 1987. Role of insulin receptor phosphorylation in the insulinomimetic efffect of hydrogen peroxide. *Proc Natl Acad Sci USA* 84: 8115.

He, Z. et al. 1995a. Mechanism-based inactivation of porcine kidney diamine oxidase by 1,4-diamino-2-butene. *Biochim Biophys Acta* 1253: 117.

He, Z., Y. Zou, and F.T. Greenaway. 1995b. Cyanide inhibition of porcine kidney diamine oxidase and bovine plasma amine oxidase: evidence for multiple interaction sites. *Arch Biochem Biophys* 319: 185.

Healy, J. and K.F. Tipton. 2007. Ceruloplasmin and what it might do. *J Neural Transm* 114: 777.

Heck, H.d'A. 1988. Toxicology of formaldehyde. *ISI Atl Sci Pharmacol* 2: 5.

Heffetz, D., W.J. Rutter, and Y. Zick. 1992. The insulinomimetic agents H_2O_2 and vanadate stimulate tyrosine phoshorylation of potential target proteins for the insulin-receptor kinase in intact cells. *Biochem J* 288: 631.

Heim, W.G. et al. 2007. Cloning and characterization of a *Nicotiana tabacum* methylputrescine oxidase transcript. *Phytochemistry* 68: 454.

Hellman, N. and J. Gitlin. 2002. Ceruloplasmin metabolism and function. *Annu Rev Nutr* 22: 439.

Heniquez, A. et al. 2003. High expression of semicarbazide-sensitive amine oxidase genes AOC2 and AOC3, but not the diamine oxidase gene AOC1 in human adipocytes. *Inflamm Res* 52: S74.

Hensley, K. et al. 1995. Brain regional correspondence between Alzheimer's disease histopathology and biomarkers of protein oxidation. *J Neurochem* 65: 2146.

Hernandez, M.D. et al. 2005. Human plasma semicarbazide sensitive amine oxidase (SSAO), β-amyloid protein and aging. *Neurosci Lett* 384: 183.

Hernandez, M. et al. 2006. Soluble semicarbazide sensitive amine oxidase (SSAO) catalysis induces apoptosis in vascular smooth muscle cells. *Biochim Biophys Acta* 1763: 164.

Hesterberg, R. et al. 1981. The start of a programme for measuring diamine oxidase activity in biopsy specimens of human rectal mucosa. *Agents Actions* 11: 33.

Hevel, J.M., S.A. Mills, and J.P. Klinman. 1999. Mutation of a strictly conserved, active-site residue alters substrate specificity and cofactor biogenesis in a copper amine oxidase. *Biochemistry* 38: 3683.

Higgins, D.F. et al. 2007. Hypoxia promotes fibrogenesis in vivo via HIF-1 stimulation of epithelial-to-mesenchymal transition. *J Clin Invest* 117: 3810.

Hill, J.M. and P.J.G. Mann. 1962. The inhibition of pea-seedling diamine oxidase by chelating agents. *Biochem J* 85: 198.

Hill, J.M. and P.J.G. Mann. 1964. Further properties of the diamine oxidase of pea seedlings. *Biochem J* 91: 171.

Hiraku, Y. et al. 1999. Oxidative DNA damage induced by aminoacetone, an amino acid metabolite. *Arch Biochem Biophys* 365: 62.

Hirota, S. et al. 2001. Spectroscopic observation of intermediates formed during the oxidative half-reaction of copper/topa quinone-containing phenylethylamine oxidase. *Biochemistry* 40: 15789.

Hirsch, J.G. 1953. Spermine oxidase: an enzyme with specificity for spermine and spermidine. *J Exp Med* 97: 337.

Hockenbery, D.M. et al. 1993. Bcl-2 functions in an antioxidant pathway to prevent apoptosis. *Cell* 75: 241.

Hofstra, C.L. et al. 2003. Histamine H4 receptor mediates chemotaxis and calcium mobilization of mast cells. *J Pharmacol Exp Therap* 305: 1212.

Høgdall, E.V. et al. 1998. Structure and tissue-specific expression of genes encoding bovine copper amine oxidases. *Eur J Biochem* 251: 320.

Holinka, C.F. and E. Gurpide. 1984. Diamine oxidase activity in human decidua and endometrium. *Am J Obstet Gynecol* 150: 359.

Holmen, C. et al. 2005. Circulating inflammatory endothelial cells contribute to endothelial progenitor cell dysfunction in patients with vasculitis and kidney involvement. *J Am Soc Nephrol* 16: 11020.

Holmen, C. et al. 2007. Anti endothelial cell autoantibodies selectively activate SAPK/JNK signalling in Wegener's granulomatosis. *J Am Soc Nephrol* 18: 2497.

Holt, A., D. Sharman, and B.A. Callingham. 1992a. Effects *in vitro* of procarbazine metabolites on some amine oxidase activities in the rat. *J Pharm Pharmacol* 44: 494.

Holt, A., D. Sharman, B.A. Callingham. 1992b. Characteristics of procarbazine as an inhibitor *in vitro* of rat semicarbazide-sensitive amine oxidase. *J Pharm Pharmacol* 44: 487.

Holt, A. and B.A. Callingham. 1994. Location of the active site of rat vascular semicarbazide-sensitive amine oxidase. *J Neural Transm Suppl* 41: 433.

Holt, A. and B.A. Callingham. 1995. Further studies on the *ex vivo* effects of procarbazine and monomethylhydrazine on rat semicarbazide-sensitive amine oxidase and monoamine oxidase activities. *J Pharm Pharmacol* 47: 837.

Holt, A. and G.B. Baker. 1995. Metabolism of agmatine (clonidine-displacing substance) by diamine oxidase and the possible implications for studies of imidazoline receptors. *Prog Brain Res* 106: 187.

Holt, A. et al. 1997. A continuous spectrophotometric assay for monoamine oxidase and related enzymes in tissue homogenates. *Anal Biochem* 244: 384.

Holt, A., K.G. Todd, and G.B. Baker. 2003. The effects of chronic administration of inhibitors of flavin and quinone amine oxidases on imidazoline I_1 receptor density in rat whole brain. *Ann NY Acad Sci* 1009: 309.

Holt, A., B. Wieland, and G.B. Baker. 2004. Allosteric modulation of semicarbazide-sensitive amine oxidase activities *in vitro* by imidazoline receptor ligands. *Br J Pharmacol* 143: 495.

Holt, A. and M.M. Palcic. 2006. A peroxidase-coupled continuous absorbance plate-reader assay for flavin monoamine oxidases, copper-containing amine oxidases and related enzymes. *Nat Prot* 1: 2498.

Holt, A. 2007. Practical enzymology. Quantifying enzyme activity and the effects of drugs thereupon. In: *Handbook of Neurochemistry and Molecular Neurobiology*, Baker, G.B., S.A. Dunn, A. Holt (eds) Vol. 18, Practical Neurochemistry Methods, 3rd ed. New York, NY: Kluwer Academic / Plenum.

Holt, A. et al. 2007. The effects of buffer cations on interactions between mammalian copper-containing amine oxidase and their substrates. *J Neural Transm* 114: 733.

Holt, A. et al. 2008. Multiple binding sites for substrates and modulators of semicarbazide-sensitive amine oxidases: kinetic consequences. *Mol Pharmacol* 73: 525.

Hong, J., H. Ghourchian, and A.A. Moosavi–Movahedi. 2006. Direct electron transfer of redox proteins on a Nafion–cysteine modified gold electrode. *Electrochem Commun* 8: 1572.

Hou, F.F. et al. 1998. Aminoguanidine inhibits advanced glycation end products formation on beta2-microglobulin. *J Am Soc Nephrol* 9: 277.

Houen, G. et al. 1993. Purification and partial characterization of mammalian Cu-dependent amine oxidases. *Acta Chem Scand* 47: 902.

Houen, G. 1999. Mammalian Cu-containing amine oxidases (CAOs): new methods of analysis, structural relationships, and possible functions. *APMIS Suppl* 96: 1.

Hubálek, F. et al. 2005. Demonstration of isoleucine 199 as a structural determinant for the selective inhibition of human monoamine oxidase B by specific reversible inhibitors. *J Biol Chem* 280: 15761.

Huheey, J.E. 1972. *Inorganic Chemistry: Principles of Structure and Reactivity*, New York: Harper & Row, p. 214.

Hunter, D.R., R.A. Haworth, and J.H. Southard. 1976. Relationship between configuration, function, and permeability in calcium-treated mitochondria. *J Biol Chem* 251: 5069.

Ibrahim, J. et al. 1995. Role of polyamines in hypertension induced by angiotensin II. *Cardiovasc Res* 29: 50.

Iffiú-Soltész, Z. et al. 2007. Influence of benzylamine acute and chronic administration of benzylamine on glucose tolerance in diabetic and obese mice fed on very high-fat diet. *J Physiol Biochem* 63: 305.

Iglesias-Osma, M.C. et al. 2005. Methylamine but not mafenide mimics insulin-like activity of the semicarbazide-sensitive amine oxidase-substrate benzylamine on glucose tolerance and on human adipocyte metabolism. *Pharmacol Res* 52: 475.

Ignesti, G. 2003. Equations of substrate-inhibition kinetics applied to pig kidney diamine oxidase (DAO, E.C. 1.4.3.6). *J Enzyme Inhib Med Chem* 18: 463.

Imamura, Y. et al. 1997. Human retina-specific amine oxidase (RAO): cDNA cloning, tissue expression, and chromosomal mapping. *Genomics* 40: 277.

Imamura, Y. et al. 1998. Human retina-specific amine oxidase: genomic structure of the gene (AOC2), alternatively spliced variant, and mRNA expression in retina. *Genomics* 51: 293.

Ionescu, G. and R. Kiehl. 1989. Cofactor levels of mono- and diamine oxidase in atopic eczema. *Allergy* 44: 298.

Irjala, H. et al. 2001. Vascular adhesion protein-1 mediates binding of immunotherapeutic effector cells to tumor endothelium. *J Immunol* 166: 6937.

Ishizaki, H. and K.T. Yasunobu. 1980. Bovine plasma amine oxidase interactions with concanavalin A in solution and with concanavalin A-Sepharose. *Biochim Biophys Acta* 611: 27.

Ishizaki, F. 1990. Plasma benzylamine oxidase activity in cerebrovascular disease. *Eur Neurol* 30: 104.

Isobe, K. et al. 1980. Determination of polyamines with immobilized beef plasma amine oxidase. *Agric Biol Chem* 44: 615.

Jaakkola, K. et al. 1999. Human vascular adhesion protein-1 in smooth muscle cells. *Am J Pathol* 155: 1953.

Jaakkola, K. et al. 2000a. *In vivo* detection of vascular adhesion protein-1 in experimental inflammation. *Am J Pathol* 157: 463.

Jaakkola, K. et al. 2000b. Vascular adhesion protein-1, intercellular adhesion molecule-1 and P-selectin mediate leukocyte binding to ischemic heart in humans. *J Am Coll Cardiol* 36: 122.

Jackson, C.V. et al. 1986. Electrolysis-induced myocardial disfunction: novel method for the study of free radical mediated tissue injury. *J Pharmacol Methods* 15: 305.

Jacotot, E. et al. 1999. Mitochondrial membrane permeabilization during the apoptotic process. *Ann NY Acad Sci* 887: 18.

Jakobsson, E. et al. 2005. Structure of human semicarbazide-sensitive amine oxidase/vascular adhesion protein-1. *Acta Cryst D* 61: 1550.

Jalkanen, S. et al. 1986. A distinct endothelial cell recognition system that controls lymphocyte traffic into inflamed synovium. *Science* 233: 556.

Jalkanen, S. et al. 2007. The oxidase activity of vascular adhesion protein-1 (VAP-1) induces endothelial E- and P-selectins and leukocyte binding. *Blood* 110: 1706.

Janes, S.M. et al. 1990. A new redox cofactor in eukaryotic enzymes: 6-hydroxydopa at the active site of bovine serum amine oxidase. *Science* 248: 981.

Janes, S.M. et al. 1992. Identification of topaquinone and its consensus sequence in copper amine oxidases. *Biochemistry* 31: 12147.

Jansen, M.K. and K. Csiszar. 2007. Intracellular localization of the matrix enzyme lysyl oxidase in polarized epithelial cells. *Matrix Biol* 26: 136.

Jeffery, C.J. 2003. Moonlighting proteins: old proteins learning new tricks. *Trends Genet* 19: 415.

Jeon, H.B. and L.M. Sayre. 2003. Highly potent propargylamine and allylamine inhibitors of bovine plasma amine oxidase. *Biochem Biophys Res Commun* 304: 788.

Jeon, H.B. et al. 2003a. Inhibition of bovine plasma amine oxidase by 1,4-diamino-2-butenes and -2-butynes. *Bioorg Med Chem* 11: 4631.

Jeon, H.B., G. Sun, and L.M. Sayre. 2003b. Inactivation of bovine plasma amine oxidase by 4-aryloxy-2-butynamines and related analogs. *Biochim Biophys Acta* 1647: 343.

Jia, L. et al. 1996. Increased activity and sensitivity of mitochondrial respiratory enzymes to tumor necrosis factor alpha-mediated inhibition is associated with increased cytotoxicity in drug-resistant leukemic cell lines. *Blood* 87: 2401.

Jia, L. et al. 1997. Mitochondrial electron transport chain activity, but not ATP synthesis, is required for drug-induced apoptosis in human leukaemic cells: a possible novel mechanism of regulating drug resistance. *Br J Haematol* 98: 686.

Jiang, Z.J., J.S. Richardson, and P.H. Yu. 2008. The contribution of cerebral vascular semicarbazide-sensitive amine oxidase to cerebral amyloid angiopathy in Alzheimer's disease. *Neuropathol Appl Neurobiol* 34: 194.

Johnson, B.J. et al. 2007. Exploring molecular oxygen pathways in *Hansenula polymorpha* copper-containing amine oxidase. *J Biol Chem* 282: 17767.

Jones, J.D. and P.C. Brunett. 1975. Creatinine metabolism and toxicity. *Kidney Int Suppl* 3: 294.

Jones, N.L. et al. 1998. Ketotifen therapy for acute ulcerative colitis in children: a pilot study. *Dig Dis Sci* 43: 609.

Joshi, B. et al. 1999. Apoptosis induction by a novel anti-prostate cancer compound, BMD188 (a fatty acid-containing hydroxamic acid), requires the mitochondrial respiratory chain. *Cancer Res* 59: 4343.

Juda, G.A. et al. 2006. A comparative study of the binding and inhibition of four copper-containing amine oxidases by azide: implications for the role of copper during the oxidative half-reaction. *Biochemistry* 45: 8788.

Juhlin, L. 1981. Factors influencing anthralin erythema. *Br J Dermatol* 105, Suppl. 20: 87.

Junn, E. and M.M. Mouradian. 2001. Apoptotic signaling in dopamine-induced cell death: role of oxidative stress, p38 mitogen-activated protein kinase, cytochrome c and caspases. *J Neurochem* 78: 374.

Jutel, M., K. Blaser, and C.A. Akdis. 2006. The role of histamine in regulation of immune responses. *Chem Immunol Allergy* 91: 174.

Kagan, H.M. et al. 1979. Purification and properties of four species of lysyl oxidase from bovine aorta. *Biochem J* 177: 203.

Kagan, H.M. 1986. Characterization and regulation of lysyl oxidase. In *Biology of Extracellular Matrix: Regulation of Matrix Accumulation*, Vol. I, Mecham, R.P., Ed. Orlando: Academic Press, p. 321.

Kagan, H.M. and W. Li. 2003. Lysyl oxidase: properties, specificity, and biological roles inside and outside of the cell. *J Cell Biochem* 88: 660.

Kalaria, R.N. 1999. The blood–brain barrier and cerebrovascular pathology in Alzheimer's disease. *Ann NY Acad Sci* 893: 113.

Kalaria, R.N. and C. Ballard. 1999. Overlap between pathology of Alzheimer's disease and vascular dementia. *Alzheimer Dis Assoc Disord* 13, Suppl. 3: S115.

Kamei, H. et al. 2005. Quick recovery of serum diamine oxidase activity in patients undergoing total gastrectomy by oral enteral nutrition. *Am J Surg* 189: 38.

Kamiya, S. et al. 2004. The value of bile replacement during external biliary drainage: analysis of intestinal permeability, integrity, and microflora. *Ann Surg* 239: 510.

Kaneko, H. et al. 1998. Inhibition of post-ischemic reperfusion injury of the kidney by diamine oxidase. *Biochim Biophys Acta* 1407: 193.

Karadi, I. et al. 2002. Serum semicarbazide-sensitive amine oxidase (SSAO) activity is an independent marker of carotid atherosclerosis. *Clin Chim Acta* 323: 139.

Karube, I. 1987. Analytical application of enzymes: enzyme sensors for clinical, process and environmental analyses. *Enzyme Technol* 7: 685.

Karvonen, E. 1987. Inhibition of diamine oxidase of rat small intestine by pentamidine and berenil (diminazene aceturate). *Biochem Pharmacol* 36: 2863.

Katoh, A., T. Shoji, and T. Hashimoto. 2007. Molecular cloning of *N*-methylputrescine oxidase from tobacco. *Plant Cell Physiol* 48: 550.

Kennedy, J.F. and C.A. White. 1985. Principles of immobilization of enzymes. In *Handbook of Enzyme Technology*, Wiseman, A., Ed. Chichester: Ellis Horwood, p. 147.

Kenten, R.H. and P.J.G. Mann. 1952. The oxidation of amines by extracts of pea seedlings. *Biochem J* 50: 360.

Kerr, J.F., A.H. Wyllie, and A.R. Currie. 1972. Apoptosis: a basic biological phenomenon with wide-ranging implications in tissue kinetics. *Br J Cancer* 26: 239.

Kim, M. et al. 2002. X-ray snapshots of quinone cofactor biogenesis in bacterial copper amine oxidase. *Nat Struct Biol* 9: 591.

Kim, J. et al. 2006. Inactivation of bovine plasma amine oxidase by haloallylamines. *Bioorg Med Chem* 14: 1444.

Kinemuchi, H. et al. 2001. Inhibition of tissue-bound semicarbazide-sensitive amine oxidase by two haloamines, 2-bromoethylamine and 3-bromopropylamine. *Arch Biochem Biophys* 385: 154.

Kinemuchi, H. et al. 2004. Selective inhibitors of membrane-bound semicarbazide-sensitive amine oxidase (SSAO) activity in mammalian tissues. *Neurotoxicology* 25: 325.

Kirkel, A.Z. et al. 1983. Decrease of placental amine oxidase activities in premature birth. *Vopr Med Khim* 29: 83.

Kirton, C.M. et al. 2005. Function-blocking antibodies to human vascular adhesion protein-1: a potential anti-inflammatory therapy. *Eur J Immunol* 35: 3119.

Kishishita, S. et al. 2003. Role of cupper ion in bacterial copper amine oxidase: spectroscopic and crystallographic studies of metal-substituted enzymes. *J Am Chem Soc* 125: 1041.

Kleywegt, G.J. and T.A. Jones. 1994. Detection, delineation, measurement and display of cavities in macromolecular structures. *Acta Crystallogr D* 50: 178.

Knowles, P.F. and K.D.S. Yadav. 1984. Amine oxidases. In *Copper Proteins and Copper Enzymes*. Boca Raton: CRC Press.

Knowles, P. et al. 2007. Hydrazine and amphetamine binding to amine oxidases: old drugs with new prospects. *J Neural Transm* 114: 743.

Kokoszka, J.E. et al. 2004. The ADP/ATP translocator is not essential for the mitochondrial permeability transition pore. *Nature* 427: 461.

Kolaříková, K. et al. 2008. Functional expression of amine oxidase from *Aspergillus niger* (AO-I) in *Saccharomyces cerevisiae*. *Mol Biol Rep* DOI: 10.1007/s11033-007-9146-7.

Kopple, J.D. et al. 1978. Diamine oxidase in renal failure. *Kidney Int* 13, Suppl. 8: S20.

Kordowiak, A.M. et al. 2005. Sodium orthovanadate exerts influence on liver Golgi complexes from control and streptozotocin-diabetic rats. *J Inorg Biochem* 99: 1083.

Koskinen, K. et al. 2004. Granulocyte transmigration through endothelium is regulated by the oxidase activity of vascular adhesion protein-1 (VAP-1). *Blood* 103: 3388.

Koskinen, K. et al. 2007. VAP-1-deficient mice display defects in mucosal immunity and antimicrobial responses: implications for antiadhesive applications. *J Immunol* 179: 6160.

Koutroubakis, I.E. et al. 2002. Circulating soluble vascular adhesion protein 1 in patients with inflammatory bowel disease. *Eur J Gastroenterol Hepatol* 14: 405.

Koyanagi, T. et al. 2000. Molecular cloning and heterologous expression of pea seedling copper amine oxidase. *Biosci Biotechnol Biochem* 64: 717.

Krieger-Brauer, H.I., P.K. Medda, and H. Kather. 1997. Insulin-induced activation of NADPH-dependent H_2O_2 generation in human adipocyte plasma membranes is mediated by $G\alpha_{i2}$. *J Biol Chem* 272: 1013.

Krissinel, E. and K. Henrick. 2004. Secondary-structure matching (SSM), a new tool for fast protein structure alignment in three dimensions. *Acta Cryst D* 60: 2256.

Kroemer, G. et al. 2005. Classification of cell death: recommendations of the nomenclature committee on cell death. *Cell Death Differ* 12: 1463.

Kumar, V. et al. 1996. Crystal structure of a eukaryotic (pea seedling) copper–containing amine oxidase at 2.2 Å resolution. *Structure* 4: 943.

Kunduzova, O.R. et al. 2002. Hydrogen peroxide production by monoamine oxidase during ischemia/reperfusion. *Eur J Pharmacol* 448: 225.

Kunkel, E.J. and E.C. Butcher. 2002. Chemokines and the tissue-specific migration of lymphocytes. *Immunity* 16: 1.

Kurkijärvi, R. et al. 1998. Circulating form of human vascular adhesion protein-1 (VAP-1): increased serum levels in inflammatory liver diseases. *J Immunol* 161: 1549.

Kurkijärvi, R. et al. 2000. Circulating soluble vascular adhesion protein 1 accounts for the increased serum monoamine oxidase activity in chronic liver disease. *Gastroenterology* 119: 1096.

Kurkijärvi, R. et al. 2001. Vascular adhesion protein-1 (VAP-1) mediates lymphocyte-endothelial interactions in chronic kidney rejection. *Eur J Immunol* 31: 2876.

Kusano, T. et al. 2007. Advances in polyaminc research in 2007. *J Plant Res* 120: 345.

Kusche, J., T. Biegański, and R. Hesterberg. 1980. The influence of carcinoma growth on diamine oxidase activity in human gastrointestinal tract. *Agents Actions* 10: 110.

Kusche, J. et al. 1986. The effect of experimental carcinogenesis on intestinal diamine oxidase, a polyamine deaminating enzyme. *Cancer Det Prev* 9: 17.

Kusche, J. et al. 1988. Large bowel tumor promotion by diamine oxidase inhibition: animal model and clinical aspects. *Adv Exp Med Biol* 250: 745.

Laidler, K. and P. Bunting. 1973. *The Chemical Kinetics of Enzyme Action.* 2nd Ed. Oxford: Clarendon Press, p. 52.

Lalor, P.F. et al. 2002. Vascular adhesion protein-1 mediates adhesion and transmigration of lymphocytes on human hepatic endothelial cells. *J Immunol* 169: 983.

Lalor, P.F. et al. 2007. Activation of vascular adhesion protein-1 on liver endothelium results in an NF-κ-B-dependent increase in lymphocyte adhesion. *Hepatology* 45: 465.

Lamplot, Z. et al. 2004. 1,5-Diamino-2-pentyne is both a substrate and inactivator of plant copper amine oxidases. *Eur J Biochem* 271: 4696.

Lamplot, Z. et al. 2005. Reactions of N^6-aminoalkyl adenines with plant amine oxidases. *J Enzym Inhib Med Chem* 20: 143.

Lange, J. and C. Wittmann. 2002. Enzyme sensor array for the determination of biogenic amines in food samples. *Anal Bioanal Chem* 372: 276.

Langford, S.D. et al. 1999. Developmental vasculotoxicity associated with inhibition of semicarbazide-sensitive amine oxidase. *Toxicol Appl Pharmacol* 155: 237.

Langford, S.D., M.B. Trent, and P.J. Boor. 2002. Semicarbazide-sensitive amine oxidase and extracellular matrix deposition by smooth-muscle cells. *Cardiovasc Toxicol* 2: 141.

Langley, D.B. et al. 2006. The copper-containing amine oxidase from *Arthrobacter globiformis*: refinement at 1.55 Å and 2.20 Å resolution in two crystal forms. *Acta Cryst F* 62: 1052.

Langley, D.B. et al. 2008a. Enantiomer-specific binding of ruthenium(II) molecular wires by the amine oxidase of *Arthrobacter globiformis*. *J Am Chem Soc* 130: 8069.

Langley, D.B. et al. 2008b. Complexes of the copper-containing amine oxidase from *Arthrobacter globiformis* with the inhibitors benzylhydrazine and tranylcypromine. *Acta Cryst F* 64: 577.

Laurenzi, M. et al. 2001. Analysis of the distribution of copper amine oxidase in cell walls of legume seedlings. *Planta* 214: 37.

Lee, Y., H. Huang, and L.M. Sayre. 1996. Model studies support pyrrolylation of the topaquinone cofactor to explain inactivation of bovine plasma amine oxidase by 3-pyrrolines: unusual processing of a secondary amine. *J Am Chem Soc* 118: 7241.

Lee, Y. and L.M. Sayre. 1998. Reaffirmation that metabolism of polyamines by bovine plasma amine oxidase occurs strictly at the primary amino termini. *J Biol Chem* 273: 19490.

Lee, Y. et al. 2001a. Catalytic turnover of substrate benzylamines by the quinone-dependent plasma amine oxidase leads to H_2O_2-dependent inactivation: evidence for generation of a cofactor-derived benzoxazole. *Biochemistry* 40: 822.

Lee, Y. et al. 2001b. Temporary inactivation of plasma amine oxidase by alkylhydrazines: a combined enzyme/model study implicates cofactor reduction/reoxidation but cofactor deoxygenation and subsequent reoxygenation in the case of hydrazine itself. *J Org Chem* 66: 1925.

Lee, Y. et al. 2002. 3-Pyrrolines are mechanism-based inactivators of the quinone-dependent amine oxidases but only substrates of the flavin-dependent amine oxidases. *J Am Chem Soc* 124: 12135.

Lefer, A.M. and D.J. Lefer. 1993. Pharmacology of the endothelium in ischemia–reperfusion and circulatory shock. *Annu Rev Pharmacol Toxicol* 33: 71.

Lewandrowski, U. et al. 2008. Identification of new tyrosine phosphorylated proteins in rat brain mitochondria. *FEBS Lett* 582: 1104.

Lewinsohn, R. 1977. Human serum amine oxidase: enzyme activity in severely burnt patients and in patients with cancer. *Clin Chim Acta* 81: 247.

Lewinsohn, R. et al. 1978. A benzylamine oxidase distinct from monoamine oxidase B: widespread distribution in man and rat. *Biochem Pharmacol* 27: 1857.

Lewinsohn, R. 1984. Mammalian monoamine-oxidizing enzyme, with special reference to benzylamine oxidase in human tissues. *Braz J Med Res* 17: 223.

Ley, K. and G.S. Kansas. 2004. Selectins in T-cell recruitment to non-lymphoid tissues and sites of inflammation. *Nat Rev Immunol* 4: 325.

Ley, K. et al. 2007. Getting to the site of inflammation: the leukocyte adhesion cascade updated. *Nat Rev Immunol* 7: 678.

Li, H. et al. 2004. Assay of plasma semicarbazide-sensitive amine oxidase and determination of its endogenous substrate methylamine by liquid chromatography. *J Chromatogr B Analyt Technol Biomed Life Sci* 810: 277.

Li, H.Y. et al. 2005. Change in vascular adhesion protein-1 and metabolic phenotypes after vertical banded gastroplasty for morbid obesity. *Obes Res* 13: 855.

Li, R., J.P. Klinman, and F.S. Mathews. 1998. Copper amine oxidase from *Hansenula polymorpha*: the crystal structure determined at 2.4 resolution reveals the active conformation. *Structure* 6: 293.

Li, W. et al. 2000. Hydrogen peroxide-mediated, lysyl oxidase-dependent chemotaxis of vascular smooth muscle cells. *J Cell Biochem* 78: 550.

Li, W. et al. 2003. 3-Aminopropanal, formed during cerebral ischaemia, is a potent lysosomotropic neurotoxin. *Biochem J* 371: 429.

Lichszteld, K.K.L. 1979. Singlet molecular oxygen in formaldehyde oxidation. *Z Phys Chem NF* 108: 167.

Lindemann, L. and M.C. Hoener. 2005. A renaissance in trace amines inspired by a novel GPCR family. *Trends Pharmacol Sci* 26: 274.

Lindley, P.F. 2001. Multi-copper oxidases. In *Handbook on Metalloproteins*, Bertini, I. et al., Eds. Basel: Marcel Dekker, p. 763.

Lindsay, G.S. and H.M. Wallace. 1999. Changes in polyamine catabolism in HL60 human promyelogenous leukaemic cells in response to etoposide-induced apoptosis. *Biochem J* 337: 83.

Lindström, A. and G. Pettersson. 1978. Active-site titration of pig plasma benzylamine oxidase. *Eur J Biochem* 83: 131.

Ling, K.Q. and L.M. Sayre. 2005. A dopaquinone model that mimics the water addition step of cofactor biogenesis in copper amine oxidases. *J Am Chem Soc* 127: 4777.

Lingueglia, E. et al. 1993. Molecular cloning and functional expression of different molecular forms of rat amiloride-binding proteins. *Eur J Biochem* 216: 679.

Liu, E. et al. 2007a. Increased expression of vascular endothelial growth factor in kidney leads to progressive impairment of glomerular functions. *J Am Soc Nephrol* 18: 2094.

Liu, D. et al. 2007b. Inactivation of *Escherichia coli* L-aspartate aminotransferase by (S)-4-amino-4,5-dihydro-2-thiophenecarboxylic acid reveals "a tale of two mechanisms." *Biochemistry* 46: 10517.

Liu, G., K. Nellaiappan, and H.M. Kagan. 1997. Irreversible inhibition of lysyl oxidase by homocysteine thiolactone and its selenium and oxygen analogues: implications for homocystinuria. *J Biol Chem* 272: 32370.

Lizcano, J.M. et al. 1990. Amine oxidase activities in bovine lung. *J Neural Transm Suppl* 32: 341.

Lizcano, J.M. et al. 1991. The oxidation of dopamine by the semicarbazide-sensitive amine oxidase (SSAO) from rat vas deferens. *Biochem Pharmacol* 41: 1107.

Lizcano, J.M. et al. 1994. Several aspects on the amine oxidation by semicarbazide-sensitive amine oxidase (SSAO) from bovine lung. *J Neural Transm Suppl* 41: 415.

Lizcano, J.M. et al. 1996. Inhibition of bovine lung semicarbazide-sensitive amine oxidase (SSAO) by some hydrazine derivatives. *Biochem Pharmacol* 52: 187.

Lizcano, J.M., K.F. Tipton, and M.Unzeta. 1998. Purification and characterization of membrane-bound semicarbazide-sensitive amine oxidase (SSAO) from bovine lung. *Biochem J* 331: 69.

Lizcano, J.M., K.F. Tipton, and M. Unzeta. 2000a. Time-dependent activation of the semicarbazide-sensitive amine oxidase (SSAO) from ox lung microsomes. *Biochem J* 351: 789.

Lizcano, J.M., M. Unzeta, and K.F. Tipton. 2000b. A spectrophotometric method for determining the oxidative deamination of methylamine by the amine oxidases. *Anal Bioch* 286: 75.

Longu, S. et al. 2005a. Mechanism-based inactivators of plant copper/quinone containing amine oxidases. *Phytochemistry* 66: 1751.

Longu, S. et al. 2005b. Nitric oxide covalently labels a 6-hydroxydopa-derived free radical intermediate in the catalytic cycle of copper/quinone-containing amine oxidase from lentil seedlings. *Biol Chem* 386: 25.

Loske, C. et al. 2000. Transition metal-mediated glycoxidation accelerates cross-linking of β-amyloid peptide. *Eur J Biochem* 267: 4171.

Lowden, D.A. et al. 1994. Renal cysts in transgenic mice expressing transforming growth factor-α. *J Lab Clin Med* 124: 386.

Lowis, S., M.A. Eastwood, and W.G. Brydon. 1985. The influence of creatinine, lecithin and choline feeding on aliphatic amine production and excretion in the rat. *Br J Nutr* 54: 43.

Luhová, L. et al. 1995. Amino oxidase from *Trigonella foenum-graecum*. *Phytochemistry* 38: 23.

Luhová, L. et al. 1996. Comparison of kinetic properties between plant and fungal amine oxidases. *J Enz Inhib* 10: 251.

Luk, G.D., L.J. Marton, and S.B. Baylin. 1980a. Ornithine decarboxylase is important in intestinal mucosal maturation and recovery from injury in rats. *Science* 10: 195.

Luk, G.D., T.M. Bayless, and S.B. Baylin. 1980b. Diamine oxidase (histaminase): a circulating marker for rat intestinal mucosal maturation and integrity. *J Clin Invest* 66: 66.

Lunelli, M. et al. 2005. Crystal structure of amine oxidase from bovine serum. *J Mol Biol* 346: 991.

Luo, B.H., C.V. Carman, and T.A. Springer. 2007. Structural basis of integrin regulation and signaling. *Annu Rev Immunol* 25: 619.

Luster, A.D., R. Alon, and U.H. von Andrian. 2005. Immune cell migration in inflammation: present and future therapeutic targets. *Nat Immunol* 6: 1182.

Lyles, G.A. and B.A. Callingham. 1975. Evidence for a clorgyline-resistant monoamine metabolizing activity in the rat heart. *J Pharm Pharmacol* 27: 682.

Lyles, G.A. and B.A. Callingham. 1981. The effect of DSP-4 (N-[2-chloroethyl]-N-ethyl-2-bromobenzylamine) on monoamine oxidase activities in tissues of the rat. *J Pharm Pharmacol* 33: 632.

Lyles, G.A. and B.A. Callingham. 1982a. Hydralazine is an irreversible inhibitor of the semicarbazide-sensitive, clorgyline-resistant amine oxidase in rat aorta homogenates. *J Pharm Pharmacol* 34: 139.

Lyles, G.A. and B.A. Callingham. 1982b. *In vitro* and *in vivo* inhibition by benserazide of clorgyline-resistant amine oxidases in rat cardiovascular tissues. *Biochem Pharmacol* 31: 1417.

Lyles, G.A. 1984. The interaction of semicarbazide-sensitive amine oxidase with MAO inhibitors. In *Monoamine Oxidase and Disease*, Tipton, K.F. et al., Eds. London: Academic Press, p. 547.

Lyles, G.A. and I. Singh. 1985. Vascular smooth muscle cells: a major source of the semicarbazide-sensitive amine oxidase of the rat aorta. *J Pharm Pharmacol* 37: 637.

Lyles, G.A. et al. 1987. Inhibition of rat aorta semicarbazide-sensitive amine oxidase by 2-phenyl-3-haloallylamines and related compounds. *Biochem Pharmacol* 36: 2847.

Lyles, G.A. and D.T. Taneja. 1987. Effects of amine oxidase inhibitors upon tryptamine metabolism and tryptamine-induced contractions of rat aorta. *Br J Pharmacol* 90: 16P.

Lyles, G.A. and S.A. McDougall. 1988. The enhanced daily excretion of urinary methlyamine in rats treated with semicarbazide or hydralazine may be related to the inhibition of semicarbazide-sensitive amine oxidase activities. *J Pharm Pharmacol* 41: 97.

Lyles, G.A., A. Holt, and C.M.S. Marshall. 1990. Further studies on the metabolism of methylamine by semicarbazide-sensitive amine oxidase activities in human plasma, umbilical artery and rat aorta. *J Pharm Pharmacol* 42: 332.

Lyles, G.A. and J. Chalmers. 1992. The metabolism of aminoacetone to methylglyoxal by semicarbazide-sensitive amine oxidase in human umbilical artery. *Biochem Pharmacol* 43: 1409.

Lyles, G.A. and J. Chalmers. 1995. Aminoacetone metabolism by semicarbazide-sensitive amine oxidase in rat aorta. *Biochem Pharmacol* 49: 416.

Lyles, G.A. 1995. Substrate-specificity of mammalian tissue-bound semicarbazide-sensitive amine oxidase. *Prog Brain Res* 106: 293.

Lyles, G.A. 1996. Mammalian plasma and tissue-bound semicarbazide-sensitive amine oxidases: biological, pharmacological and toxicological aspects. *Int J Biochem Cell Biol* 28: 259.

Ma, L.J. and A.B. Fogo. 2007. Modulation of glomerulosclerosis. *Semin Immunopathol* 29: 385.

Maccarrone, M. et al. 1997. Involvement of 5-lipoxygenase in programmed cell death of cancer cells. *Cell Death Differ* 4: 396.

Maccarrone, M. et al. 2001. Oxidation products of polyamines induce mitochondrial uncoupling and cytochrome c release. *FEBS Lett* 507: 30.

Macholán, L., L. Rozprimová, and E. Sedláčková. 1967. 1,4-Diamino-2-butanon (2-ketoputrescine) as strong and short acting competitive inhibitor of diamine oxidase. *Biochim Biophys Acta* 132: 505.

Macholán, L. 1969. Substrate-like inhibitors of diamine oxidase: some relations between the structure of aliphatic aminoketones and their inhibitory effect. *Arch Biochem Biophys* 134: 302.

Macholán, L. 1974. Selective and reversible inhibition of diamine oxidase by 1,5-diamino-3-pentanone. *Coll Czech Chem Commun* 39: 653.

Macholán, L., F. Hubálek, and H. Šubová. 1975. Oxidation of 1,4-diamino-2-butene to pyrrole, a sensitive test of diamine oxidase activity. *Coll Czech Chem Commun* 40: 1247.

Macholán, L. and J. Slanina. 1991. Use of inhibited enzyme electrode for the estimation of pea diamine oxidase inhibitors. *Coll Czech Chem Commun* 56: 1545.

Male, B.K. et al. 1996. Amperometric biosensor for total histamine, putrescine and cadaverine using diamine oxidase. *J Food Sci* 61: 1012.

Mallick, I.H. et al. 2004. Ischemia–reperfusion injury of the intestine and protective strategies against injury. *Dig Dis Sci* 49: 1359.

Malorni, W. et al. 1998. Protection against apoptosis by monoamine oxidase A inhibitors. *FEBS Lett* 426: 155.

Mamhidir, A.G. et al. 2007. Weight increase in patients with dementia, and alteration in meal routines and meal environment after integrity-promoting care. *J Clin Nurs* 16: 987.

Mann, P.J.G. 1955. Purification and properties of the amine oxidase of pea seedlings. *Biochem J* 59: 609.

Mann, P.J.G. 1961. Further purification and properties of the amine oxidase of pea seedlings. *Biochem J* 79: 623.

Manna, F. et al. 2002. Inhibition of amine oxidases activity by 1-acetyl-3,5-diphenyl-4,5-dihydro-(1H)-pyrazole derivatives. *Bioorg Med Chem Lett* 12: 3629.

Mannaioni, P.F. et al. 1991. Mast cells as a source of superoxide anions and nitric oxide-like factor: relevance to histamine release. *Int J Tissue React* 13: 271.

Mansueto, P. et al. 2006. Food allergy in gastroenterologic diseases: review of literature. *World J Gastroenterol* 12: 7744.

Mantle, T.J., K.F. Tipton, and N.J. Garrett. 1976. Inhibition of monoamine oxidase by amphetamine and related compounds. *Biochem Pharmacol* 25: 2073.

Mantyh, P.W. et al. 1991. Alterations in receptors for sensory neuropeptides in human inflammatory bowel disease. *Adv Exp Med Biol* 298: 253.

Marini, F. et al. 2001. Polyamine metabolism is upregulated in response to tobacco mosaic virus in hypersensitive, but not susceptible, tobacco. *New Phytol* 149: 301.

Martelius, T. et al. 2000. Induction of vascular adhesion protein-1 during liver allograft rejection and concomitant cytomegalovirus infection in rats. *Am J Pathol* 157: 1229.

Martelius, T. et al. 2004. Blockade of vascular adhesion protein 1 inhibits lymphocyte infiltration in rat liver allograft rejection. *Am J Pathol* 165: 1993.

Marti, L. et al. 1998. Tyramine and vanadate synergistically stimulate glucose transport in rat adipocytes by amine oxidase-dependent generation of hydrogen peroxide. *J Pharmacol Exp Ther* 285: 342.

Marti, L. et al. 2001. Combined treatment with benzylamine and low dosages of vanadate enhances glucose tolerance and reduces hyperglycemia in streptozotocin-induced diabetic rats. *Diabetes* 50: 2061.

Marttila-Ichihara, F. et al. 2006. Vascular amine oxidases are needed for leukocyte extravasation into inflamed joints *in vivo*. *Arthritis Rheum* 54: 2852.

Maruyama, W. et al. 2000. Mechanism underlying anti-apoptotic activity of a (-)deprenyl-related propargylamine, rasagiline. *Mech Ageing Dev* 116: 181.

Marzo, I. et al. 1998a. BAX and adenine nucleotide translocator cooperate in the mitochondrial control of apoptosis. *Science* 281: 2027.

Marzo, I. et al.1998b. The permeability transition pore complex: a target for apoptosis regulation by caspases and Bcl-2-related proteins. *J Exp Med* 187: 1261.

Masini, E. et al. 1999. Cardioprotective activity of endogenous and exogenous nitric oxide on ischaemia reperfusion injury in isolated guinea pig hearts. *Inflamm Res* 48: 561.

Masini, E. et al. 2002. A plant histaminase modulates cardiac anaphylactic response in guinea pig. *Biochem Biophys Res Commun* 296: 840.

Masini E. et al. 2003. Protective effect of a plant histaminase in myocardial ischaemia and reperfusion *in vivo*. *Biochem Biophys Res Commun* 309: 432.

Masini, E. et al. 2004. Effect of a plant histaminase on asthmalike reaction induced by inhaled antigen in sensitized guinea pig. *Eur J Pharmacol* 502: 253.

Masini, E. et al. 2005a. Histamine and histidine decarboxylase up-regulation in colorectal cancer: correlation with tumor stage. *Inflamm Res* 54: S80.

Masini, E. et al. 2005b. Reduction of antigen-induced respiratory abnormalities and airway inflammation in sensitized guinea pigs by a superoxide dismutase mimetic. *Free Radic Biol Med* 39: 520.

Masini, E. et al. 2007. Beneficial effects of a plant histaminase in a rat model of splanchnic artery occlusion and reperfusion. *Shock* 27: 409.

Mateescu, M.A. et al. 1995. Protection of myocardial tissue against deleterious effects of oxygen free radicals by ceruloplasmin. *Arzneim Forsch Drug Res* 45: 476.

Mateescu, M.A. et al. 1997. A new physiological role of copper amine oxidases: cardioprotection against reactive oxygen intermediates. *J Physiol Pharmacol* 48: 110.

Matsumoto, T. et al. 1981a. A new enzymatic assay for total diamines and polyamines in urine of cancer patients. *J Cancer Res Clin Oncol* 100: 73.

Matsumoto, T. et al. 1981b. A fluorometric assay for total diamines in human urine using human placental diamine oxidase. *Clin Chim Acta* 112: 141.

Matsumoto, S. et al. 2005. Probiotic Lactobacillus-induced improvement in murine chronic inflammatory bowel disease is associated with the down-regulation of pro-inflammatory cytokines in lamina propria mononuclear cells. *Clin Exp Immunol* 140: 417.

Matsunami, H. et al. 2004. Chemical rescue of a site-specific mutant of bacterial copper amine oxidase for generation of the topa quinone cofactor. *Biochemistry* 43: 2178.

Matsuzaki, R. et al. 1994. Generation of the topa quinone cofactor in bacterial monoamine oxidase by cupric ion-dependent autooxidation of a specific tyrosyl residue. *FEBS Lett* 351: 360.

Matsuzaki, R. et al. 1995. Spectroscopic studies on the mechanism of the topa quinone generation in bacterial monoamine oxidase. *Biochemistry* 34: 4524.

Matsuzaki, R. and K. Tanizawa. 1998. Exploring a channel to the active site of copper/topaquinone-containing phenylethylamine oxidase by chemical modification and site-specific mutagenesis. *Biochemistry* 37: 13947.

Maula, S.M. et al. 2005 Carbohydrates located on the top of the "cap" contribute to the adhesive and enzymatic functions of vascular adhesion protein-1. *Eur J Immunol* 35: 2718.

May, J.M. and C. de Haën. 1979. The insulin-like effect of hydrogen peroxide on pathways of lipid synthesis in rat adipocytes. *J Biol Chem* 254: 9017.

McDonald, I.A. et al. 1985. Enzyme-activated irreversible inhibitors of monoamine oxidase: phenylallylamine structure-activity relationships. *J Med Chem* 28: 186.

McDonald, A.G. et al. 2007. Modelling the roles of MAO and SSAO in glucose transport. *J Neural Transm* 114: 783.

McEwen, C. and J.D. Cohen. 1963. An amine oxidase in normal human serum. *J Lab Clin Med* 62: 766.

McEwen, C.M. Jr. 1965a. Human plasma monoamine oxidase I. Purification and identification. *J Biol Chem* 240: 2003.

McEwen, C.M. Jr. 1965b. Human plasma monoamine oxidase II. Kinetic studies. *J Biol Chem* 240: 2011.

McEwen, C.M., K. Cullen Jr. and A.J. Sober. 1966. Rabbit serum monoamine oxidase. *J Biol Chem* 241: 4544.

McEwen, C. and D.O. Castell. 1967. Abnormalities of serum monoamine oxidase in chronic liver disease. *J Lab Clin Med* 70: 36.

McGuirl, M.A., D.E. Brown, and D.M. Dooley. 1997. Cyanide as a copper-directed inhibitor of amine oxidases: implications for the mechanism of amine oxidation. *J Biol Inorg Chem* 2: 336.

McIntire, W.S. and C. Hartmann. 1992. Copper-containing amine oxidases. In *Principles and Applications of Quinoproteins*, Davidson, V.L., Ed. New York: Marcel Dekker, p. 97.

McKhann, G. et al. 1984. Clinical diagnosis of Alzheimer's disease: report of NINCDS-ADRDA work group under the auspices of Department of Health and Human Services Task Force on Alzheimer's disease. *Neurology* 34: 939.

Medda, R., A. Padiglia, and G. Floris. 1995a. Plant copper-amine oxidases. *Phytochemistry* 39: 1.

Medda, R. et al. 1995b. The reaction mechanism of copper amine oxidase: detection of intermediates by the use of substrates and inhibitors. *Biochemistry* 34: 16375.

Medda, R. et al. 1997a. Inhibition of copper amine oxidase by haloamines: a killer product mechanism. *Biochemistry* 36: 2595.

Medda, R. et al. 1997b. Tryptamine as substrate and inhibitor of lentil seedling copper amine oxidase. *Eur J Biochem* 250: 377.

Medda, R. et al. 1998. Intermediate in the catalytic cycle of lentil (*Lens esculenta*) seedling copper-containing amine oxidase. *Biochem J* 332: 431.

Mennigen, R. et al. 1988. Large bowel tumors and diamine oxidase (DAO) activity in patients: a new approach for risk group identification. *Agents Actions* 23: 351.

Mennigen, R. et al. 1990. Diamine oxidase activities in the large bowel mucosa of ulcerative colitis patients. *Agents Actions* 30: 264.

Mercier, N. et al. 2001. Semicarbazide-sensitive amine oxidase activation promotes adipose conversion of 3T3-L1 cells. *Biochem J* 358: 335.

Mercier, N. et al. 2003. Regulation of semicarbazide-sensitive amine oxidase expression by tumor necrosis factor-α in adipocytes: functional consequences on glucose transport. *J Pharmacol Exp Ther* 304: 1197.

Mercier, N. et al. 2006. Carotid arterial stiffness, elastic fibre network and vasoreactivity in semicarbazide-sensitive amine-oxidase null mouse. *Cardiovasc Res* 72: 349.

Mercier, N. et al. 2007. Modifications of arterial phenotype in response to amine oxidase inhibition by semicarbazide. *Hypertension* 50: 234.

Meredith, S.C. 2005. Protein denaturation and aggregation: cellular responses to denatured and aggregated proteins. *Ann NY Acad Sci* 1066: 181.

Merinen, M. et al. 2005. Vascular adhesion protein-1 is involved in both acute and chronic inflammation in the mouse. *Am J Pathol* 166: 793.

Mészáros, Z. et al. 1999a. Determination of human serum semicarbazide-sensitive amine oxidase activity: a possible clinical marker of atherosclerosis. *Eur J Drug Metab Pharmacokinet* 24: 299.

Mészáros, Z. et al. 1999b. Elevated serum semicarbazide-sensitive amine oxidase activity in non-insulin-dependent diabetes mellitus: correlation with body mass index and serum triglyceride. *Metabolism* 48: 113.

Mills, S.A. and J.P. Klinman. 2000. Evidence against reduction of Cu^{2+} to Cu^{+} during dioxygen activation in a copper amine oxidase from yeast. *J Am Chem Soc* 122: 9897.

Mills, S.A. et al. 2002. Mechanistic comparison of the cobalt-substituted and wild-type copper amine oxidase from *Hansenula polymorpha*. *Biochemistry* 41: 10577.

Milovic, V. 2001. Polyamines in the gut lumen: bioavailability and biodistribution. *Eur J Gastroenterol Hepatol* 13: 1021.

Mitchell, J.L. et al. 1992. Abnormal accumulation and toxicity of polyamines in a difluoro methylornithine-resistant HTC cell variant. *Biochim Biophys Acta* 1136: 136.

Mlíčková, K. et al. 2001. Inhibition of copper amine oxidases by pyridine-derived aldoximes and ketoximes. *Biochimie* 83: 995.

Moenne-Loccoz, P. et al. 1995. Characterization of the topaquinone cofactor in amine oxidase from *Escherichia coli* by resonance Raman spectroscopy. *Biochemistry* 34: 7020.

Moldes, M., B. Fève, and J. Pairault. 1999. Molecular cloning of a major mRNA species in murine 3T3 adipocyte lineage differentiation-dependent expression, regulation, and identification as semicarbazide-sensitive amine oxidase. *J Biol Chem* 274: 9515.

Møller, S.G. and M.J. McPherson. 1998. Developmental expression and biochemical analysis of *Arabidopsis atao1* gene encoding an H_2O_2–generating diamine oxidase. *Plant J* 13: 781.

Møller, S.G. et al. 1998. Nematode-induced expression of *atao1*, a gene encoding an extracellular diamine oxidase associated with developing vascular tissue. *Physiol Mol Plant Pathol* 53: 73.

Molnar, J. et al. 2003. Structural and functional diversity of lysyl oxidase and the LOX-like proteins. *Biochim Biophys Acta* 1647: 220.

Mondovì, B. et al. 1964. Purification of pig-kidney diamine oxidase and its identity with histaminase. *Biochem J* 91: 408.

Mondovì, B. et al. 1967a. Diamine oxidase from pig kidney. *J Biol Chem* 242: 1160.

Mondovì, B. et al. 1967b. Diamine oxidase inactivation by hydrogen peroxide. *Biochim Biophys Acta* 132: 521.

Mondovì, B. et al. 1969. Copper reduction by substrate in diamine oxidase. *FEBS Lett* 2: 182.

Mondovì, B. et al. 1975. *In vivo* anti-histaminic activity of histaminase. *Agents Actions* 5: 460.

Mondovì, B., P. Gerosa, and R. Cavaliere. 1982. Studies on the effect of polyamines and their products on Ehrlich ascites tumours. *Agents Actions* 12: 450.

Mondovì, B. et al. 1983. Amine oxidase activity in malignant human brain tumors. In *Advances in Polyamine Research,* Bachrach, U. et al., Eds. New York: Raven, p. 183.

Mondovì, B., Ed. 1985. *Structure and Functions of Amine Oxidases*. Boca Raton: CRC Press.

Mondovì, B. and P. Riccio. 1990. Copper amine oxidases: current state of knowledge. *Biol Metals* 3: 110.

Mondovì, B. et al. 1992. Specific temperature dependance of diamine oxidase activity and its thermal stability in the presence of polyvinylalcohol. *Agents Actions* 37: 220.

Mondovì, B. et al. 1994. Amine oxidase as possible antineoplastic drugs. In *Proceedings of the 6th European Congress on Biotechnology*, Alberghina, L., Frontali, L., and Sensi, P., Eds. Elsevier Science, p. 755.

Mondovì, B. et al. 1997. New aspects on the physiological role of copper amineoxidases. *Curr Top Med Chem* 2: 31.

Moore, R.H. et al. 2007. Trapping of a dopaquinone intermediate in the TPQ cofactor biogenesis in a copper-containing amine oxidase from *Arthrobacter globiformis. J Am Chem Soc* 129: 11524.

Morgan, M.D. et al. 2006. Anti-neutrophil cytoplasm-associated glomerulonephritis. *J Am Soc Nephrol* 17: 1224.

Morimoto, T., Y. Yamamoto, and A. Yamatodani. 2001. Brain histamine and feeding behavior. *Behav Brain Res* 124: 45.

Morin, N. et al. 2001. Semicarbazide-sensitive amine oxidase substrates stimulate glucose transport and inhibit lipolysis in human adipocytes. *J Pharmacol Exp Ther* 297: 563.

Morin, N. et al. 2002. Tyramine stimulates glucose uptake in insulin-sensitive tissues *in vitro* and *in vivo* via its oxidation by amine oxidases. *J Pharmacol Exp Ther* 303: 1238.

Morpurgo, L. et al. 1987. Reactions of bovine serum amine oxidase with *NN*-diethyldithiocarbamate: selective removal of one copper ion. *Biochem J* 248: 865.

Morpurgo, L. et al. 1988. Spectroscopic studies of the reaction between bovine serum amine oxidase (copper-containing) and some hydrazides and hydrazines. *Biochem J* 256: 565.

Morpurgo, L. et al. 1989. Benzylhydrazine as a pseudo-substrate of bovine serum amine oxidase. *Biochem J* 260: 19.

Morpurgo, L. et al. 1990. The role of copper in bovine serum amine oxidase. *Biol Metals* 3: 114.

Morpurgo, L. et al. 1992. Bovine serum amine oxidase: half-site reactivity with phenylhydrazine, semicarbazide, and aromatic hydrazides. *Biochemistry* 31: 2615.

Morpurgo, L. et al. 1994. Properties of cobalt-substituted bovine serum amine oxidase. *Eur J Biochem* 222: 727.

Morris, N.J. et al. 1997. Membrane amine oxidase cloning and identification as a major protein in the adipocyte plasma membrane. *J Biol Chem* 272: 9388.

Mu, D. et al. 1992. Tyrosine codon corresponds to topa quinone at the active site of copper amine oxidases. *J Biol Chem* 267: 7979.

Mu, D. et al. 1994. Primary structures for a mammalian cellular and serum copper amine oxidase. *J Biol Chem* 269: 9926.

Mukherjee, A. et al. 2008. An inner-sphere mechanism for molecular oxygen reduction catalyzed by copper amine oxidases. *J Amer Chem Soc* 130: 9459.

Mukherjee, B. et al. 2004. Vanadium: element of atypical biological significance. *Toxicol Lett* 150: 135.

Mukherjee, S.P. 1980. Mediation of the antilipolytic and lipogenic effects of insulin in adipocytes by intracellular accumulation of hydrogen peroxide. *Biochem Pharmacol* 29: 1239.

Munch, G. et al. 1997. Influence of advanced glycation end products and AGE inhibitors on nucleation-dependent polymerization of β-amyloid peptide. *Biochim Biophys Acta* 1360: 17.

Murakawa, T. et al. 2006. Quantum mechanical hydrogen tunneling in bacterial copper amine oxidase reaction. *Biochem Biophys Res Commun* 342: 414.

Murata, Y. et al. 2005. Granulocyte macrophage-colony stimulating factor increases the expression of histamine and histamine receptors in monocytes/macrophages in relation to arteriosclerosis. *Arterioscler Thromb Vasc Biol* 25: 430.

Mure, M. and J.P. Klinman. 1993. Synthesis and spectroscopic characterization of model compounds for the active site cofactor in copper amine oxidases. *J Am Chem Soc* 115: 7117.

Mure, M. and J.P. Klinman. 1995a. Model studies of topaquinone-dependent amine oxidases 1. Oxidation of benzylamine by topaquinone analogs. *J Am Chem Soc* 117: 8698.

Mure, M. and J.P. Klinman. 1995b. Model studies of topaquinone-dependent amine oxidases 2. Characterization of reaction intermediates and mechanism. *J Am Chem Soc* 117: 8707.

Mure, M., S.A. Mills, and J.P. Klinman. 2002. Catalytic mechanism of the topaquinone-containing copper amine oxidases. *Biochemistry* 41: 9269.

Mure, M., S.X. Wang, and J.P. Klinman. 2003. Synthesis and characterization of model compounds of the lysyl tyrosine quinone (LTQ) cofactor of lysyl oxidase. *J Am Chem Soc* 125: 6113.

Mure, M. 2004. Tyrosine-derived quinone cofactors. *Acc Chem Res* 37: 131.

Mure, M. et al. 2005a. Active site rearrangement of the 2-hydrazinopyridine adduct in *Escherichia coli* amine oxidase to an azo copper (II) chelate form: a key role for tyrosine 369 in controlling the mobility of the TPQ-2HP adduct. *Biochemistry* 44: 1583.

Mure, M. et al. 2005b. Role of interactions between the active site base and the substrate schiff base in amine oxidase catalysis: evidence from structural and spectroscopic studies of the 2-hydrazinopyridine adduct of *Escherichia coli* amine oxidase. *Biochemistry* 44: 1568.

Murooka, Y., H. Azakami, and M. Yamashita. 1996. The monoamine regulon including syntheses of arylsulfatase and monoamine oxidase in bacteria. *Biosci Biotechnol Biochem* 60: 935.

Murray, J.M. et al. 1999. The active site base controls cofactor reactivity in *Escherichia coli* amine oxidase: x-ray crystallographic studies with mutational variants. *Biochemistry* 38: 8217.

Murray, J.M. et al. 2001. Conserved tyrosine-369 in the active site of *Escherichia coli* copper amine oxidase is not essential. *Biochemistry* 40: 12808.

Nagan, N., P.S. Callery, and H.M. Kagan. 1998. Aminoalkylaziridines as substrates and inhibitors of lysyl oxidase: specific inactivation of the enzyme by N-(5-aminopentyl)aziridine. *Front Biosci* 3: A23.

Nakamura, N. et al. 1996. Biosynthesis of topaquinone cofactor in bacterial amine oxidases. Solvent origin of C2 oxygen determined by Raman spectroscopy. *J Biol Chem* 271: 4718.

Narang, D. et al. 2008. On the hormetic behaviour of drugs binding to different redox states of amine oxidase enzymes. *Am J Pharmacol Toxicol* 3: 122.

Nelson, D.R. and A.K. Huggins. 1974. Interference of 5-hydroxytryptamine in the assay of glucose by glucose oxidase: peroxidase: chromogen based methods. *Anal Biochem* 59: 46.

Nemcsik, J. et al. 2007. Alteration of serum semicarbazide-sensitive amine oxidase activity in chronic renal failure. *J Neural Transm* 114: 841.

Niculescu, M. et al. 2000a. Redox hydrogel-based amperometric bienzyme electrodes for fish freshness monitoring. *Anal Chem* 72: 1591.

Niculescu, M. et al. 2000b. Electro-oxidation mechanism of biogenic amines at amine oxidase modified graphite electrode. *Anal Chem* 72: 5988.

Niculescu, M. et al. 2000c. Amine oxidase based amperometric biosensors for histamine detection. *Electroanalysis* 12: 369.

Niculescu, M. et al. 2001. Detection of histamine and other biogenic amines using biosensors based on amine oxidase. *Inflamm Res* 50: 146.

Nilsson, S.E., N. Tryding, and G. Tufvesson. 1968. Serum monoamine oxidase (MAO) in diabetes mellitus and some other internal diseases. *Acta Med Scand* 184: 105.

Nitta, T., K. Igarashi, and N. Yamamoto. 2002. Polyamine depletion induces apoptosis through mitochondria-mediated pathway. *Exp Cell Res* 276: 120.

Noda, K. et al. 2008. Inhibition of vascular adhesion protein-1 suppresses endotoxin-induced uveitis. *FASEB J* 22: 1094.

Nordquist, J.E., C. Gokturk, and L. Oreland. 2002. Semicarbazide-sensitive amine oxidase (SSAO) gene expression in alloxan-induced diabetes in mice. *Mol Med* 8: 824.

Nordqvist, A., L. Oreland, and C.J. Fowler. 1982. Some properties of monoamine oxidase and a semicarbazide sensitive amine oxidase capable of the deamination of 5-hydroxytryptamine from porcine dental pulp. *Biochem Pharmacol* 31: 2739.

Nori, M. et al. 2003. Ebastine inhibits T cell migration, production of Th2-type cytokines and proinflammatory cytokines. *Clin Exp Allergy* 33: 1544.

Novotny, W.F. et al. 1994. Diamine oxidase is the amiloride-binding protein and is inhibited by amiloride analogues. *J Biol Chem* 269: 9921.

Obata, T. and Y. Yamanaka. 2000a. Evidence for existence of immobilization stress-inducible semicarbazide-sensitive amine oxidase inhibitor in rat brain cytosol. *Neurosci Lett* 296: 58.

Obata, T. and Y. Yamanaka. 2000b. Inhibition of monkey brain semicarbazide-sensitive amine oxidase by various antidepressants. *Neurosci Lett* 286: 131.

Obata, T. 2006. Diabetes and semicarbazide-sensitive amine oxidase (SSAO) activity: a review. *Life Sci* 79: 417.

Ochiai, Y. et al. 2005. Molecular cloning and characterization of rat semicarbazide-sensitive amine oxidase. *Biol Pharm Bull* 28: 413.

Ochiai, Y. et al. 2006. Substrate selectivity of monoamine oxidase A, monoamine oxidase B, diamine oxidase, and semicarbazide-sensitive amine oxidase in COS-1 expression systems. *Biol Pharm Bull* 29: 2362.

Ohashi, M. et al. 1994. Oxygen-sensor-based simple assay of histamine in fish using purified amine oxidase. *J Food Sci* 59: 519.

Okajima, T. et al. 2005. Re-investigation of metal ion specificity for quinone cofactor biogenesis in bacterial copper amine oxidase. *Biochemistry* 44: 12041.

Okeley, N.M. and W.A. van der Donk. 2001. Novel cofactors via post-translational modifications of enzyme active sites. *Chem Biol* 7: 159.

Okuyama, T. and Y. Kobayashi. 1961. Determination of diamine oxidase activity by liquid scintillation counting. *Arch Biochem Biophys* 95: 242.

Olivieri, A., K.F. Tipton, and J. O'Sullivan. 2007. L-lysine as a recognition molecule for the VAP-1 function of SSAO. *J Neural Transm* 114: 747.

Omar, B.A. et al. 1990. Cardioprotection by Cu,Zn-superoxide dismutase is lost at high doses in the reoxygenated heart. *Free Radical Biol Med* 9: 465.

O'Brien, P.J., A.G. Siraki, and N. Shangari. 2005. Aldehyde sources, metabolism, molecular toxicity mechanisms, and possible effects on human health. *Crit Rev Toxicol* 35: 609.

O'Connell, K.M. et al. 2004. Differential inhibition of six copper amine oxidases by a family of 4-(aryloxy)-2-butynamines: evidence for a new mode of inactivation. *Biochemistry* 43: 10965.

O'Rourke, A.M. et al. 2007. Benefit of inhibiting SSAO in relapsing experimental autoimmune encephalomyelitis. *J Neural Transm.* 114: 845.

O'Rourke, A.M. et al. 2008. Anti-inflammatory effects of LJP 1586 [Z-3-fluoro-2-(4-methoxybenzyl)allylamine hydrochloride], an amine-based inhibitor of semicarbazide-sensitive amine oxidase activity. *J Pharmacol Exp Ther* 324: 867.

O'Sullivan, J., K.F. Tipton, and W.E. McDevitt. 2002. Immunolocalization of semicarbazide-sensitive amine oxidase in human dental pulp and its activity toward serotonin. *Arch Oral Biol* 47: 399.

O'Sullivan, J. et al. 2003a. The inhibition of semicarbazide-sensitive amine oxidase by aminohexoses. *Biochim Biophys Acta* 1647: 367.

O'Sullivan, J. et al. 2003b. Semicarbazide-sensitive amine oxidases in pig dental pulp. *Biochim Biophys Acta* 1647: 333.

O'Sullivan, J. et al. 2004. Semicarbazide-sensitive amine oxidases: enzymes with quite a lot to do. *Neurotoxicology* 25: 303.

O'Sullivan, J. et al. 2006. Inhibition of amine oxidases by the histamine-1 receptor antagonist hydroxyzine. *J Neural Transm.* Suppl. 71: 105.

O'Sullivan, J. et al. 2007. Hydrogen peroxide derived from amine oxidation mediates the interaction between amino sugars and semicarbazide-sensitive amine oxidase. *J Neural Transm* 114: 751.

Ozaita, A. et al. 1997. Inhibition of monoamine oxidase A and B activities by imidazol(ine)/guanidine drugs, nature of the interaction and distinction from I2-imidazoline receptors in rat liver. *Br J Pharmacol* 121: 901.

Padiglia, A., R. Medda, and G. Floris. 1992. Lentil seedling amine oxidase: interaction with carbonyl reagents. *Biochem Int* 28: 1097.

Padiglia, A. et al. 1998a. Characterization of *Euphorbia characias* latex amine oxidase. *Plant Physiol* 117: 1363.

Padiglia, A. et al. 1998b. Inhibitors of plant copper amine oxidases. *J Enzym Inhib* 13: 311.

Padiglia, A. et al. 1999a. Effect of metal substitution in copper amine oxidase from lentil seedlings. *J Biol Inorg Chem* 4: 608.

Padiglia, A. et al. 1999b. Interaction of pig kidney and lentil seedling copper-containing amine oxidases with guanidinium compounds. *J Enzym Inhib* 15: 91.

Padiglia, A. et al. 2001a. The reductive and oxidative half-reactions and the role of copper ions in plant and mammalian copper-amine oxidases. *Eur J Inorg Chem* 1: 35.

Padiglia, A. et al. 2001b. Irreversible inhibition of pig kidney copper-containing amine oxidase by sodium and lithium ions. *Eur J Biochem* 268: 4686.

Padiglia, A. et al. 2004. Inhibition of lentil copper/TPQ amine oxidase by the mechanism-based inhibitor derived from tyramine. *Biol Chem* 385: 323.

Page, R.D.M. 1996. TreeView: an application to display phylogenetic trees on personal computers. *Comput Appl Biosci* 12: 357.

Palcic, M. and J.P. Klinman. 1983. Isotopic probes yield microscopic constants: separation of binding energy from catalytic efficiency in the bovine plasma amine oxidase reaction. *Biochemistry* 22: 5957.

Palfreyman, M.G. et al. 1986 The rational design of suicide substrates of amine oxidases. *Biochem Soc Trans* 14: 410.

Palfreyman, M.G. et al. 1994. Haloallylamine inhibitors of MAO and SSAO and their therapeutic potential. *J Neural Transm* Suppl. 41: 407.

Parchment, R.E. et al. 1990. Serum amine oxidase activity contributes to crisis in mouse embryo cell lines. *Proc Natl Acad Sci USA* 87: 4340.

Parsons, M.R. et al. 1995. Crystal structure of a quinoenzyme: copper amine oxidase of *Escherichia coli* at 2 Å resolution. *Structure* 3: 1171.

Paschalidis, K.A. and K.A. Roubelakis-Angelakis. 2005. Sites and regulation of polyamine catabolism in the tobacco plant: correlations with cell division/expansion, cell cycle progression, and vascular development. *Plant Physiol* 138: 2174.

Payne, S.L., M.J.C. Hendrix, and D.A. Kirschmann. 2007. Paradoxical roles for lysyl oxidases in cancer: a prospect. *J Cell Biochem* 101: 1338.

Paz, M.A. et al. 1991. Specific detection of quinoproteins by redox-cycling staining. *J Biol Chem* 266: 689.

Pearce, F.L. 1991. Biological effects of histamine: an overview. *Agents Actions* 33: 4.

Peč, P. and L. Macholán. 1976. Inhibition of diamine oxidase by cinchona alkaloids and lobeline. *Collect Czech Chem Commun* 41: 3474.

Peč, P. 1985. Inhibition of pea diamine oxidase by alkaloids with piperidine skeleton and by piperidine derivatives. *Biológia (Bratislava)* 40: 1209.

Peč, P. and E. Hlídková. 1987. Inhibition of pea amine oxidase by some derivatives of 4,5-dihydroimidazole. *Acta Univ Palacky* 88: 199.

Peč, P. and A. Haviger. 1988. Inhibition of diamine oxidase from pea cotyledons by hydrazides and azide. *Acta Univ Palacky* 88: 245.

Peč, P. and T. Šimůnková. 1988. Interaction of diamine oxidase with some alicyclic amines. *Acta Univ Palacki Olomuc Fac Rer Nat* 91: 235.

Peč, P., J. Chudý, and L. Macholán. 1991. Determination of the activity of diamine oxidase from pea with Z-1,4-diamino-2-butene as a substrate. *Biológia (Bratislava)* 46: 665.

Peč, P. and I. Frébort. 1992a. Some amines as inhibitors of pea diamine oxidase. *J Enzym Inhib* 5: 323.

Peč, P. and I. Frébort. 1992b. 1,4-Diamino-2-butyne as the mechanism-based pea diamine oxidase inhibitor. *Eur J Biochem* 209: 661.

Peč, P. et al. 1992a. Time dependent inhibitions of pea cotyledons diamine oxidase by some hydrazides. *J Enz Inhib* 6: 243.

Peč, P., A. Haviger, and I. Frébort. 1992b. Determination of the dissociation constants of pea diamine oxidase. *Biochem Int* 26: 87.

Pegg, A.E. and H. Hibasami. 1980. Polyamine metabolism during cardiac hypertrophy. *Am J Physiol* 239: E372.

Pel, H.J. et al. 2007. Genome sequencing and analysis of the versatile cell factory *Aspergillus niger* CBS 513.88. *Nat Biotechnol* 25: 221.

Penning, L.C. et al. 1998. Sensitization of TNF-induced apoptosis with polyamine synthesis inhibitors in different human and murine tumor cell lines. *Cytokine* 10: 423.

Peppler, H.J., and G. Reed. 1987. Enzymes in food and feed processing. *Enzyme Technol* 7: 547.

Perin, A., A. Sessa, and M.A. Desiderio. 1985. Diamine oxidase in regenerating and hypertrophic tissues. In *Structure and Functions of Amine Oxidases*, Mondovì, B., Ed. Boca Raton: CRC Press, p. 179.

Petřivalský, M. et al. 2007. Aminoaldehyde dehydrogenase activity during wound healing of mechanically injured pea seedlings. *J Plant Physiol* 164: 1410.

Pettersson, G.A. 1985. Plasma amine oxidase. In *Structure and Functions of Amine Oxidases*, Mondovì, B., Ed. Boca Raton: CRC Press, p. 105.

Pietrangeli, P. et al. 2000. Modulation of bovine serum amine oxidase activity by hydrogen peroxide. *Biochem Biophys Res Commun* 267: 174.

Pietrangeli, P. et al. 2004. Inactivation of copper-containing amine oxidases by turnover products. *Eur J Biochem* 271: 146.

Pietrangeli, P. and B. Mondovì. 2004. Amine oxidases and tumors. *Neurotoxicology* 25: 317.

Pietrangeli, P. et al. 2007. Substrate specificity of copper-containing plant amine oxidases. *J Inorg Biochem* 101: 997.

Pinnell, S.R. and G.R. Martin. 1968. The cross-linking of collagen and elastin: enzymatic conversion of lysine in peptide linkage to α -aminoadipic- δ -semialdehyde (allysine) by an extract from bone. *Proc Natl Acad Sci USA* 61: 708.

Pirisino, R. et al. 2001. Methylamine and benzylamine induced hypophagia in mice: modulation by semicarbazide-sensitive benzylamine oxidase inhibitors and aODN towards Kv1.1 channels. *Br J Pharmaco*l 134: 880.

Pirisino, R. et al. 2004. Methylamine, but not ammonia, is hypophagic in mouse by interaction with brain Kv1.6 channel subtype. *Br J Pharmacol* 142: 381.

Plastino, J. et al. 1999. Unexpected role for the active site base in cofactor orientation and flexibility in the copper amine oxidase from *Hansenula polymorpha*. *Biochemistry* 38: 8204.

Poli, E., C. Pozzoli, and G. Coruzzi. 2001. Role of histamine H(3) receptors in the control of gastrointestinal motility: an overview. *J Physiologie* 95: 67.

Poulin, R. et al. 1993. Enhancement of the spermidine uptake system and lethal effects of spermidine overaccumulation in ornithine decarboxylase-overproducing L1210 cells under hyposmotic stress. *J Biol Chem* 268: 4690.

Poulin, R., G. Pelletier, and A.E. Pegg. 1995. Induction of apoptosis by excessive polyamine accumulation in ornithine decarboxylase-overproducing L1210 cells. *Biochem J* 311: 723.

Prabhakaram, M. and B.J. Ortwerth. 1994. Determination of glycation crosslinking by the sugar dependent incorporation of [14C]lysine into protein. *Anal Biochem* 216: 305.

Precious, E. and G.A. Lyles. 1988. Properties of a semicarbazide-sensitive amine oxidase in human umbilical artery. *J Pharm Pharmacol* 40: 627.

Precious, E., C.E. Gunn, and G.A. Lyles. 1988. Deamination of methylamine by semicarbazide-sensitive amine oxidase in human umbilical artery and rat aorta. *Biochem Pharmacol* 37: 707.

Premkumar, A. and R. Simantov. 2002. Mitochondrial voltage-dependent anion channel is involved in dopamine-induced apoptosis. *J Neurochem* 82: 345.

Prévot, D. et al. 2007. Prolonged treatment with aminoguanidine strongly inhibits adipocyte semicarbazide-sensitive amine oxidase and slightly reduces fat deposition in obese Zucker rats. *Pharmacol Res* 56: 70.

Qiao, C. 2004. New copper amine oxidase probes: synthesis, mechanism, and enzymology. Ph. D. thesis, Case Western Reserve University, Cleveland, OH, August 2004.

Qiao, C., H.B. Jeon, and L.M. Sayre. 2004. Selective inhibition of bovine plasma amine oxidase by homopropargylamine, a new inactivator motif. *J Am Chem Soc* 126: 8038.

Qiao, C. et al. 2006. Mechanism-based cofactor derivatization of a copper amine oxidase by a branched primary amine recruits the oxidase activity of the enzyme to turn inactivator into substrate. *J Am Chem Soc* 128: 6206.

Raasch, W. et al. 1995. Agmatine, the bacterial amine, is widely distributed in mammalian tissues. *Life Sci* 56: 2319.

Raddatz, R., A. Parini, and S.M. Lanier. 1997. Imidazoline/guanidinium binding domains on monoamine oxidases: relationship to subtypes of imidazoline-binding proteins and tissue-specific interaction of imidazoline ligands with monoamine oxidase B. *Mol Pharmacol* 52: 549.

Raimondi, L. et al. 1990. Cultured preadipocytes produce a semicarbazide-sensitive amine oxidase (SSAO) activity. *J Neural Transm* 32: 331.

Raimondi, L. et al. 1991. Semicarbazide-sensitive amine oxidase activity (SSAO) of rat epididymal white adipose tissue. *Biochem Pharmacol* 41: 467.

Raimondi, L. et al. 1993. Histamine lipolytic activity and semicarbazide-sensitive amine oxidase (SSAO) of rat white adipose tissue (WAT). *Biochem Pharmacol* 46: 1369.

Raimondi, L. et al. 2007. Methylamine-dependent release of nitric oxide and dopamine in the CNS modulates food intake in fasting rats. *Br J Pharmacol* 150: 1003.

Raingeaud, J. et al. 1995. Pro-inflammatory cytokines and environmental stress cause p38 mitogen-activated protein kinase activation by dual phosphorylation on tyrosine and threonine. *J Biol Chem* 270: 7420.

Raithel, M. et al. 1995. Mucosal histamine content and histamine secretion in Crohn's disease, ulcerative colitis and allergic enteropathy. *Int Arch Allergy Immunol* 108: 127.

Rando, R.R. and J. De Mairena. 1974. Propargyl amine-induced irreversible inhibition of non-flavin-linked amine. *Biochem Pharmacol* 23: 463.

Ray, S. and M. Ray. 1983. Formation of methylglyoxal from aminoacetone by amine oxidase from goat plasma. *J Biol Chem* 258: 3461.

Rea, G. et al. 1998. Developmentally and wound-regulated expression of the gene encoding a cell wall copper amine oxidase in chickpea seedlings. *FEBS Lett* 437: 177.

Rea, G. et al. 2002. Copper amine oxidase expression in defense responses to wounding and *Ascochyta rabiei* invasion. *Plant Physiol* 128: 865.

Reed, J.C. 2002. Apoptosis-based therapies. *Nat Rev Drug Discov* 1: 111.

Reese, I. et al. 2008. Diagnostic approach for suspected pseudoallergic reaction to food ingredients. *Journal der Deutschen Dermatologischen Gesellschaft* 7: 1, 70–77.

Remy, L. et al. 1998. Legume lectin structure. *Biochim Biophys Acta* 1383: 9.

Rhee, S.G. 1999. Redox signaling: hydrogen peroxide as intracellular messenger. *Exp Mol Med* 31: 53.

Rinaldi, A., P. Vecchini, and G. Floris. 1982. Diamine oxidase from pig kidney: new purification method and amino acid composition. *Prep Biochem* 12: 11.

Rinaldi, A. et al. 1984. Lentil seedlings amine oxidase: preparation and properties of the copper-free enzyme. *Biochem Biophys Commun* 120: 242.

Rinaldi, A., G. Floris, and A. Giartosio. 1985. Plant amine oxidases. In *Structure and Functions of Amine Oxidases*, Mondovì, B., Ed. Boca Raton: CRC Press, p. 51.

Robinson-White, A. et al. 1985. Binding of diamine oxidase activity to rat and guinea pig microvascular endothelial cells: comparisons with lipoprotein lipase binding. *J Clin Invest* 76: 93.

Roessner, V. et al. 2006a. Decreased serum semicarbazide sensitive aminooxidase (SSAO) activity in patients with major depression. *Prog Neuropsychopharmacol Biol Psychiatr* 30: 906.

Roessner, V. et al. 2006b. Serum level of semicarbazide-sensitive amine oxidase in children with ADHD. *Behav Brain Funct* 2: 5; http://www.behavioralandbrainfunctions.com/content/2/1/5.

Rokkas, T. et al. 1990a. Is the intestine the sole source of heparin-stimulated plasma diamine oxidase? Acute effects of jejunectomy, ileectomy and total enterectomy. *Digestion* 46, Suppl. 2: 439.

Rokkas, T. et al. 1990b. Effect of intestinal diamine oxidase (DAO) depletion by heparin on mucosal polyamine metabolism. *Digestion* 46, Suppl. 2: 378.

Rokkas, T. et al. 1990c. Postheparin plasma diamine oxidase in health and intestinal disease. *Gastroenterology* 98: 1493.

Rossi, A., R. Petruzelli, and A. Finazzi-Agrò. 1992. cDNA-derived amino-acid sequence of lentil seedlings' amine oxidase. *FEBS Lett* 301: 253.

Roth, J.A. and C.N. Gillis. 1975. Multiple forms of amine oxidase in perfused rabbit lung. *J Pharmacol* 194: 537.

Rotilio, G. 1985. Spectroscopic and chemical properties of the amine oxidase copper. In *Structure and Functions of Amine Oxidases,* Mondovì, B., Ed. Boca Raton: CRC Press, p. 127.

Rucker, R.B. and B.L. O'Dell. 1971. Connective tissue amine oxidase I. Purification of bovine aorta amine oxidase and its comparison with plasma amine oxidase. *Biochim Biophys Acta* 235: 32.

Rucker, R.B. and W. Goettlich-Riemann. 1972. Properties of rabbit aorta amine oxidase. *Proc Soc Exp Biol Med* 139: 286.

Rucker, R.B. et al. 1996. Modulation of lysyl oxidase by dietary copper in rats. *J Nutr* 126: 51.

Rudich, A. et al. 1997. Oxidant stress reduces insulin responsiveness in 3T3-L1 adipocytes. *Am J Physiol* 272: E261.

Ruggiero, C.E. et al. 1997. Mechanistic studies of topa quinone biogenesis in phenylethylamine oxidase. *Biochemistry* 36: 1953.

Ruggiero, C.E. and D.M. Dooley. 1999. Stoichiometry of the topa quinone biogenesis reaction in copper amine oxidases. *Biochemistry* 38: 2892.

Ruoslahti, E. and M.D. Pierschbacher. 1987. New perspectives in cell adhesion: RGD and integrins. *Science* 238: 491.

Russel, D.H. 1917. Increased polyamine concentration in the urine of human cancer patients. *Nature* 233: 144.

Sagrinati, C. et al. 2006. Isolation and characterization of multipotent progenitor cells from the Bowman's capsule of adult human kidneys. *J Am Soc Nephrol* 17: 2443.

Sakata, K. et al. 2003. Acrolein produced from polyamines as one of the uraemic toxins. *Biochem Soc Trans* 31: 371.

Salmi, M. and S. Jalkanen. 1992. A 90-kilodalton endothelial cell molecule mediating lymphocyte binding in humans. *Science* 257: 1407.

Salmi, M., K. Kalimo, and S. Jalkanen. 1993. Induction and function of vascular adhesion protein-1 at sites of inflammation. *J Exp Med* 178: 2255.

Salmi, M. and S. Jalkanen. 1995. Different forms of human vascular adhesion protein-1 (VAP-1) in blood vessels *in vivo* and in cultured endothelial cells: implications for lymphocyte-endothelial cell adhesion models. *Eur J Immunol* 25: 2803.

Salmi, M. and S. Jalkanen. 1996. Human vascular adhesion protein 1 (VAP-1) is a unique sialoglycoprotein that mediates carbohydrate-dependent binding of lymphocytes to endothelial cells. *J Exp Med* 183: 569.

Salmi, M. et al. 1997. Vascular adhesion protein 1 (VAP-1) mediates lymphocyte subtype-specific, selectin-independent recognition of vascular endothelium in human lymph nodes. *J Exp Med* 186: 589.

Salmi, M., S. Tohka, and S. Jalkanen. 2000. Human vascular adhesion protein-1 (VAP-1) plays a critical role in lymphocyte-endothelial cell adhesion cascade under shear. *Circ Res* 86: 1245.

Salmi, M. and S. Jalkanen. 2001. VAP-1: an adhesin and an enzyme. *Trends Immunol* 22: 211.

Salmi, M. et al. 2001. A cell surface amine oxidase directly controls lymphocyte migration. *Immunity* 14: 265.

Salmi, M. et al. 2002. Insulin-regulated increase of soluble vascular adhesion protein-1 in diabetes. *Am J Pathol* 161: 2255.

Salmi, M. and S. Jalkanen. 2005. Cell-surface enzymes in control of leukocyte trafficking. *Nat Rev Immunol* 5: 760.

Salmi, M. and S. Jalkanen. 2006. Developmental regulation of the adhesive and enzymatic activity of vascular adhesion protein-1 (VAP-1) in humans. *Blood* 108: 1555.

Salminen, A.T. et al. 1998. Structural model of the catalytic domain of an enzyme with cell adhesion activity: human vascular adhesion protein-1 (HVAP-1) D4 domain is an amine oxidase. *Prot Eng* 11: 1195.

Salter-Cid, L.M. et al. 2005. Anti-inflammatory effects of inhibiting the amine oxidase activity of semicarbazide-sensitive amine oxidase. *J Pharmacol Exp Ther* 315: 553.

Samuels, N.M. and J.P. Klinman. 2005. 2,4,5-Trihydroxyphenylalanine quinone biogenesis in the copper amine oxidase from *Hansenula polymorpha* with the alternate metal nickel. *Biochemistry* 44: 14308.

Sander, L.E. et al. 2006. Selective expression of histamine receptors H1R, H2R and H4R, but not H3R, in the human intestinal tract. *Gut* 55: 498.

Sattler, J. et al. 1985. Inhibition of human and canine diamine oxidase by drugs used in an intensive care unit: relevance for clinical side effects? *Agents Actions* 16: 91.

Sattler, J. et al. 1988. Food-induced histaminosis as an epidemiological problem: plasma histamine elevation and haemodynamic alterations after oral histamine administration and blockade of diamine oxidase (DAO). *Agents Actions* 23: 361.

Sattler, J. et al. 1989. Food-induced histaminosis under diamine oxidase (DAO) blockade in pigs: further evidence of the key role of elevated plasma histamine levels as demonstrated by successful prophylaxis with antihistamines. *Agents Actions* 27: 212.

Sattler, J. and W. Lorenz. 1990. Intestinal diamine oxidases and enteral-induced histaminosis: studies on three prognostic variables in an epidemiological model. *J Neural Transm* Suppl. 32: 291.

Sava, I.G. et al. 2006. Free radical scavenging action of the natural polyamine spermine in rat liver mitochondria. *Free Radic Biol Med* 41: 1272.

Saysell, C.G. et al. 2000. Investigation into the mechanism of λ_{max} shifts and their dependence on pH for the 2-hydrazinopyridine derivatives of two copper amine oxidases. *J Mol Cat B* 8: 17.

Saysell, C.G. et al. 2002. Probing the catalytic mechanism of *E. coli* amine oxidase using mutational variants and a reversible inhibitor as a substrate analogue. *Biochem J* 365: 809.

Schafer, T. et al. 2007. Industrial enzymes. *Adv Biochem Eng Biotechnol* 105: 59.

Schayer, R.W., R.L. Smiley, and E.H. Kaplan. 1952. The metabolism of epinephrine containing isotopic carbon II. *J Biol Chem* 198: 545.

Schilling, B. and K. Lerch. 1995. Cloning, sequencing and heterologous expression of the monoamine oxidase gene from *Aspergillus niger*. *Mol Gen Genet* 247: 430.

Schipper, R.G. et al. 2000. Involvement of polyamines in apoptosis. Facts and controversies: effectors or protectors? *Sem Cancer Biol* 10: 55.

Schmidt, W.U. et al. 1990. Human intestinal diamine oxidase (DAO) activity in Crohn's disease: a new marker for disease assessment? *Agents Actions* 30: 267.

Schulze-Osthoff, K. et al. 1993. Depletion of the mitochondrial electron transport abrogates the cytotoxic and gene-inductive effects of TNF. *EMBO J* 12: 3095.

Schwartz, B. et al. 1998. Relationship between conserved consensus site residues and the productive conformation for the TPQ cofactor in a copper-containing amine oxidase from yeast. *Biochemistry* 37: 16591.

Schwartz, B., J.E. Dove, and J.P. Klinman. 2000. Kinetic analysis of oxygen utilization during cofactor biogenesis in a copper-containing amine oxidase from yeast. *Biochemistry* 39: 3699.

Schwelberger, H.G. and E. Bodner. 1997. Purification and characterization of diamine oxidase from porcine kidney and intestine. *Biochim Biophys Acta* 1340: 152.

Schwelberger, H.G., A. Hittmair, and S.D. Kohlwein. 1998. Analysis of tissue and subcellular localization of mammalian diamine oxidase by confocal laser scanning fluorescence microscopy. *Inflamm Res* 47: S60.

Schwelberger, H.G. 1999. Molecular cloning of mammalian diamine oxidase genes and cDNAs. *Inflamm Res* 48 Suppl 1: S79–80.

Schwelberger, H.G., H. Dieplinger, and S.D. Kohlwein. 1999. Diamine oxidase and catalase are expressed in the same cells but are present in different subcellular compartments in porcine kidney. *Inflamm Res* 48: S81.

Schwelberger, H.G., A. Drasche, and E. Hütter. 2000. Analysis of mammalian diamine oxidase genes. *Inflamm Res* 49: S51.

Schwelberger, H.G. 2004. Diamine oxidase (DAO) enzyme and gene. In *Histamine: Biology and Medical Aspects*, Falus, A., Ed. Budapest: SpringMed, p. 43.

Schwelberger, H.G. 2006. Origins of plasma amine oxidases in different mammalian species. *Inflamm Res* 55: S57.

Schwelberger, H.G. 2007. The origin of mammalian plasma amine oxidases. *J Neural Transm* 114: 757.

Schwelberger, H.G. and J. Feurle. 2007. Luminometric determination of amine oxidase activity. *Inflamm Res* 56: S53.

Schwelberger, H.G. and J. Feurle. 2008. Characterization of human plasma amine oxidase. *Inflamm Res* 57: S51.

Šebela, M. et al. 1997. Confirmation of the presence of a Cu(II)/topa quinone active site in the amine oxidase from fenugreek seedlings. *J Exp Bot* 48: 1897.

Šebela, M. et al. 2001. FAD-containing polyamine oxidases: a timely challenge for researchers in biochemistry and physiology of plants. *Plant Sci* 160: 197.

Šebela, M. et al. 2003. Recent news related to substrates and inhibitors of plant amine oxidases. *Biochim Biophys Acta* 1647: 355.

Šebela, M. et al. 2005. Thermostable β-cyclodextrin conjugates of two similar plant amine oxidases and their properties. *Biotechnol Appl Biochem* 41: 77.

Šebela, M. et al. 2007. Interaction of plant amine oxidases with diaminoethers. *Arkivoc* 2007: 222.

Segel, I.H. 1993. *Enzyme Kinetics: Behavior and Analysis of Rapid Equilibrium and Steady State Enzyme Systems*. New York: John Wiley & Sons, p. 100.

Seiler, N. 2000. Oxidation of polyamines and brain injury. *Neurochem Res* 25: 471.

Seiler, N. 2004. Catabolism of polyamines. *Amino Acids* 26: 217.

Selkoe, D.J. 2001. Alzheimer's disease: genes, proteins, and therapy. *Physiol Rev* 81: 741.

Sessa, A., M.A. Desiderio, and A. Perin. 1984. Stimulation of hepatic and renal diamine oxidase activity after acute ethanol administration. *Biochim Biophys Acta* 801: 285.

Shakir, K.M., S. Margolis, and S.B. Baylin. 1977. Localization of histaminase (diamine oxidase) in rat small intestinal mucosa: site of release by heparin. *Biochem Pharmacol* 26: 2343.

Shepard, E.M. et al. 2002. Toward the development of selective amine oxidase inhibitors. Mechanism-based inhibition of six copper containing amine oxidases. *Eur J Biochem* 269: 3645.

Shepard, E.M. et al. 2003. Inhibition of six copper-containing amine oxidases by the anti depressant drug tranylcypromine. *Biochim Biophys Acta* 1647: 252.

Shepard, E.M. et al. 2004. Cyanide as a copper and quinone-directed inhibitor of amine oxidases from pea seedlings (*Pisum sativum*) and *Arthrobacter globiformis*: evidence for both copper coordination and cyanohydrin derivatization of the quinone cofactor. *J Biol Inorg Chem* 9: 256.

Shepard, E.M. and D.M. Dooley. 2006. Intramolecular electron transfer rate between active-site copper and TPQ in *Arthrobacter globiformis* amine oxidase. *J Biol Inorg Chem* 11: 1039.

Shepard, E.M, K.M. Okonski, and D.M. Dooley. 2008. Kinetics and spectroscopic evidence that the Cu(I) semi-quinone intermediate reduces molecular oxygen in the oxidative half-reaction of *Arthrobacter globiformis* amine oxidase. *Biochemistry* 47: 13907.

Shih, J.C. 2004. Cloning, after cloning, knock-out mice, and physiological functions of MAO A and B. *NeuroToxicology* 25: 21.

Shimizu, S. et al. 2001. Essential role of voltage-dependent anion channel in various forms of apoptosis in mammalian cells. *J Cell Biol* 152: 237.

Shindler, J.S. and W.G. Bardsley. 1976. Human kidney diamine oxidase: inhibition studies. *Biochem Pharmacol* 25: 2689.

Sibon, I. et al. 2008. Association between semicarbazide-sensitive amine oxidase, a regulator of the glucose transporter, and elastic lamellae thinning during experimental cerebral aneurysm development: laboratory investigation. *J Neurosurg* 108: 558.

Sikkema, J.M. et al. 2002. Semicarbazide-sensitive amine oxidase in pre-eclampsia: no relation with markers of endothelial cell activation. *Clin Chim Acta* 324: 31.

Sims, A.H. et al. 2005. Transcriptome analysis of recombinant protein secretion by *Aspergillus nidulans* and the unfolded-protein response *in vivo*. *Appl Environ Microbiol* 71: 2737.

Slorach, S.A. 1991. Histamine in food. In *Histamine and Histamine Antagonists*, Uvnas, B., Ed. Berlin: Springer, p. 511.

Smith, D.J. et al. 1998. Cloning of vascular adhesion protein 1 reveals a novel multifunctional adhesion molecule. *J Exp Med* 188: 17.

Smith, M.A. et al. 1996. Trypsin interaction with the senile plaques of Alzheimer disease is mediated by beta-protein precursor. *Mol Chem Neuropathol* 27: 145.

Soda, K. and K. Yonaha. 1987. Application of free enzymes in pharmaceutical and chemical industries. In *Enzyme Technology,* Kennedy, J.F., Ed. Weinheim: VCH, p. 605.

Sohal, R.S. 1997. Mitochondria generate superoxide anion radicals and hydrogen peroxide. *FASEB J* 11: 1269.

Solé, M. et al. 2008. p53 phosphorylation is involved in vascular cell death induced by the catalytic activity of membrane-bound SSAO/VAP-1. *Biochim Biophys Acta* 1783: 1085.

Soltesz, Z. et al. 2007. Studies on the insulinomimetic effects of benzylamine, exogenous substrate of semicarbazide-sensitive amine oxidase enzyme in streptozotocin-induced diabetic rats. *J Neural Transm* 114: 851.

Somfai, G.M. et al. 2006. Soluble semicarbazide-sensitive amine oxidase (SSAO) activity is related to oxidative stress and subchronic inflammation in streptozotocin-induced diabetic rats. *Neurochem Int* 48: 746.

Souto, R.P. et al. 2003. Immunopurification and characterization of rat adipocyte caveolae suggest their dissociation from insulin signaling. *J Biol Chem* 278: 18321.

Spencer, J.P. et al. 1998. Conjugates of catecholamines with cysteine and GSH in Parkinson's disease: possible mechanisms of formation involving reactive oxygen species. *J Neurochem* 71: 2112.

Spina, M.B. and G. Cohen. 1989. Dopamine turnover and glutathione oxidation: implications for Parkinson disease. *Proc Natl Acad Sci USA* 86: 1398.

Stefanelli, C. et al. 1998. Spermine causes caspase activation in leukaemia cells. *FEBS Lett* 437: 233.

Stefanelli, C. et al. 1999. Spermine triggers the activation of caspase-3 in a cell-free model of apoptosis. *FEBS Lett* 451: 95.

Stefanelli, C. et al. 2000. Polyamines directly induce release of cytochrome c from heart mitochondria. *Biochem J* 347: 875.

Stein, J. et al. 1994. Reduced postheparin plasma diamine oxidase activity in patients with chronic renal failure. *Z Gastroenterol* 32: 236.

Steinebach, V. et al. 1996. Cloning of the maoA gene that encodes aromatic amine oxidase of *Escherichia coli* W3350 and characterization of overexpressed enzyme. *Eur J Biochem* 237: 584.

Stevanato, R. et al. 1989. Characterization of free and immobilized amine oxidases. *Biotechnol Appl Biochem* 11: 266.

Stevanato, R. et al. 1990. Spectrophotometric assay for total polyamines by immobilized amine oxidase. *Anal Chim Acta* 237: 391.

Stevanato, R. et al. 1994. Electrostatic control of oxidative deamination catalysed by bovine serum amine oxidase. *Biochem J* 299: 317.

Stevanato, R., F. Vianello, and A. Rigo. 1995. Thermodynamic analysis of the oxidative deamination of polyamines by bovine serum amine oxidase. *Arch Biochem Biophys* 324: 374.

Stolen, C.M. et al. 2004a. Semicarbazide sensitive amine oxidase overexpression has dual consequences: insulin mimicry and diabetes-like complications. *FASEB J* 18: 702.

Stolen, C.M. et al. 2004b. Origins of serum semicarbazide-sensitive amine oxidase. *Circ Res* 95: 50.

Stolen, C.M. et al. 2005. Absence of the endothelial oxidase AOC3 leads to abnormal leukocyte traffic in vivo. *Immunity* 22: 105.

Stránská, J. et al. 2007. Inhibition of plant amine oxidases by a novel series of diamine derivatives. *Biochimie* 89: 135.

Strolin Benedetti, M. 2001. Biotransformation of xenobiotics by amine oxidases. *Fundam Clin Pharmacol* 15: 75.

Strolin Benedetti, M., K.F. Tipton, and R. Whomsley. 2007. Amine oxidases and monooxygenases in the *in vivo* metabolism of xenobiotic amines in humans: has the involvement of amine oxidases been neglected? *Fundam Clin Pharmacol* 21: 467.

Su, Q.J. and J.P. Klinman. 1998. Probing the mechanism of proton coupled electron transfer to dioxygen: the oxidative half-reaction of bovine serum amine oxidase. *Biochemistry* 37: 12513.

Suda, S.A., K. Dolmer, and P.G. Gettins. 1997. Critical role of asparagine 1065 of human α-2-macroglobulin in formation and reactivity of the thiol ester. *J Biol Chem* 272: 31107.

Sugino, H. et al. 1991. Gene cloning of the maoA gene and overproduction of a soluble monoamine oxidase from *Klebsiella aerogenes*. *Appl Microbiol Biotechnol* 35: 606.

Susin, S.A., N. Zamzami, and G. Kroemer. 1998. Mitochondria as regulators of apoptosis: doubt no more. *Biochim Biophys Acta* 1366: 151.

Suzuki, S. et al. 1980. Spectroscopic aspects of copper binding site in bovine serum amine oxidase. *FEBS Lett* 116: 17.

Suzuki, S. et al. 1981. Preparation and characterization of Cobalt(II)-substituted bovine serum amine oxidase. *J Biochem* 90: 905.

Suzuki, S. et al. 1982. A new enzymatic method for quantitation of spermine in human semen. *Z Rechtsmed* 88: 67.

Suzuki, S. et al. 1983. Effect of metal substitution on the chromophore of bovine serum amine oxidase. *Biochemistry* 22: 1630.

Suzuki, S. et al. 1984. Roles of two copper ions in bovine serum amine oxidase. *Biochemistry* 25: 338.

Svenson, A. and P.A. Hynning. 1981. Preparation of amine oxidase from bovine serum by affinity chromatography on aminohexyl–Sepharose. *Prep Biochem* 11: 99.

Tabata, A. et al. 2003. Further evidence for suicide inhibition of semicarbazide-sensitive amine oxidase in guinea pig lung by 2-bromoethylamine. *Methods Fund Exp Clin Pharmacol* 25: 785.

Tabor, C.W., H. Tabor, and S.M. Rosenthal. 1954. Purification of amine oxidase from beef plasma. *J Biol Chem* 208: 645.

Tabor, H. and C.W. Tabor. 1964. Spermidine, spermine and related amines. *Pharmacol Rev* 16: 245.

Takabe, W. et al. 2001. Oxidative stress promotes the development of transformation: involvement of a potent mutagenic lipid peroxidation product, acrolein. *Carcinogenesis* 22: 935.

Tam, C.F. et al. 1979. Diamine oxidase activity in plasma and urine in uremia. *Nephron* 23: 23.

Tamura, H. et al. 1989. Kinetic studies on the inhibition mechanism of diamine oxidase from porcine kidney by aminoguanidine. *J Biochem* 105: 299.

Tamura, B.K., K.H. Masaki, and P. Blanchette. 2007. Weight loss in patients with Alzheimer's disease. *J Nutr Elder* 26: 21.

Tanaka,Y. et al. 2003. Clinical significance of plasma diamine oxidase activity in pediatric patients: influence of nutritional therapy and chemotherapy. *Kurume Med J* 50: 131.

Tang, S.S., P.C. Trackman, and H.M. Kagan. 1983. Reaction of aortic lysyl oxidase with β-aminopropionitrile. *J Biol Chem* 258: 4331.

Tang, S.S., D.E. Simpson, and H.M. Kagan. 1984. β-substituted ethylamine derivatives as suicide inhibitors of lysyl oxidase. *J Biol Chem* 259: 975.

Tang, S.S., C.O. Chichester, and H.M. Kagan. 1989. Comparative sensitivities of purified preparations of lysyl oxidase and other amine oxidases to active site-directed enzyme inhibitors. *Connect Tissue Res* 19: 933.

Tanizawa, K. et al. 1994. Cloning and sequencing of phenylethylamine oxidase from *Arthrobacter globiformis* and implication of Tyr 382 as the precursor to its covalently bound quinone cofactor. *Biochem Biophys Res Commun* 199: 1096.

Tanizawa, K. 1995. Biogenesis of novel quinone coenzymes. *J Biochem* 118: 671.

Taylor, S.L. 1986. Histamine food poisoning: toxicology and clinical aspects. *Crit Rev Toxicol* 17: 91.

Thomas, T. et al. 1999. Polyamine biosynthesis inhibitors alter protein-protein interactions involving estrogen receptor in MCF-7 breast cancer cells. *J Mol Endocrinol* 22: 131.

Thompson, J.S. et al. 1988a. Intestinal mucosa diamine oxidase activity reflects intestinal involvement in Crohn's disease. *Am J Gastroenterol* 83: 756.

Thompson, J.S. et al. 1988b. Plasma postheparin diamine oxidase activity. Development of a simple technique of assessing Crohn's disease. *Dis Colon Rectum* 31: 529.

Thompson, J.S., D.A. Burnett, and W.P. Vaughan. 1991. Factors affecting plasma postheparin diamine oxidase activity. *Dig Dis Sci* 36: 1582.

Thurmond, R.L. et al. 2004. A potent and selective histamine H4 receptor antagonist with anti-inflammatory properties. *J Pharmacol Exp Ther* 309: 404.

Tibbles, L.A. et al. 1996. MLK-3 activates the SAPK/JNK and p38/RK pathways via SEK1 and MKK3/6. *EMBO J* 15: 7026.

Tipping, A.J. and M.J. McPherson. 1995. Cloning and molecular analysis of the pea seedling copper amine oxidase. *J Biol Chem* 270: 16939.

Tipton, K.F., G. Davey, and M. Motherway. 2000. Monoamine oxidase assays. In *Current Protocols in Pharmacology*, Enna, S.J. et al., Eds. New York: John Wiley & Sons, p. 3.6.1.

Tipton, K.F. et al. 2003. It can be a complicated life being an enzyme. *Biochem Soc Trans* 31: 711.

Tohka, S. et al. 2001. Vascular adhesion protein 1 (VAP-1) functions as a molecular brake during granulocyte rolling and mediates their recruitment *in vivo*. *FASEB J* 15: 373.

Tombelli, S. and M. Mascini. 1998. Electrochemical biosensors for biogenic amines: a comparison between different approaches. *Anal Chim Acta* 358: 277.

Toninello, A. et al. 2000. Kinetics and free energy profiles of spermine transport in liver mitochondria. *Biochemistry* 39: 324.

Toninello, A., M. Salvi, and B. Mondovì. 2004. Interaction of biologically active amines with mitochondria and their role in the mitochondrial-mediated pathway of apoptosis. *Curr Med Chem* 11: 2349.

Toninello, A. et al. 2006. Amine oxidases in apoptosis and cancer. *Biochim Biophys Acta Rev Cancer* 1765: 1.

Toraya, T. et al. 1976. Affinity chromatography of amine oxidase from *Aspergillus Niger. Biochim Biophys Acta* 420: 316.

Torrigiani, P., D. Serafini-Fracassini, and A. Fara. 1989. Diamine oxidase activity in different physiological stages of *Helianthus tuberosus* tuber. *Plant Physiol* 89: 69.

Toul, Z. and L. Macholan. 1975. Enzyme electrode for rapid determination of biogenic polyamines. *Collect Czech Chem Commun* 40: 2208.

Trackman, P.C. et al. 1992. Post-translational glycosylation and proteolytic processing of a lysyl oxidase precursor. *J Biol Chem* 267: 8666.

Tressel, T. et al. 1986. Interaction between L-threonine dehydrogenase and aminoacetone synthetase and mechanism of aminoacetone production. *J Biol Chem* 261: 16428.

Tryding, N. et al. 1969. Physiological and pathological influences on serum monoamine oxidase level. *Scand J Clin Lab Invest* 23: 79.

Tsujikawa, T. et al. 1999. Changes in serum diamine oxidase activity during chemotherapy in patients with hematological malignancies. *Cancer Lett* 147: 195.

Tsunooka, N. et al. 2004. Bacterial translocation secondary to small intestinal mucosal ischemia during cardiopulmonary bypass: measurement by diamine oxidase and peptidoglycan. *Eur J Cardiothorac Surg* 25: 275.

Turini, P. et al. 1982. Purification of bovine plasma amine oxidase. *Anal Biochem* 125: 294.

Turowski, P.N., M.A. McGuirl, and D.M. Dooley. 1993. Intramolecular electron transfer rate between active-site copper and topa quinone in pea seedling amine oxidase. *J Biol Chem* 268: 17680.

Ucar, G. et al. 2005. Elevated semicarbazide-sensitive amine oxidase (SSAO) activity in lung with ischemia-reperfusion injury: protective effect of ischemic preconditioning plus SSAO inhibition. *Life Sci* 78: 421.

Uchida, M. et al. 2003. Stereochemistry of 2-phenylethylamine oxidation catalyzed by bacterial copper amine oxidase. *Biol Biotech Biochem* 67: 2664.

Unzeta, M. et al. 2007. Semicarbazide-sensitive amine oxidase (SSAO) and its possible contribution to vascular damage in Alzheimer's disease. *J Neural Transm* 114: 857.

Uragoda, C.G. and S.R. Kottegoda. 1977. Adverse reactions to isoniazid on ingestion of fish with a high histamine content. *Tubercle* 58: 83.

Uragoda, C.G. and S.C. Lodha. 1979. Histamine intoxication in a tuberculous patient after ingestion of cheese. *Tubercle* 60: 59.

Uzbekov, M.G. et al. 2006. Biochemical profile in patients with anxious depression under the treatment with serotonergic antidepressants with different mechanisms of action. *Hum Psychopharmacol Clin Exp* 21: 109.

Vainio, P.J. et al. 2005. Safety of blocking vascular adhesion protein-1 in patients with contact dermatitis. *Basic Clin Pharmacol Toxicol* 96: 429.

Valen, G. et al. 1996. Activity of histamine metabolizing and catabolizing enzymes during reperfusion of isolated, globally ischemic rat hearts. *Inflamm Res* 45: 145.

Van der Kaay, J. et al. 1999. Distinct phosphatidylinositol 3-kinase lipid products accumulate upon oxidative and osmotic stress and lead to different cellular responses. *J Biol Chem* 274: 35963.

van Dijk, J. et al. 1995. Determination of semicarbazide-sensitive amine oxidase activity in human plasma by high-performance liquid chromatography with fluorimetric detection. *J Chromatogr B Biomed Appl* 663: 43.

Varga, C. et al. 2005. Inhibitory effects of histamine H4 receptor antagonists on experimental colitis in the rat. *Eur J Pharmacol* 522: 130.

Vercesi, A.E. et al. 1997. The role of reactive oxygen species in mitochondrial permeability transition. *Biosci Rep* 17: 43.

Vergona, R.A. et al. 1987. Protective effects of hydralazine in a renal ischemia model in the rat. *Life Sci* 41: 563.

Vermillion, D.L. et al. 1993. Altered small intestinal smooth muscle function in Crohn's disease. *Gastroenterology* 104: 1692.

Vestweber, D. 2007. Adhesion and signaling molecules controlling the transmigration of leukocytes through endothelium. *Immunol Rev* 218: 178.

Vianello, F. et al. 1992. Isolation of amine oxidase from bovine plasma by a two-step procedure. *Protein Exp Purif* 3: 362.

Vianello, F. et al. 1993. Purification and characterization of amine oxidase from soybean seedlings. *Arch Biochem Biophys* 307: 35.

Vianello, F. et al. 1999. Purification and characterization of amine oxidase from pea seedlings. *Protein Exp Purif* 15: 196.

Vidrio, H. et al. 2000. Enhancement of hydralazine hypotension by low doses of isoniazid. Possible role of semicarbazide-sensitive amine oxidase inhibition. *Gen Pharmacol* 35: 195.

Vidrio, H. 2003. Semicarbazide-sensitive amine oxidase: role in the vasculature and vasodilation after *in situ* inhibition. *Auton Autacoid Pharmacol* 23: 275.

Vidrio, H. and M. Medina. 2005. 2-Bromoethylamine, a suicide inhibitor of semicarbazide-sensitive amine oxidase, increases hydralazine hypotension in rats. *J Cardiovasc Pharmacol* 46: 316.

Vidrio, H. and M. Medina. 2007. Hypotensive effect of hydroxylamine, an endogenous nitric oxide donor and SSAO inhibitor. *J Neural Transm* 114: 863.

Vinters, H.V., Z.Z. Wang, and D.L. Secor. 1996. Brain parenchymal and microvascular amyloid in Alzheimer's disease. *Brain Pathol* 6: 179.

Visentin, V. et al. 2003. Inhibition of rat fat cell lipolysis by monoamine oxidase and semicarbazide-sensitive amine oxidase substrates. *Eur J Pharmacol* 466: 235.

Visentin, V. et al. 2004. Alteration of amine oxidase activity in the adipose tissue of obese subjects. *Obes Res* 12: 547.

Visentin, V. et al. 2005a. Influence of high-fat diet on amine oxidase activity in white adipose tissue of mice prone or resistant to diet-induced obesity. *J Physiol Biochem* 61: 343.

Visentin, V. et al. 2005b. Glucose handling in streptozotocin-induced diabetic rats is improved by tyramine but not by the amine oxidase inhibitor semicarbazide. *Eur J Pharmacol* 522: 139.

von Andrian, U.H. and T.R. Mempel. 2003. Homing and cellular traffic in lymph nodes. *Nat Rev Immunol* 3: 867.

Vujcic, S. et al. 2002. Identification and characterization of a novel flavin-containing spermine oxidase of mammalian cell origin. *Biochem J* 367: 665.

Vujcic, S. et al. 2003. Genomic identification and biochemical characterization of the mammalian polyamine oxidase involved in polyamine back-conversion. *Biochem J* 370: 19.

Walsh, D.M. and D.J. Selkoe. 2004. Oligomers on the brain: the emerging role of soluble protein aggregates in neurodegeneration. *Protein Pept Lett* 11: 213.

Walters, D.R. 2003. Polyamines and plant disease. *Phytochemistry* 64: 97.

Wanecq, E. et al. 2006. Increased monoamine oxidase and semicarbazide-sensitive amine oxidase activities in white adipose tissue of obese dogs fed a high-fat diet. *J Physiol Biochem* 62: 113.

Wang, E.Y. et al. 2006. Design, synthesis, and biological evaluation of semicarbazide-sensitive amine oxidase (SSAO) inhibitors with anti-inflammatory activity. *J Med Chem* 49: 2166.

Wang, L. et al. 2000. Altered gene expression in kidneys of mice with 2,8-dihydroxyadenine nephrolithiasis. *Kidney Int* 58: 528.

Wang, X. et al. 1996a. Extended substrate specificity of serum amine oxidase: possible involvement in protein post-translational modification. *Biochem Biophys Res Commun* 223: 91.

Wang, X. et al. 1996b. A crosslinked cofactor in lysyl oxidase: redox function for amino acid side chains. *Science* 273: 1078.

Wantke, F. et al. 1993. Inhibition of diamine oxidase is a risk in specific immunotherapy. *Allergy* 48: 552.

Watson, G.S. and S. Craft. 2006. Insulin resistance, inflammation, and cognition in Alzheimer's disease: lessons for multiple sclerosis. *J Neurol Sci* 245: 21.

Weiss, H.G. et al. 2003. Plasma amine oxidase: a postulated cardiovascular risk factor in non-diabetic obese patients. *Metabolism* 52: 688.

Wilce, M.C.J. et al. 1997. Crystal structures of the copper-containing amine oxidase from *Arthrobacter globiformis* in the holo and apo forms: implications for the biogenesis of topaquinone. *Biochemistry* 36: 16116.

Williamson, J.D. et al. 2007. The action to control cardiovascular risk in diabetes memory in diabetes study (ACCORD-MIND): rationale, design, and methods. *Am J Cardiol* 99: 112.

Wilmot, C.M. et al. 1997. Catalytic mechanism of the quinoenzyme amine oxidase from *Escherichia coli*: exploring the reductive half-reaction. *Biochemistry* 36: 1608.

Wilmot, C.M. et al. 1999. Visualization of dioxygen bound to copper during enzyme catalysis. *Science* 286: 1724.

Wilmot, C.M. et al. 2004. Medical implications from the crystal structure of a copper-containing amine oxidase complexed with the antidepressant drug tranylcypromine. *FEBS Lett* 576: 301.

Wimmerová, M. and L. Macholán. 1999. Sensitive amperometric biosensor for the determination of biogenic and synthetic amines using pea seedlings amine oxidase: a novel approach for enzyme immobilisation. *Biosens Bioelectron* 14: 695.

Winter, R. et al. 1973. Aminoacetone as a substrate for bovine serum monoaminooxidase. *Bull Acad Pol Sci Biol* 21: 321.

Winterkamp, S. et al. 2002. Urinary excretion of N-methylhistamine as a marker of disease activity in inflammatory bowel disease. *Am J Gastroenterol* 97: 3071.

Wisniewski, J.P. et al. 2000. Involvement of diamine oxidase and peroxidase in insolubilization of the extracellular matrix: implications for pea nodule initiation by *Rhizobium leguminosarum*. *Mol Plant Microbe Interact* 13: 413.

Xavier, R.J. and D.K. Podolsky. 2007. Unravelling the pathogenesis of inflammatory bowel disease. *Nature* 448: 427.

Xiao, S. and P.H. Yu. 2008. A fluorometric high performance liquid chromatography procedure for simultaneous determination of methylamine and aminoacetone in blood and tissues. *Anal Biochem*. doi: 10.1016/j.ab.2008.09.029.

Xie, Z. and G.M. Miller. 2008. Beta-phenylethylamine alters monoamine transporter function via trace amine-associated receptor 1: implication for modulatory roles of trace amines in brain. *J Pharmacol Exp Ther* 325: 617.

Xing, S.G. et al. 2007. Higher accumulation of γ-aminobutyric acid induced by salt stress through stimulating the activity of diamine oxidases in *Glycine max* (L.) Merr. roots. *Plant Physiol Biochem* 45: 560.

Xu, H.L. et al. 2005. Vascular adhesion protein-1 plays an important role in postischemic inflammation and neuropathology in diabetic, estrogen-treated ovariectomized female rats subjected to transient forebrain ischemia *J Pharmacol Exp Ther* 317: 19.

Yamada, H., O. Adachi, and K. Ogata. 1965. Amine oxidase of microorganisms. II. Further properties of amine oxidase of *Aspergillus niger*. *Agric Biol Chem* 29: 912.

Yamashita, M. et al. 1993. Purification and characterization of monoamine oxidase from *Klebsiella aerogenes*. *J Ferment Bioeng* 76: 289.

Yamazaki, R.K. et al. 2005. The effects of peroxovanadate and peroxovanadyl on glucose metabolism in vivo and identification of signal transduction proteins involved in the mechanism of action in isolated soleus muscle. *Mol Cell Biochem* 273: 145.

Yanagisawa, H., E. Hirasawa, and Y. Suzuki. 1981. Purification and properties of diamine oxidase from pea epicotyls. *Phytochemistry* 20: 2105.

Yegutkin, G.G. et al. 2004. A peptide inhibitor of vascular adhesion protein-1 (VAP-1) blocks leukocyte–endothelium interactions under shear stress. *Eur J Immunol* 34: 2276.

Yoong, K.F. et al. 1998. Vascular adhesion protein-1 and ICAM-1 support the adhesion of tumor-infiltrating lymphocytes to tumor endothelium in human hepatocellular carcinoma. *J Immunol* 160: 3978.

Yoshida, A. et al. 1997. Vanillin formation by microbial amine oxidases from vanillylamine. *J Ferm Bioeng* 84: 603.

Yraola, F. et al. 2006. New efficient substrates for semicarbazide-sensitive amine oxidase/VAP-1 enzyme: analysis by SARs and computational docking. *J Med Chem* 49: 6197.

Yraola, F., F. Albericio, and M. Royo. 2007a. Inhibition of VAP-1: quickly gaining ground as an anti-inflammatory therapy. *Chem Med Chem* 2: 173.

Yraola, F. et al. 2007b. Understanding the mechanism of action of the novel SSAO substrate $(C_7NH_{10})_6(V_{10}O_{28}).2H_2O$, a prodrug of peroxovanadate insulin mimetics. *Chem Biol Drug Des* 69: 423.

Yu, P.H. 1990. Oxidative deamination of aliphatic amines by rat aorta semicarbazide-sensitive amine oxidase. *J Pharm Pharmacol* 42: 882.

Yu, P.H., C.Y. Fang, and C.M. Yang. 1992. Semicarbazide-sensitive amine oxidase from the smooth muscles of dog aorta and trachea: activation by the MAO-A inhibitor clorgyline. *J Pharm Pharmacol* 44: 981.

Yu, P.H. and D.M. Zuo. 1993. Oxidative deamination of methylamine by semicarbazide-sensitive amine oxidase leads to cytotoxic damage in endothelial cells: possible consequences for diabetes. *Diabetes* 42: 594.

Yu, P.H., D.M. Zuo, and B.A. Davis. 1994. Characterization of human serum and umbilical artery semicarbazide-sensitive amine oxidase (SSAO): species heterogeneity and stereoisomeric specificity. *Biochem Pharmacol* 47: 1055.

Yu, P.H. and D.M. Zuo. 1996. Formaldehyde produced endogenously via deamination of methylamine: a potential risk factor for initiation of endothelial injury. *Atherosclerosis* 120: 189.

Yu, P.H. and D.M. Zuo. 1997. Aminoguanidine inhibits semicarbazide-sensitive amine oxidase activity: implications for advanced glycation and diabetic complications. *Diabetologia* 40: 1243.

Yu, P.H., C.T. Lai, and D.M. Zuo. 1997. Formation of formaldehyde from adrenaline *in vivo*: a potential risk factor for stress-related angiopathy. *Neurochem Res* 22: 615.

Yu, P.H. and R.F. Dyck. 1998. Impairment of methylamine clearance in uremic patients and its nephropathological implications. *Clin Nephrol* 49: 299.

Yu, P.H. and Y.L. Deng. 1998. Endogenous formaldehyde as a potential factor of vulnerability of atherosclerosis: involvement of semicarbazide-sensitive amine oxidase-mediated methylamine turnover. *Atherosclerosis* 140: 357.

Yu, P.H. 2001. Involvement of cerebrovascular semicarbazide-sensitive amine oxidase in the pathogenesis of Alzheimer's disease and vascular dementia. *Med Hypotheses* 57: 175.

Yu, P.H., B.A. Davis, and Y. Deng. 2001. 2-Bromoethylamine as a potent selective suicide inhibitor for semicarbazide-sensitive amine oxidase. *Biochem Pharmacol* 61: 741.

Yu, P.H. et al. 2003a. A novel sensitive high-performance liquid chromatography–electrochemical procedure for measuring formaldehyde produced from oxidative deamination of methylamine and in biological samples. *Anal Biochem* 318: 285.

Yu, P.H. et al. 2003b. Physiological and pathological implications of semicarbazide-sensitive amine oxidase. *Biochim Biophys Acta* 1647: 193.

Yu, P.H. et al. 2004. Involvement of SSAO-mediated deamination in adipose glucose transport and weight gain in obese diabetic KKAy mice. *Am J Physiol Endocrinol Metab* 286: E634.

Yu, P.H. et al. 2006. Involvement of semicarbazide-sensitive amine oxidase-mediated deamination in lipopolysaccharide-induced pulmonary inflammation. *Am J Physiol Endocrinol Metab* 168: 718.

Yuan, Q., R.M. Ray, and L.R. Johnson. 2002. Polyamine depletion prevents camptothecin-induced apoptosis by inhibiting the release of cytochrome c. *Am J Physiol Cell Physiol* 282: C1290.

Zeisel, S.H., J.S. Wishnok, and J.K. Blusztajn. 1983. Formation of methylamines from ingested choline and lecithin. *J Pharmacol Exp Ther* 225: 320.

Zeller, E.A. 1938. Über den enzymatischen abbau von histamine and diaminen. *Helv Chim Acta* 21: 880.

Zeller, E.A. 1940. Über die beeinflussung der diamin-oxydase durch kaliumcyanid. 7. mitteilung über den enzymatischen abbau von poly-aminen. *Helv Chim Acta* 23: 1418.

Zeller, E.A. 1959. The role of amine oxidases in the destruction of catecholamines. *Pharmacol Rev* 11: 387.

Zeller, E.A. 1963. Diamine oxidse. In *The Enzymes*, Summer, J.B. and Myebäck, K., Eds. New York: Academic Press, p. 313.

Zhang, X., J.H. Fuller, and W.S. McIntire. 1993. Cloning, sequencing, expression, and regulation of the structural gene for the copper/topa quinone-containing methylamine oxidase from Arthrobacter strain P1, a Gram-positive facultative methylotroph. *J Bacteriol* 175: 5617.

Zhang, Y. et al. 2007. Highly potent 3-pyrroline mechanism-based inhibitors of bovine plasma amine oxidase and mass spectrometric confirmation of cofactor derivatization. *Bioorg Med Chem* 15: 1868.

Zhou, M. and N. Panchuk-Voloshina. 1997. A one-step fluorimetric method for the continuous measurement of monoamine oxidase activity. *Anal Biochem* 237: 169.

Zimmermann, P. et al. 2004. GENEVESTIGATOR: Arabidopsis microarray database and analysis toolbox. *Plant Physiol* 136: 2621.

Zlokovic, B.V. 2005. Neurovascular mechanisms of Alzheimer's neurodegeneration. *Trends Neurosci* 28: 202.

Zorzano, A. et al. 2003. Semicarbazide-sensitive amine oxidase activity exerts insulin-like effects on glucose metabolism and insulin-signaling pathways in adipose cells. *Biochim Biophys Acta* 1647: 3.

Zuo, D.M. and P.H. Yu. 1994. Semicarbazide-sensitive amine oxidase and monoamine oxidase in rat brain microvessels, meninges, retina and eye sclera. *Brain Res Bull* 33: 307.

Index

A

B

J

K

L

M

T

U

V

Printed and bound by CPI Group (UK) Ltd, Croydon, CR0 4YY
18/10/2024
01776262-0015